TOPOLOGICAL FIELDS

PURE AND APPLIED MATHEMATICS

A Program of Monographs, Textbooks, and Lecture Notes

MONOGRAPHS AND TEXTBOOKS IN PURE AND APPLIED MATHEMATICS

1. *K. Yano,* Integral Formulas in Riemannian Geometry (1970)*(out of print)*
2. *S. Kobayashi,* Hyperbolic Manifolds and Holomorphic Mappings (1970) *(out of print)*
3. *V. S. Vladimirov,* Equations of Mathematical Physics (A. Jeffrey, editor; A. Littlewood, translator) (1970) *(out of print)*
4. *B. N. Pshenichnyi,* Necessary Conditions for an Extremum (L. Neustadt, translation editor; K. Makowski, translator) (1971)
5. *L. Narici, E. Beckenstein, and G. Bachman,* Functional Analysis and Valuation Theory (1971)
6. *D. S. Passman,* Infinite Group Rings (1971)
7. *L. Dornhoff,* Group Representation Theory (in two parts). Part A: Ordinary Representation Theory. Part B: Modular Representation Theory (1971, 1972)
8. *W. Boothby and G. L. Weiss (eds.),* Symmetric Spaces: Short Courses Presented at Washington University (1972)
9. *Y. Matsushima,* Differentiable Manifolds (E. T. Kobayashi, translator) (1972)
10. *L. E. Ward, Jr.,* Topology: An Outline for a First Course (1972) *(out of print)*
11. *A. Babakhanian,* Cohomological Methods in Group Theory (1972)
12. *R. Gilmer,* Multiplicative Ideal Theory (1972)
13. *J. Yeh,* Stochastic Processes and the Wiener Integral (1973) *(out of print)*
14. *J. Barros-Neto,* Introduction to the Theory of Distributions (1973) *(out of print)*
15. *R. Larsen,* Functional Analysis: An Introduction (1973) *(out of print)*
16. *K. Yano and S. Ishihara,* Tangent and Cotangent Bundles: Differential Geometry (1973) *(out of print)*
17. *C. Procesi,* Rings with Polynomial Identities (1973)
18. *R. Hermann,* Geometry, Physics, and Systems (1973)
19. *N. R. Wallach,* Harmonic Analysis on Homogeneous Spaces (1973) *(out of print)*
20. *J. Dieudonné,* Introduction to the Theory of Formal Groups (1973)
21. *I. Vaisman,* Cohomology and Differential Forms (1973)
22. *B.-Y. Chen,* Geometry of Submanifolds (1973)
23. *M. Marcus,* Finite Dimensional Multilinear Algebra (in two parts) (1973, 1975)
24. *R. Larsen,* Banach Algebras: An Introduction (1973)
25. *R. O. Kujala and A. L. Vitter (eds.),* Value Distribution Theory: Part A; Part B: Deficit and Bezout Estimates by Wilhelm Stoll (1973)
26. *K. B. Stolarsky,* Algebraic Numbers and Diophantine Approximation (1974)
27. *A. R. Magid,* The Separable Galois Theory of Commutative Rings (1974)
28. *B. R. McDonald,* Finite Rings with Identity (1974)
29. *J. Satake,* Linear Algebra (S. Koh, T. A. Akiba, and S. Ihara, translators) (1975)

30. *J. S. Golan,* Localization of Noncommutative Rings (1975)
31. *G. Klambauer,* Mathematical Analysis (1975)
32. *M. K. Agoston,* Algebraic Topology: A First Course (1976)
33. *K. R. Goodearl,* Ring Theory: Nonsingular Rings and Modules (1976)
34. *L. E. Mansfield,* Linear Algebra with Geometric Applications: Selected Topics (1976)
35. *N. J. Pullman,* Matrix Theory and Its Applications (1976)
36. *B. R. McDonald,* Geometric Algebra Over Local Rings (1976)
37. *C. W. Groetsch,* Generalized Inverses of Linear Operators: Representation and Approximation (1977)
38. *J. E. Kuczkowski and J. L. Gersting,* Abstract Algebra: A First Look (1977)
39. *C. O. Christenson and W. L. Voxman,* Aspects of Topology (1977)
40. *M. Nagata,* Field Theory (1977)
41. *R. L. Long,* Algebraic Number Theory (1977)
42. *W. F. Pfeffer,* Integrals and Measures (1977)
43. *R. L. Wheeden and A. Zygmund,* Measure and Integral: An Introduction to Real Analysis (1977)
44. *J. H. Curtiss,* Introduction to Functions of a Complex Variable (1978)
45. *K. Hrbacek and T. Jech,* Introduction to Set Theory (1978)
46. *W. S. Massey,* Homology and Cohomology Theory (1978)
47. *M. Marcus,* Introduction to Modern Algebra (1978)
48. *E. C. Young,* Vector and Tensor Analysis (1978)
49. *S. B. Nadler, Jr.,* Hyperspaces of Sets (1978)
50. *S. K. Segal,* Topics in Group Rings (1978)
51. *A. C. M. van Rooij,* Non-Archimedean Functional Analysis (1978)
54. *L. Corwin and R. Szczarba,* Calculus in Vector Spaces (1979)
53. *C. Sadosky,* Interpolation of Operators and Singular Integrals: An Introduction to Harmonic Analysis (1979)
54. *J. Cronin,* Differential Equations: Introduction and Quantitative Theory (1980)
55. *C. W. Groetsch,* Elements of Applicable Functional Analysis (1980)
56. *I. Vaisman,* Foundations of Three-Dimensional Euclidean Geometry (1980)
57. *H. I. Freedman,* Deterministic Mathematical Models in Population Ecology (1980)
58. *S. B. Chae,* Lebesgue Integration (1980)
59. *C. S. Rees, S. M. Shah, and C. V. Stanojević,* Theory and Applications of Fourier Analysis (1981)
60. *L. Nachbin,* Introduction to Functional Analysis: Banach Spaces and Differential Calculus (R. M. Aron, translator) (1981)
61. *G. Orzech and M. Orzech,* Plane Algebraic Curves: An Introduction Via Valuations (1981)
62. *R. Johnsonbaugh and W. E. Pfaffenberger,* Foundations of Mathematical Analysis (1981)
63. *W. L. Voxman and R. H. Goetschel,* Advanced Calculus: An Introduction to Modern Analysis (1981)
64. *L. J. Corwin and R. H. Szcarba,* Multivariable Calculus (1982)
65. *V. I. Istrătescu,* Introduction to Linear Operator Theory (1981)
66. *R. D. Järvinen,* Finite and Infinite Dimensional Linear Spaces: A Comparative Study in Algebraic and Analytic Settings (1981)

67. *J. K. Beem and P. E. Ehrlich,* Global Lorentzian Geometry (1981)
68. *D. L. Armacost,* The Structure of Locally Compact Abelian Groups (1981)
69. *J. W. Brewer and M. K. Smith, eds.,* Emmy Noether: A Tribute to Her Life and Work (1981)
70. *K. H. Kim,* Boolean Matrix Theory and Applications (1982)
71. *T. W. Wieting,* The Mathematical Theory of Chromatic Plane Ornaments (1982)
72. *D. B. Gauld,* Differential Topology: An Introduction (1982)
73. *R. L. Faber,* Foundations of Euclidean and Non-Euclidean Geometry (1983)
74. *M. Carmeli,* Statistical Theory and Random Matrices (1983)
75. *J. H. Carruth, J. A. Hildebrant, and R. J. Koch,* The Theory of Topological Semigroups (1983)
76. *R. L. Faber,* Differential Geometry and Relativity Theory: An Introduction (1983)
77. *S. Barnett,* Polynomials and Linear Control Systems (1983)
78. *G. Karpilovsky,* Commutative Group Algebras (1983)
79. *F. Van Oystaeyen and A. Verschoren,* Relative Invariants of Rings: The Commutative Theory (1983)
80. *I. Vaisman,* A First Course in Differential Geometry (1984)
81. *G. W. Swan,* Applications of Optimal Control Theory in Biomedicine (1984)
82. *T. Petrie and J. D. Randall,* Transformation Groups on Manifolds (1984)
83. *K. Goebel and S. Reich,* Uniform Convexity, Hyperbolic Geometry, and Nonexpansive Mappings (1984)
84. *T. Albu and C. Năstăsescu,* Relative Finiteness in Module Theory (1984)
85. *K. Hrbacek and T. Jech,* Introduction to Set Theory, Second Edition, Revised and Expanded (1984)
86. *F. Van Oystaeyen and A. Verschoren,* Relative Invariants of Rings: The Noncommutative Theory (1984)
87. *B. R. McDonald,* Linear Algebra Over Commutative Rings (1984)
88. *M. Namba,* Geometry of Projective Algebraic Curves (1984)
89. *G. F. Webb,* Theory of Nonlinear Age-Dependent Population Dynamics (1985)
90. *M. R. Bremner, R. V. Moody, and J. Patera,* Tables of Dominant Weight Multiplicities for Representations of Simple Lie Algebras (1985)
91. *A. E. Fekete,* Real Linear Algebra (1985)
92. *S. B. Chae,* Holomorphy and Calculus in Normed Spaces (1985)
93. *A. J. Jerri,* Introduction to Integral Equations with Applications (1985)
94. *G. Karpilovsky,* Projective Representations of Finite Groups (1985)
95. *L. Narici and E. Beckenstein,* Topological Vector Spaces (1985)
96. *J. Weeks,* The Shape of Space: How to Visualize Surfaces and Three-Dimensional Manifolds (1985)
97. *P. R. Gribik and K. O. Kortanek,* Extremal Methods of Operations Research (1985)
98. *J.-A. Chao and W. A. Woyczynski, eds.,* Probability Theory and Harmonic Analysis (1986)
99. *G. D. Crown, M. H. Fenrick, and R. J. Valenza,* Abstract Algebra (1986)
100. *J. H. Carruth, J. A. Hildebrant, and R. J. Koch,* The Theory of Topological Semigroups, Volume 2 (1986)

101. *R. S. Doran and V. A. Belfi,* Characterizations of C*-Algebras: The Gelfand-Naimark Theorems (1986)
102. *M. W. Jeter,* Mathematical Programming: An Introduction to Optimization (1986)
103. *M. Altman,* A Unified Theory of Nonlinear Operator and Evolution Equations with Applications: A New Approach to Nonlinear Partial Differential Equations (1986)
104. *A. Verschoren,* Relative Invariants of Sheaves (1987)
105. *R. A. Usmani,* Applied Linear Algebra (1987)
106. *P. Blass and J. Lang,* Zariski Surfaces and Differential Equations in Characteristic $p > 0$ (1987)
107. *J. A. Reneke, R. E. Fennell, and R. B. Minton.* Structured Hereditary Systems (1987)
108. *H. Busemann and B. B. Phadke,* Spaces with Distinguished Geodesics (1987)
109. *R. Harte,* Invertibility and Singularity for Bounded Linear Operators (1988).
110. *G. S. Ladde, V. Lakshmikantham, and B. G. Zhang,* Oscillation Theory of Differential Equations with Deviating Arguments (1987)
111. *L. Dudkin, I. Rabinovich, and I. Vakhutinsky,* Iterative Aggregation Theory: Mathematical Methods of Coordinating Detailed and Aggregate Problems in Large Control Systems (1987)
112. *T. Okubo,* Differential Geometry (1987)
113. *D. L. Stancl and M. L. Stancl,* Real Analysis with Point-Set Topology (1987)
114. *T. C. Gard,* Introduction to Stochastic Differential Equations (1988)
115. *S. S. Abhyankar,* Enumerative Combinatorics of Young Tableaux (1988)
116. *H. Strade and R. Farnsteiner,* Modular Lie Algebras and Their Representations (1988)
117. *J. A. Huckaba,* Commutative Rings with Zero Divisors (1988)
118. *W. D. Wallis,* Combinatorial Designs (1988)
119. *W. Więsław,* Topological Fields (1988)

Other Volumes in Preparation

TOPOLOGICAL FIELDS

WITOLD WIĘSŁAW

Wrocław University
Wrocław, Poland

MARCEL DEKKER, INC. New York and Basel

Więsław, Witold.
Topological fields.

(Pure and applied mathematics ; v. 119)
Bibliography: p.
Includes indexes.
1. Topological fields. I. Title. II. Series:
Monographs and textbooks in pure and applied mathematics ;
v. 119.
QA247.W54 1988 512'.74 88-7102
ISBN 0-8247-7731-X

MARCEL DEKKER, INC.
270 Madison Avenue, New York, New York 10016

Current printing (last digit):
10 9 8 7 6 5 4 3 2 1

PRINTED IN THE UNITED STATES OF AMERICA

Preface

This book is based on lectures that I gave at Wrocław University in the period 1975-1983. My main purpose was to present most of the known results dealing with topological fields. It was, of course, impossible to prove all results stated. The theorems on valuations and absolute values are standard and are included in this book only for completeness. Other results require many technical means, thus making impossible the presentation of complete proofs. In principle, however, the most fundamental theorems are given with proofs.

The contents of this book are as follows. In Chapter 1 the fundamental notions are introduced and various examples are presented. Chapter 2 presents fundamental properties of absolute-valued fields. The culminating point here is a theorem of Šafarevič-Kaplansky. In Chapter 3 we discuss normed fields and their properties and characterization, due to P. M. Cohn. The following chapters deal with Krull valuations, topologies of type V, and locally compact fields. The final five chapters contain different results and examples of topological fields, such as connected fields, connections of arithmetic and topology (Chapter 12), and extensions of topologies. Most results are presented here for the first time in book form.

The idea of writing this book was suggested to me by Professor Hans-Joachim Kowalsky. When nine years ago I presented a lecture at Braunschweig Technische Universität, I remarked to Professor Kowalsky that it would be a good idea for someone to write a book on topological fields. "You can write it," said Professor Kowalsky; and so I did.

A preliminary edition of this book appeared in 1982 in the *Acta Universitatis Wratislaviensis*, and I received useful comments from Professors Jerzy Browkin, Władysław Narkiewicz, Niel Shell, Seth L. Warner, Hans Weber, and Heinrich Wefelscheid. I would like to thank them cordially for their help.

Witold Więsław

Contents

TOPOLOGICAL FIELDS

1

Definitions and Examples

In this chapter we shall be concerned with the definition and fundamental properties of topological fields. All rings considered are assumed to be associative and with unity. We consider commutative fields only.

1. BASES OF ZERO NEIGHBORHOODS

We start with a general

DEFINITION 1 Let R be a ring. A pair $(R,\mathcal{T})$ is said to be a *topological ring* if $\mathcal{T}$ is a topology on R and the functions $(x,y) \mapsto x \pm y$, $(x,y) \mapsto xy$ are continuous. If $\mathcal{T}$ is neither discrete nor antidiscrete, we say that $\mathcal{T}$ is a *proper ring topology* on R and $(R,\mathcal{T})$ is a *proper topological ring*.

DEFINITION 2 Assume now that K is a field. By a *topological field* we shall mean a topological ring $(K,\mathcal{T})$, where, moreover, division $x \mapsto x^{-1}$ $(x \neq 0)$ is continuous. Then we say that $\mathcal{T}$ is a field topology. In the case where the field topology $\mathcal{T}$ is neither discrete nor antidiscrete (trivial) we shall say that $\mathcal{T}$ is a *proper field topology* and $(K,\mathcal{T})$ is a *proper topological field*.

Remark 1 Recall (Kohn and Newman [1]) that if a field K has characteristic $\mathrm{char}(K) \neq 2$, then the equality

$$xy = [(x + y - 2)^{-1} - (x + y + 2)^{-1}]^{-1} - [(x - y - 2)^{-1} - (x - y + 2)^{-1}]^{-1}$$

holds. Thus, if $\mathcal{T}$ is a topology on K such that addition, subtraction, and inverse are continuous, then multiplication is also continuous.

Remark 2 Let K be a field and $\mathcal{T}$ a proper metrizable ring topology on K. Then $\mathcal{T}$ is a field topology if and only if the set $K^\times$ of all nonzero elements of K is a G_δ (i.e., is an intersection of countably many open sets (see Banach [1, Th. II])).

Notation Let A, B be two subsets of a field K. Then A + B denotes the set of all sums a + b, where a ∈ A and b ∈ B. Similarly, AB is the set of all products ab with a ∈ A and b ∈ B. If 0 ∉ A, then A^{-1} denotes the set of all elements a^{-1}, a ∈ A, and B/A is defined as BA^{-1}. If b ∈ K, then we write b + A (resp. bA) instead of {b} + A (resp. {b}A). Moreover, $K^\times = K \setminus \{0\}$ is the multiplicative group of a field K and K^+ is the additive group of K. A subset A is called symmetric if A = -A, where -A is the set of all -a with a ∈ A.

Let K be a field and $\mathcal{T}$ a topology on K. By a *neighborhood* of an element a ∈ K we shall always mean a set whose interior contains a. Let $\mathcal{B} = \mathcal{B}(\mathcal{T})$ be a base of symmetric neighborhoods of zero. It is clear that $\mathcal{B}$ is a filter base of sets. Since $\mathcal{T}$ is a group topology on the additive group K^+ of K, then $\mathcal{B}_a(\mathcal{T}) = a + \mathcal{B} = \{a + B : B \in \mathcal{B}\}$ is a base of the neighborhoods of an element a ∈ K.

LEMMA 1 Let K be a field and $\mathcal{T}$ a ring topology on K. If $\mathcal{T}$ is nontrivial, then $(K,\mathcal{T})$ is a Hausdorff space.

Proof: By our assumption, there exists a nonempty open set U in K, U ≠ K. Let a ∈ K \ U. Then for arbitrary b ∈ U, -b + U is an open set and a - b ∉ -b + U. Therefore, $V = (a - b)^{-1}(-b + U)$ is an open neighborhood of zero not containing the identity element 1. Let $c \in K^\times$ be an arbitrary element. Then cV is an open neighborhood of zero and c ∉ cV. This implies that $\mathcal{T}$ satisfies the T_0 separation axiom. Since $(K,\mathcal{T})$ is a topological group (with respect to the addition in K), and $(K,\mathcal{T})$ is T_0, it must be Hausdorff.

THEOREM 1 Let $(K,\mathcal{T})$ be a proper topological field. Then $\mathcal{T}$ is uniquely determined by a base $\mathcal{B} = \mathcal{B}(\mathcal{T})$ of the symmetric neighborhoods of zero satisfying the following conditions:

(B_1) $\bigcap \mathcal{B} = \{0\}$

(B_2) $\forall\, U,V \in \mathcal{B}\ \exists\, W \in \mathcal{B}$ $\quad W \subset U \cap V$

(B_3) $\forall\, U \in \mathcal{B}\ \exists\, V \in \mathcal{B}$ $\quad V + V \subset U$

(B_4) $\forall\, U \in \mathcal{B}\ \exists\, V \in \mathcal{B}$ $\quad VV \subset U$

(B_5) $\forall\, U \in \mathcal{B}\ \forall\, x \in K\ \exists\, V \in \mathcal{B}$ $\quad xV \subset U$

(B_6) $\forall\, U \in \mathcal{B}\ \exists\, V \in \mathcal{B}$ $\quad (1 + V)^{-1} \subset 1 + U$

Conversely, every family $\mathcal{B}$ satisfying (B_1)-(B_6) defines a field topology.

Proof: First, let $(K,\mathcal{T})$ be a proper topological field. Let $\mathcal{A}$ be the family of open sets in $(K,\mathcal{T})$. Define $\mathcal{B}$ to be the set of all sets $B \in \mathcal{A}$ that contain zero. Since $\mathcal{T}$ is Hausdorff, (B_1) holds. Moreover, $V = -V$ for $V \in \mathcal{B}$, by definition. The continuity of addition at $(0,0)$, multiplication [at $(0,0)$], and inversion (at 1) implies (B_3), (B_4), and (B_6), respectively. Condition (B_2) is obvious. Finally, let $x \in K$. For $x = 0$ condition (B_5) is clear, and if $x \neq 0$, then the continuity of the function $y \mapsto yx$ at 0 implies (B_5).

One readily sees that specifying $\{x + B : B \in \mathcal{B}\}$ as a neighborhood base at x determines a topology.

Assume now that a family of sets $\mathcal{B} \subset 2^K$ satisfies (B_1)-(B_6). We are going to prove that there is a topology $\mathcal{T}$ on K such that $\mathcal{B}$ is a base of zero neighborhoods of $\mathcal{T}$ and for arbitrary $x \in K$, the sets $\{x + B : B \in \mathcal{B}\}$ form a base of neighborhoods of x.

Continuity of addition follows from (B_3). For arbitrary $x,y \in K$ we have $(x + V) + (y + V) \subset (x + y) + U$. Continuity of subtraction follows from the symmetry of the family $\mathcal{B}$ ($V = -V$ for every $V \in \mathcal{B}$). We shall prove the continuity of multiplication,

(B_4') $\forall\, U \in \mathcal{B}\ \forall\, x,y \in K\ \exists\, V \in \mathcal{B}$ $\quad (x + V) \subset xy + U$

and continuity of division,

(B_5') $\forall\, U \in \mathcal{B}\ \forall\, x \in K^\times\ \exists\, V \in \mathcal{B}$ $\quad (x + V)^{-1} \subset x^{-1} + U$

Let $U \in \mathcal{B}$ be arbitrary. There exist $U_1 \in \mathcal{B}$ satisfying $U_1 + U_1 + U_1 \subset U$ [by (B_3)]. Condition (B_5) implies the existence of $V_1,V_2 \in \mathcal{B}$ with $xV_1 \subset U_1$, $yV_2 \subset U_1$. By (B_4), there exists $V_3 \in \mathcal{B}$ such that $V_3V_3 \subset U_1$. Let $V = V_1 \cap V_2 \cap V_3$. Then

$$(x + V)(y + V) = xy + xV + yV + VV \subset xy + xV_1 + yV_2$$
$$+ V_3V_3 \subset xy + U_1 + U_1 + U_1 \subset U + xy$$

proving (B_4').

Let $x \in K^\times$ and $U \in \mathcal{B}$ be arbitrary. Then there exists $U_1 \in \mathcal{B}$ satisfying $U_1x^{-1} \subset U$, by (B_5). We can find $V_1 \in \mathcal{B}$ for U_1 such that $(1 + V_1)^{-1} \subset 1 + U_1$ [see (B_6)]. Finally, by (B_5) there exists $V \in \mathcal{B}$, $x^{-1}V \subset V_1$. This implies that

$$(1 + x^{-1}V)^{-1} \subset (1 + V_1)^{-1} \subset 1 + U_1 \subset 1 + xU$$

i.e.,

$$x^{-1}(1 + x^{-1}V)^{-1} = (x + V)^{-1} \subset x^{-1} + U$$

proving (B_5').

We shall sometimes need an equivalent form of (B_6), namely

(B_6') $\forall\, U \in \mathcal{B}\ \exists\, V \in \mathcal{B} \quad \dfrac{V}{1 + V} \subset U$

LEMMA 2 Conditions (B_1)-(B_6), and (B_1)-(B_5) and (B_6') are equivalent.

Proof: Suppose that conditions (B_1)-(B_6) hold. Let $U \in \mathcal{B}$ be arbitrary. Then there are $U_1, U_2 \in \mathcal{B}$ satisfying $U_1 + U_1 \subset U$, $U_2 \subset U_1$, and $U_2U_2 \subset U_1$. By (B_6), for every $U_2 \in \mathcal{B}$ there exists $V \subset U_2$, $V \in \mathcal{B}$ with $(1 + V)^{-1} \subset 1 + U_2$. Then

$$\frac{V}{1 + V} \subset V(1 + U_2) \subset U_2 + U_2U_2 \subset U_1 + U_1 \subset U$$

Now suppose that (B_1)-(B_5) and (B_6') hold and take $U \in \mathcal{B}$ arbitrary. Let $V = -V$ satisfy (B_6'). Then

$$(1 + V)^{-1} = \left\{\frac{1}{1 + v} : v \in V\right\} = \left\{1 + \frac{-v}{1 + v} : v \in V\right\}$$

i.e.,

$$(1 + V)^{-1} \subset 1 + \frac{V}{1 + V} \subset 1 + U$$

We shall need the following condition:

(B_5'') $\forall\, U \in \mathcal{B}\ \forall\, x \in K\ \exists\, V \in \mathcal{B} \quad Vx \subset U$

THEOREM 1' Let $(K,\mathcal{T})$ be a proper topological ring. Then $\mathcal{T}$ is uniquely determined by a base $\mathcal{B}(\mathcal{T})$ of symmetric neighborhoods of zero satisfying (B_1)-(B_5) and (B_5''). Conversely, every family $\mathcal{B} \subset 2^K$ with (B_1)-(B_5) and (B_5'') defines a ring topology.

The proof is similar to that of Theorem 1 and we shall omit it. As an easy corollary from Theorem 1' we have

LEMMA 1' The closure $\overline{\{0\}}$ of zero set $\{0\}$ is an ideal in any topological ring. In particular, in a field with a ring topology, $\overline{\{0\}} = \{0\}$ (i.e., the topology is Hausdorff) or $\overline{\{0\}}$ is the whole field (i.e., thc topology is trivial).

DEFINITION 3 Let V be a vector space equipped with a topology $\mathcal{S}$ and let $(K,\mathcal{T})$ be a topological field. We shall say that $(V,\mathcal{S})$ is a *topological vector space* over $(K,\mathcal{T})$ if the function $(x,y) \mapsto x + y$ from $(V,\mathcal{S}) \times (V,\mathcal{S})$ to $(V,\mathcal{S})$ and the functions $(a,x) \mapsto ax$ from $(K,\mathcal{T}) \times (V,\mathcal{S})$ to $(V,\mathcal{S})$ are continuous.

One can easily formulate appropriate conditions for bases of zero neighborhoods for any topological vector space $(V,\mathcal{S})$ over a topological field $(K,\mathcal{T})$. We shall often apply this notion in the case when $(V,\mathcal{S})$ is a field (or a ring) extension of $(K,\mathcal{T})$.

THEOREM 2 (Tamari [1]) Let K be a field with a topology $\mathcal{T}$, which is neither discrete nor antidiscrete. If K is a topological group in $\mathcal{T}$ with respect to the addition $(x,y) \mapsto x + y$ and $K^\times = K \setminus \{0\}$ is a topological group in $\mathcal{T}$ with respect to the multiplication, then $(K,\mathcal{T})$ is a topological field.

Proof: As the inversion $x \mapsto x^{-1}$ is continuous in $K^\times$, and the addition and the subtraction $(x,y) \mapsto x \pm y$ are continuous by assumption, it is sufficient to show the continuity of the multiplication with respect to the topology $\mathcal{T}$.

Let $\mathcal{B}(\mathcal{T})$ be a base of the symmetric neighborhoods of zero in the additive group K^+ of K and $\mathcal{B}_a(\mathcal{T}) = \{a + U : U \in \mathcal{B}(\mathcal{T}),\ 0 \notin a + U\}$ be a base of neighborhoods of an element a of the multiplicative group $K^\times$ of K. We have to show that

(1) $\forall\ x,y \in K\ \forall\ U \in \mathcal{B}(\mathcal{T})\ \exists\ V \in \mathcal{B}(\mathcal{T}) \qquad (V + x)(V + y) \subset U + xy$

Condition (1) is satisfied for $x,y \in K^\times$. Indeed,

(2) $\forall\ x,y \in K^\times\ \forall\ U_{xy} \in \mathcal{B}_{xy}(\mathcal{T})\ \exists\ W_x \in \mathcal{B}_x(\mathcal{T})\ \exists\ W_y \in \mathcal{B}_y(\mathcal{T}) \qquad W_xW_y \subset U_{xy}$

Indeed, put $U_{xy} = U + xy$, with $U \in \mathcal{B}(\mathcal{T})$, $-xy \notin U$. We may assume that $W_x = x + W$, $W_y = y + W'$ with $W,W' \in \mathcal{B}(\mathcal{T})$, $-x \notin W$, $-y \notin W'$. Put $V = W \cap W'$. Then

$$(V + x)(V + y) \subset W_xW_y \subset U + xy$$

as desired.

It remains to show that if $xy = 0$, then (1) is fulfilled. First let $x = y = 0$. Let $U \in \mathcal{B}(\mathcal{T})$ be arbitrary and $-1 \notin U$, i.e., $0 \notin 1 + U$. Then $1 + U \in \mathcal{B}_1(\mathcal{T})$. There exists $W \in \mathcal{B}(\mathcal{T})$ with $W + W + W \subset U$. Consequently, as multiplication is continuous at $(1,1)$, there exists $V_1 \in \mathcal{B}_1(\mathcal{T})$ with $V_1V_1 \subset 1 + W$. Since $1 \in V_1$, then $V_1 - 1 \in \mathcal{B}(\mathcal{T})$. Let $V = V_1 - 1$. We may assume that $V \subset W$, taking, if necessary, $V \cap W$ instead of V. Thus

$$\begin{aligned} VV &= (V_1 - 1)(V_1 - 1) \subset V_1V_1 - V_1 - V_1 + 1 \subset (1 + W) - V_1 \\ &\quad - V_1 + 1 \\ &= W + (1 - V_1) + (1 - V_1) \subset W + W + W \subset U \end{aligned}$$

We have proved that

$$\forall\ U \in \mathcal{B}(\mathcal{T})\ \exists\ V \in \mathcal{B}(\mathcal{T}) \qquad VV \subset U$$

Now let $x \neq 0 = y$. Since multiplication is continuous at the point $(1,x)$, then (2) implies that

$$\forall\ U_x \in \mathcal{B}_x(\mathcal{T})\ \exists\ V_1 \in \mathcal{B}_1(\mathcal{T})\ \exists\ V_x \in \mathcal{B}_x(\mathcal{T}) \qquad V_1V_x \subset U_x$$

We can take $U_x = U + x$ with $U \in \mathcal{B}(\mathcal{T})$, $-x \notin U$, $V_1 = 1 + W$, $V_x = x + W'$ with $W,W' \in \mathcal{B}(\mathcal{T})$, and $-1 \notin W$, $-x \notin W'$. Put $V = W \cap W'$. Then

$$(1 + V)(x + V) \subset U + x$$

implying that $V + xV + VV \subset U$, and consequently, we obtain

$$\forall\ U \in \mathcal{B}(\mathcal{T})\ \forall\ x \in K^\times\ \exists\ V \in \mathcal{B}(\mathcal{T}) \qquad (V + x)V \subset U$$

Theorem 2 is proved.

2. EXAMPLES OF TOPOLOGICAL FIELDS

We shall start with

Example 1 Let $K = \mathbb{R}$ (or $\mathbb{C}$); a topology $\mathcal{T}$ is induced by ordinary absolute value, i.e., by the metric $d(x,y) = |x - y|$.

Continuity of algebraic operations: Let $d(x,x_0) < \varepsilon$, $d(y,y_0) < \varepsilon$. Then

$$d(x \pm y, x_0 \pm y_0) = |x \pm y - x_0 \mp y_0| \le |x - x_0| + |y - y_0| < 2\varepsilon$$

which implies continuity of addition and subtraction at (x_0,y_0). Moreover,

$$d(xy,x_0y_0) = |xy - x_0y_0| = |x(y - y_0) - (x_0 - x)y_0| \le |x||y - y_0| + |y_0||x - x_0| < 2M\varepsilon$$

where $M = \max(|x_0| + \varepsilon, |y_0|)$. Thus the multiplication is continuous.

Finally, let $x \neq 0$, $x_0 \neq 0$, and $d(x,x_0) < \varepsilon$, where $\varepsilon < |x_0|$. We have

$$d(x^{-1},x_0^{-1}) = |x^{-1} - x_0^{-1}| = \frac{|x - x_0|}{|x||x_0|} < \delta^2\varepsilon$$

where $\delta = \min(|x_0| - \varepsilon, |x_0|)^{-1}$, so division in $(K,\mathcal{T})$ is also continuous.

Example 2 Now let us introduce a topology not equivalent to the ordinary topology of the complex numbers $\mathbb{C}$. With this topology $\mathbb{C}$ will be again a complete metric space.

It is well known (Charnow [1]) that every algebraically closed field K has $2^{|K|}$ automorphisms. In particular, $|\mathrm{Aut}(\mathbb{C})| = 2^c$.

Let us first note

LEMMA 3 The mappings $z \mapsto z$ and $z \mapsto \bar{z}$ are the only continuous automorphisms (with respect to the ordinary topology) of the complex number field.

Proof: Let $f \in \mathrm{Aut}(\mathbb{C})$ be continuous. Since f restricted to $\mathbb{Q}$ is an isomorphism, it is obvious that $f|\mathbb{Q}$ equals the identity, and thus $f|\mathbb{R}$ maps $\mathbb{R}$ onto $\mathbb{R}$, by continuity. But the only automorphism of $\mathbb{R}$ is the identity (compare, e.g., Chapter 2, Section 6). Consequently, if $z \in \mathbb{C}$, $z = x + iy$, then $f(z) = f(x) + f(i)f(y) = x + f(i)y = z$ if $f(i) = i$, and $f(z) = \bar{z}$ if $f(i) = -i$.

Let $g \in \mathrm{Aut}(\mathbb{C})$ be neither the identity nor complex conjugation. It follows, by Lemma 3, that g is not continuous on $\mathbb{C}$ [or even, g is nonmeasurable; this follows from the well-known result that every measurable solution of the Cauchy functional equation $f(x + y) = f(x) + f(y)$ is continuous; for the case of arbitrary locally compact topological fields, see Chapter 6, Theorem 5].

Put $|x|_g = |g(x)|$ and define a metric by $d_g(x,y) = |x - y|_g$. It is easy to see that the metric d_g defines a field topology on $\mathbb{C}$. Moreover, for every such g, $(\mathbb{C},d_g)$ is a complete metric space and the topology induced by the metric d_g is different from the ordinary topology on $\mathbb{C}$.

Remark It is easy to see that every homeomorphism of finite order of the reals is the identity if it preserves the order, and has order 2 otherwise (McShane [1]).

Example 3 shows that a wide class of fields possess the property that every field from the class admits a proper field topology.

Example 3 Let K be the quotient field of a unique factorization domain A, $\mathbb{P}$ be the set of pairwise nonassociated irreducible elements of A, and U(A) be the group of units of A (invertible elements). Then every element $x \in K^{\times}$ can be uniquely written as

$$x = \varepsilon \prod_{p \in \mathbb{P}} p^{v_p(x)} \tag{3}$$

where $v_p(x) \in \mathbb{Z}$ and all $v_p(x)$ but a finite number are zero and $\varepsilon \in U(A)$.

Note that the function $v_p : K \to \mathbb{Z} \cup \{\infty\}$, defined formally at 0 as ∞, has the following properties:

$$v_p(x \pm y) \geq \min(v_p(x), v_p(y)) \tag{4}$$

$$v_p(xy) = v_p(x) + v_p(y) \tag{5}$$

Indeed, let $x,y \in K^{\times}$, $x = \varepsilon \prod_{p \in \mathbf{P}} p^{v_p(x)}$, $y = \varepsilon' \prod_{p \in \mathbf{P}} p^{v_p(y)}$; $\varepsilon, \varepsilon' \in U(A)$. Then $xy = \varepsilon\varepsilon' \prod_{p \in \mathbf{P}} p^{v_p(x)+v_p(y)}$. Since, on the other hand, $xy = \varepsilon'' \prod_{p \in \mathbf{P}} p^{v_p(xy)}$, then by uniqueness of the factorization, we obtain that $v_p(xy) = v_p(x) + v_p(y)$. This proves (5).

Assume, moreover, that $x + y \neq 0$. Let $v_p(x) = N$, $v_p(y) = M$, $M \leq N$. Then $x = p^N x_1$, $y = p^M y_1$ with $x_1, y_1 \in K^{\times}$ and $v_p(x_1) = v_p(y_1) = 0$. Thus $x + y = p^N x_1 + p^M y_1 = p^M(p^{N-M} x_1 + y_1)$, i.e., $v_p(x + y) \geq M = \min(N,M) = \min(v_p(x), v_p(y))$, proving (4). Moreover, if $v_p(x) \neq v_p(y)$, then in (4) the equality holds. For a fixed prime element $p \in \mathbf{P}$, define a function $|\cdot|_p : K \to \mathbb{R}_+$ by the formula

$$|x|_p = c^{v_p(x)} \qquad \text{for } x \neq 0 \text{ and } |0|_p = 0 \tag{6}$$

where c is a fixed constant, $0 < c < 1$. The function $|\cdot|_p$, called the *p-adic absolute value* on K, has the following properties:

$$|x|_p \geq 0, \ |x|_p = 0 \Leftrightarrow x = 0 \tag{7}$$

$$|x \pm y|_p \leq \max(|x|_p, |y|_p) \tag{8}$$

$$|xy|_p = |x|_p |y|_p \tag{9}$$

which follow immediately from (4) and (5). The family of sets

$$U_\varepsilon = \{a \in K : |a|_p < \varepsilon\} \qquad \varepsilon \in \mathbf{R},\ \varepsilon > 0$$

defines a field topology $\mathcal{T}_p$ on K. If $p \neq q$, then $\mathcal{T}_p \neq \mathcal{T}_q$, since for for $n \to \infty$ we have $p^n \to 0$ in $\mathcal{T}_p$ but $p^n \not\to 0$ in $\mathcal{T}_q$. We call $\mathcal{T}_p$ the *p-adic topology*.

Example 4 Let M be a field of meromorphic functions on a connected and open subset $D \subset \mathbb{C}$ (in $\mathbb{C}$ we take the ordinary topology). For a fixed $x_0 \in D$ and f not identically zero, define $v_{x_0}(f)$ by

$$f(x) = (x - x_0)^{v_{x_0}(f)} g(x) \qquad g(x_0) \neq 0$$

In other words, if $f(x) = \Sigma_{n=m}^{\infty} a_n(x - x_0)^n$ is the Laurent expansion of f at x_0 and $a_m \neq 0$, then $v_{x_0}(f) = m$ by definition. The function

$$\varphi_{x_0}(f) = e^{-v_{x_0}(f)} \quad (f \neq 0) \qquad \varphi_{x_0}(0) = 0$$

defines a metric of M by the formula $d(f,g) = \varphi(f - g)$. The topology defined by d on M is a field topology.

It is known from the theory of functions of complex variables that if $D = \mathbb{C} \cup \{\infty\}$ is the one-point compactification of complex numbers, then $M = \mathbb{C}(x)$. In this case Example 4 is contained in Example 3: The rational function field $\mathbb{C}(x)$ of a variable x is the quotient field of the Euclidean domain $\mathbb{C}[x]$ and $x - x_0$'s are unique (up to units) prime elements of $\mathbb{C}[x]$.

Example 5 Let k be an arbitrary field and K a set of sequences $(\ldots,0,\ldots,0,a_N,a_{N+1},\ldots)$ of elements of k $(N \in \mathbb{Z})$, which we shall also write in the form

$$\alpha = a_0X^N + a_1X^{N+1} + \cdots + a_nX^{N+n} + \cdots \tag{10}$$

and call a formal Laurent series of one variable X with coefficients in k. Addition in K is coordinatewise. If

$$\begin{aligned} \alpha &= X^N(a_0 + a_1X + a_2X^2 + \cdots) \qquad a_0 \neq 0 \\ \beta &= X^M(b_0 + b_1X + b_2X^2 + \cdots) \qquad b_0 \neq 0 \end{aligned} \tag{11}$$

then the product $\alpha\beta$ is defined by the formula

$$\alpha\beta = X^{N+M}(c_0 + c_1X + c_2X^2 + \cdots)$$

where

$$c_n = \sum_{j=0}^{n} a_jb_{n-j} \qquad n = 0, 1, 2, \ldots$$

It is easy to see that K is an integral domain. Moreover, K is a field: If $\alpha \neq 0$ (i.e., not every $a_i = 0$), then there exists its inverse β. Indeed, since α has the form (11), it is sufficient to prove that $\alpha' = a_0 + a_1X + \cdots$ has an inverse.

To prove it, let us consider an infinite system of equations with respect to $b_0, b_1, \ldots, b_n, \ldots$:

$$\begin{aligned} &a_0b_0 = 1 \\ &a_0b_1 + a_1b_0 = 0 \\ &\cdots \\ &a_0b_n + \cdots + a_{n-1}b_1 + a_nb_0 = 0 \\ &\cdots \end{aligned} \tag{12}$$

which is easily seen to be solvable. It follows directly that $b_0 + b_1X + b_2X^2 + \cdots$ is the inverse of α'. Define a metric in K: For α given by (11) put $\varphi(\alpha) = e^{-N}$; $d(\alpha,\beta) = \varphi(\alpha - \beta)$ is the required metric; $\varphi(0) = 0$. The metric d induces a field topology on K.

Example 6 Let $K = \mathbb{Q}$ be the rational number field. Let

$$U_\varepsilon = \{x \in \mathbb{Q} : |x| < \varepsilon \text{ and } |x|_p < \varepsilon\} \qquad \varepsilon > 0 \tag{13}$$

where p is a fixed prime number and $|\cdot|_p$ denotes the p-adic absolute value (compare Example 3). It is left to the reader to show that the family (13) can be taken as a base of zero neighborhoods for a suitable field topology on $\mathbb{Q}$.

Example 7 Let K be a field linearly ordered by the relation $\leq$. Recall that by the order absolute value of K we mean a function defined as follows:

$$|a| = \begin{cases} a & \text{if } a \geq 0 \\ -a & \text{otherwise} \end{cases}$$

Now, if ε runs over all positive elements of K, then the sets $U_\varepsilon = \{a \in K : |a| < \varepsilon\}$ can be taken for a base of the zero neighborhoods for some field topology on K, called the *order topology*.

Example 8 (Williamson [1, Ex. I]) Let M(0,1) be the ring of Lebesgue measurable functions $f: I = (0,1) \to \mathbb{R}$. For arbitrary ε, $k > 0$, put

$$N(k,\varepsilon) = \{f \in M(0,1) : m(\{t : |f(t)| \geq k\}) < \varepsilon\} \tag{14}$$

(m is the Lebesgue measure on the real line). Let $\mathcal{B}$ be the family of all sets of the form (14). We shall prove that $\mathcal{B}$ is a base of zero neighborhoods for a ring topology on M(0,1).

To prove that conditions (B_1)-(B_5) are satisfied, we prove that

$$N\left(\frac{k}{2},\frac{\varepsilon}{2}\right) + N\left(\frac{k}{2},\frac{\varepsilon}{2}\right) \subset N(k,\varepsilon) \tag{15}$$

$$\forall\ f \in M(0,1)\ \exists\ k' \qquad fN\left(\frac{k}{k'},\frac{\varepsilon}{2}\right) \subset N(k,\varepsilon) \tag{16}$$

$$N\left(\sqrt{k},\frac{\varepsilon}{2}\right)N\left(\sqrt{k},\frac{\varepsilon}{2}\right) \subset N(k,\varepsilon) \tag{17}$$

To prove (15), note that

$$\{t : |f(t) + g(t)| \geq k\} \subset \left\{t : |f(t)| \geq \frac{k}{2}\right\} \cup \left\{t : |g(t)| \geq \frac{k}{2}\right\}$$

which implies that

$$m(\{t : |f(t) + g(t)| \geq k\}) \leq m\left(\left\{t : |f(t)| > \frac{k}{2}\right\}\right) + m\left(\left\{t : |g(t)| \geq \frac{k}{2}\right\}\right)$$

Thus $f,g \in N(k/2,\varepsilon/2) \Rightarrow f + g \in N(k,\varepsilon)$.

Now let $f \in M(0,1)$. Then there exists k' satisfying

$$m(\{t : |f(t)| \geq k'\}) < \frac{\varepsilon}{2}$$

If $g \in N(k/k',\varepsilon/2)$, then $m(\{t : |g(t)| \geq k/k'\}) < \varepsilon/2$. Since $\{t : |g(t)f(t)| \geq k\} \subset \{t : |f(t)| \geq k'\} \cup \{t : |g(t)| \geq k/k'\}$, then $m(\{t : |g(t)f(t)| \geq k\}) < \varepsilon$, which proves (16). Finally, (17) follows from the inclusion

$$\{t : |f(t)g(t)| \geq k\} \subset \{t : |f(t)| \geq \sqrt{k}\} \cup \{t : |g(t)| \geq \sqrt{k}\}$$

since for $f,g \in N(\sqrt{k},\varepsilon/2)$,

$$\begin{aligned} m(\{t : |f(t)g(t)| \geq k\}) &\leq m(\{t : |f(t)| \geq \sqrt{k}\}) \\ &\quad + m(\{t : |g(t)| \geq \sqrt{k}\}) < \frac{\varepsilon}{2} + \frac{\varepsilon}{2} = \varepsilon \end{aligned}$$

The field $K = \mathbb{R}(t)$ is a subfield of M(0,1) [every function $f: (0,1) \to \mathbb{R}$ defined by a rational function of a variable t is measurable]. We are going to prove that K is a topological field in the topology induced from M(0,1). One need only prove continuity of division in K. Let k, $\varepsilon > 0$, be arbitrary and $f \in K$, $f \neq 0$. Assume (without restricting the generality) that $f(1) = 1$. Put $f(x) = 1 + g(x)$, where $g(1) = 0$. There is an open set $U \subset \mathbb{R}$ such that f is continuous on $I \setminus U$ and $m(U) < \varepsilon/2$. Thus the function $x \longrightarrow x^{-1}$, continuous on the compact set $I \setminus U$ $(0 \notin I \setminus U)$, is uniformly continuous. This implies that there exists $\ell > 0$ such that

$$|1 - f(x)| < \ell \text{ implies } |1 - f(x)^{-1}| < \frac{k}{2}$$

for every $x \in I \setminus U$. Hence

$$|g(x)| < \ell \text{ implies } \left|\frac{g(x)}{1 + g(x)}\right| < \frac{k}{2} \text{ on } I \setminus U$$

or

$$\left\{x \in I \setminus U : \left|\frac{g(x)}{1 + g(x)}\right| \geq \frac{k}{2}\right\} \subset \{x \in I \setminus U : |g(x)| \geq \ell\}$$

We thus have

$$g \in N\left(\ell, \frac{\varepsilon}{2}\right) \Rightarrow m\left(\left\{x \in I : \left|\frac{g(x)}{1 + g(x)}\right| \geq \frac{k}{2}\right\}\right) < \varepsilon$$

or

$$g \in N\left(\ell, \frac{\varepsilon}{2}\right) \Rightarrow \frac{g}{1 + g} \in N(k, \varepsilon) \tag{18}$$

Since $N(\ell/2, \varepsilon/4) + N(\ell/2, \varepsilon/4) \subset N(\ell, \varepsilon/2)$ by (15), putting $v = N(\ell/2, \varepsilon/4)$, $U = N(k, \varepsilon)$, we obtain [applying (18)] that

$$\frac{V}{1 + V} \subset U$$

Hence the division in K is continuous [see (B_6') and Lemma 2].

Example 9 Let A be a Dedekind domain and let $p \neq (0)$ be an arbitrarily fixed prime ideal of A. Let K be the quotient field of A.

Let $a \in A$, $a \neq 0$. Write $(a) = p^N b$, where b is an ideal prime to p, i.e., $(p, b) = (1)$. Put $v_p(a) = N$, $v_p(0) = \infty$. Thus v_p: $A \to \mathbb{Z} \cup \{\infty\}$. It is easily seen that v_p, called the *p-adic exponential valuation*, satisfies (4) and (5). Extending v_p to the quotient field K of A by the formula

$$v_p\left(\frac{a}{b}\right) = v_p(a) - v_p(b) \tag{19}$$

we obtain the *p*-adic exponential valuation on K, which enables us to define the *p-adic absolute value* on K in the same way as in Example 3.

Note, finally, that the definition (19) is independent of the choice of representatives of the fraction a/b: If a/b = c/d, then ad = bc, which implies that $v_p(a) + v_p(d) = v_p(b) + v_p(c)$ and thus

$$v_p\left(\frac{a}{b}\right) = v_p(a) - v_p(b) = v_p(c) - v_p(d) = v_p\left(\frac{c}{d}\right)$$

Example 10 (Gould [1]) Let $K = \mathbb{R}(X)$ be the rational function field in one variable X. At first, we take the sets

$$U_\varepsilon = \left\{f \in \mathbb{R}[X] : f(X) = \sum_{n\geq 0} a_n X^n, \ \forall\, n\ |a_n| < \varepsilon_n\right\}$$

where $\varepsilon = (\varepsilon_n)$ is any sequence of positive numbers as a base of the zero neighborhoods for a certain ring topology of $\mathbb{R}[X]$. Then the sets of the form

$$V(\varepsilon,g) = \frac{gU_\varepsilon}{1 + gU_\varepsilon} \qquad g \in \mathbb{R}[X]$$

define a proper field topology on K. Note that this topology does not extend the ordinary topology of $\mathbb{R}$ to $\mathbb{R}(X)$; indeed, it induces the discrete topology on $\mathbb{R}$.

Example 11 (Mutylin [3]) Let $K = \mathbb{Q}$. We shall show how to construct a field topology on rationals, different from the usual ones (i.e., the p-adic or the ordinary topology on $\mathbb{Q}$).

Let (P_n) be an infinite sequence of prime numbers and (a_{rn}) $(r,\ n \in \mathbb{N},\ n \geq r)$ be any sequence of natural numbers. Both sequences are fixed arbitrarily, but in the sequel will be suitably specified. Put

$$U_n = \left\{\sum_{r=n}^{\infty} \frac{k_r}{\ell_r} P_r \in \mathbb{Q} : k_r, \ell_r \in \mathbb{Z},\ k_r = 0 \text{ for sufficiently} \right.$$
$$\left. \text{large } r\text{'s},\ \left|\frac{k_r}{\ell_r}\right| \leq a_{rn},\ |\ell_r| \leq 2^{3^{r-n}}\right\} \tag{20}$$

It is obvious that all U_n's are symmetric. We shall need the U_n's to satisfy stronger conditions than (B_1)-(B_5), namely,

$$\bigcap_n U_n = \{0\} \tag{21}$$

$$U_{n+1} + U_{n+1} \subset U_n \tag{22}$$

$$U_{n+1}U_{n+1} \subset U_n \tag{23}$$

$$\frac{1}{m} U_{n+a} \subset U_n \qquad \text{for every } a,m \in \mathbb{N} \text{ satisfying } 2^a \geq m \tag{24}$$

It can be shown (Mutylin [3]) that condition

$$a_{rn} \geq \max\left\{2a_{r\ n+1},\ a_{r\ n+1}\left(2 \sum_{s=n+1}^{r-1} a_{s\ n+1}|P_s| + a_{r\ n+1}|P_r|\right)\right\}$$

guarantees satisfying (22) and (23). To obtain a Hausdorff topology, one has to assume that

$$\frac{|P_{r+1}|}{2\cdot 2^{3^r}} \geq \sum_{s=1}^{r} a_{s1}|P_s|$$

In this way we have obtained a proper ring topology on $\mathbb{Q}$, not equal to either the ordinary order topology or to a p-adic topology. This topology can be weakened to a field topology, saving the above-mentioned properties (see Theorem 4).

For a more general construction, see Zobel [1].

Example 12 (Endo [4]; Kowalsky [2]) Let $K = \mathbb{Q}$ and let (p_n) be an infinite sequence of prime numbers. Denote by $\mathcal{T}_p$ the *p-adic topology* on $\mathbb{Q}$ (see Example 3) and by $\mathcal{B}_p$ a corresponding base of neighborhoods of zero. Let $\mathcal{B}$ be the set of all intersections $U_1 \cap \cdots \cap U_n$, where $U_i \in \mathcal{B}_{p_i}$. The family $\mathcal{B}$ defines a field topology on $\mathbb{Q}$.

Other examples of topological fields will appear in subsequent chapters.

3. THE LATTICE $\mathcal{L}_r(K)$ OF RING TOPOLOGIES FOR A FIELD K

For two topologies $\mathcal{T}_1$ and $\mathcal{T}_2$ on a set X we write $\mathcal{T}_1 \leq \mathcal{T}_2$ if and only if every $\mathcal{T}_1$-open set is $\mathcal{T}_2$-open and say that $\mathcal{T}_2$ is *stronger* than $\mathcal{T}_1$ (or $\mathcal{T}_1$ is *weaker* than $\mathcal{T}_2$). If, moreover, $\mathcal{T}_1 \neq \mathcal{T}_2$, we say that $\mathcal{T}_2$ is *properly stronger* than $\mathcal{T}_1$ (or $\mathcal{T}_1$ is *properly weaker* than $\mathcal{T}_2$) and write $\mathcal{T}_1 < \mathcal{T}_2$. Thus $\mathcal{T}_1 \leq \mathcal{T}_2$ if and only if the identity mapping $(X,\mathcal{T}_2) \to (X,\mathcal{T}_1)$ is continuous. Obviously, the discrete topology $\mathbf{1}_X$ is the strongest and the trivial (or antidiscrete) topology $\mathbf{0}_X$ is the weakest among all topologies on X.

Let X = K be a field. Let $\mathcal{L}_f(K)$ [resp. $\mathcal{L}_r(K)$] denote the set of field (resp. ring) topologies on the field K. Obviously, $\mathcal{L}_f(K) \subset \mathcal{L}_r(K)$. If $\mathcal{T}_1,\mathcal{T}_2 \in \mathcal{L}_f(K)$ [$\mathcal{L}_r(K)$], there exists a uniquely determined weakest field [ring] topology on K, stronger than $\mathcal{T}_1$ and $\mathcal{T}_2$. We shall denote it by $\mathcal{T}_1 \vee \mathcal{T}_2$ [or $\sup(\mathcal{T}_1,\mathcal{T}_2)$] and call it the *supremum* of $\mathcal{T}_1$ and $\mathcal{T}_2$. Similarly, there exists the strongest ring topology among topologies on K, which is weaker than $\mathcal{T}_1$ and $\mathcal{T}_2$. We shall call it the *infimum* of $\mathcal{T}_1$ and $\mathcal{T}_2$ and denote it by $\mathcal{T}_1 \wedge \mathcal{T}_2$ [or $\inf(\mathcal{T}_1,\mathcal{T}_2)$]. If we talk simultaneously about $\mathcal{L}_f(K)$ and $\mathcal{L}_r(K)$, we shall add to the signs $\vee$ and $\wedge$ indices f (resp. r).

Similarly, $\mathcal{L}_r(A)$ denotes the lattice of all ring topologies on a ring A.

DEFINITION 4 A *field (ring) topology* $\mathcal{T}$ on K is said to be *minimal* if the trivial topology on K is the only ring topology properly weaker than $\mathcal{T}$. Similarly, we say that a field (ring) topology $\mathcal{T}$ on K is maximal if the only field (ring) topology properly stronger than $\mathcal{T}$ is the discrete topology.

It can easily be proved, applying the Kuratowski-Zorn lemma,*

*This theorem was proved for the first time by K. Kuratowski (Une méthode d'élimination des nombres transfinis des raisonnements mathematiques, *Fund. Math.* 3 (1922), 77-108; especially p. 89). The importance of this theorem and its applications was shown by M. Zorn (A remark on method in transfinite algebra, *Bull. Amer. Math. Soc.* 41 (1935), 667-670).

that for every nondiscrete field (ring) topology $\mathcal{T}$ on a field K there exists a maximal field (ring) topology $\mathcal{S}$ on K, stronger than $\mathcal{T}$.

It is not known whether for every proper field (ring) topology $\mathcal{T}$ on a field K there exists a minimal field (ring) topology $\mathcal{S}$ on K, weaker than $\mathcal{T}$.

THEOREM 3 Let K be any field. Then $\mathcal{L}_r(K)$ is a complete lattice and $\mathcal{L}_f(K)$ is a complete sublattice of $\mathcal{L}_r(K)$.

If X is a set and $(\mathcal{T}_i)_{i\in I}$ a family of topologies on X, then $\sup_{i\in I} \mathcal{T}_i$ and $\inf_{i\in I} \mathcal{T}_i$ (in the lattice of all topologies on X) always exist (see Bourbaki [1, Chap. I, Secs. 2-4]).

In particular, when X = K is a field and $\mathcal{T}_i \in \mathcal{L}_r(K)$ we must define bases of zero neighborhoods only for $\sup_{r_{i\in I}} \mathcal{T}_i$ and $\inf_{r_{i\in I}} \mathcal{T}_i$. First, let

$$\mathcal{B} = \{\bigcap_{i\in F} U_i : F \subset I \text{ runs over all finite subsets, } U_i \in \mathcal{B}_i\} \tag{25}$$

where $\mathcal{B}_i$ is a base of zero neighborhoods for $\mathcal{T}_i$ $(i \in I)$.

LEMMA 4 (Endo [4, Th. 3, p. 13]) Let K be a field.

(i) If $\mathcal{T}_i \in \mathcal{L}_r(K)$ for every $i \in I$, then $\sup_{r_{i\in I}} \mathcal{T}_i$ exists and $\mathcal{B}$, defined by (25), is its base of the neighborhoods of zero.

(ii) If $\mathcal{T}_i \in \mathcal{L}_f(K)$ for every $i \in I$, $\sup_{f_{i\in I}} \mathcal{T}_i$ exists and $\mathcal{B}$ [see (25)] is its base of zero neighborhoods.

Remark Although "sup" in $\mathcal{L}_r$ coincides with "sup" in the lattice of all topologies, "inf" does not.

Proof: It is clear that every set of $\mathcal{B}$ is open in $\sup_{i\in I} \mathcal{T}_i$ [the supremum is taken in the lattice $\mathcal{L}(K)$ of all topologies on K]. We shall prove that $\sup \mathcal{T}_i \in \mathcal{L}_r(K)$ in case (i) and $\sup \mathcal{T}_i \in \mathcal{L}_f(K)$ in case (ii). Thus it remains to show that $\mathcal{B}$ satisfies conditions (B_1)-(B_5) if $\mathcal{T}_i \in \mathcal{L}_r(K)$. For condition (B_1) this is obvious. If $U = \bigcap_{i\in F} U_i$, $V = \bigcap_{j\in G} V_j$, where F and G are finite subsets of I, then taking $W_i \subset V_i \cap V_i$ for $i \in F \cap G$ and $W_i = U_i$ for $i \in F \setminus G$, $W_j = V_j$ for $j \in G \setminus F$ and putting $W = \bigcap_{i\in F\cap G} W_i$, we obtain (B_2).

Let $U = \bigcap_{i\in F} U_i$. Take $V_i \in \mathcal{B}_{T_i}$, satisfying $V_i + V_i \subset U_i$. Then $V = \bigcap_{i\in F} V_i$ satisfies (B_3). One proves similarly that (B_4) and (B_5) hold for $\mathcal{B}$.

Now let $T_i \in L_f(K)$ for every $i \in I$. We shall prove the validity of (B_6'). Let $U = \bigcap_{i\in F} U_i$, $F \subset I$ finite, $U_i \neq K$. Take V_i satisfying (B_6') for every $i \in F$ and let $V = \bigcap_{i\in F} V_i$. Then for $a,b \in V$,

$$\frac{a}{1+b} \in \frac{V_i}{1+V_i} \subset U_i \quad \text{for every } i \in F, \text{ i.e., } \frac{V}{1+V} \subset U \text{ and } V \in \mathcal{B}$$

Proof of Theorem 3: We recall a standard result: If a partially ordered set X has a least (greatest) element and for every A, $\emptyset \neq A \subset X$, sup A (inf A) exists, then X is a complete lattice (L. S. Skornjakov [1]). Indeed, if $\emptyset \neq A \subset X$ and sup A exists, then $\inf A = \sup\{x \in X : \forall\, a \in A,\ x \leq a\}$. This and Lemma 4 prove Theorem 3.

THEOREM 4 (Gelbaum, Kalisch, and Olmstead [1, Th. 9]) Let K be a field and $T \in L_r(K)$ be a proper topology. Then there exists $T' \in L_f(K)$ satisfying $\mathbf{0}_K \neq T' \leq T$. Moreover, $T' = \sup\{S \in L_f(K) : S \leq T\}$ is a field topology.

Proof: Let $\mathcal{B} = \mathcal{B}(T)$ be a base of symmetric neighborhoods of zero. Let S be the topology defined by the family $\mathcal{B}(S)$ of all sets of the form $U/[1 + (U \setminus \{-1\})]$, where $U \in \mathcal{B}$. We shall prove that $\mathcal{B}(S)$ satisfies conditions (B_1)-(B_6) of Theorem 1, i.e., $S \in L_f(K)$. Moreover, since $U \subset U/[1 + (U\setminus\{-1\})]$ for every $U \in \mathcal{B}$, $S \leq T$. First note that $\mathbf{0}_K \neq S$, i.e., S is nontrivial (see Lemma 1). Indeed, let $x \in K^\times$ be arbitrary. Take $U \in \mathcal{B}$ with $x \notin U$. There exist $V,W \in \mathcal{B}$ satisfying $V + V \subset U$ and $xW \subset V$. We can suppose that $W \subset V$, taking $W \cap V$ instead of W, if necessary. Then $x \notin xW + W$, i.e., $(x + xW) \cap W \neq \emptyset$, since $W = -W$. It means that $x \notin W/[1 + (W\setminus\{-1\})]$ and this proves that S is a Hausdorff topology. Put $\tilde{U} = U/[1 + (U\setminus\{-1\})]$ for $U \in \mathcal{B}$. To prove (B_2), take arbitrary $U,V \in \mathcal{B}$ and $W = U \cap V$. Then $\tilde{W} \subset \tilde{U}$ and $\tilde{W} \subset \tilde{V}$, i.e., $\tilde{W} \subset \tilde{U} \cap \tilde{V}$.

Let $U \in \mathcal{B}$ be arbitrary. Then, applying Theorem 1, one can find $V,W \in \mathcal{B}$ satisfying $V + V + V + V \subset U$, $WW \subset V$, and $W \subset V$. This implies that

$$\tilde{W} + \tilde{W} \subset \frac{W + W + W(W \setminus \{-1\}) + W(W \setminus \{-1\})}{1 + (W \setminus \{-1\}) + (W \setminus \{-1\}) + (W \setminus \{-1\})(W \setminus \{-1\})}$$

$$\subset \frac{U}{1 + (U \setminus \{-1\})} = \tilde{U}$$

giving (B_3). To prove (B_4) it is necessary to find for every $U \in \mathcal{B}$ the neighborhoods $V, W \in \mathcal{B}$ satisfying $V + V + V \subset U$, $W \subset V$, and $WW \subset V$. It is easily seen that $\tilde{W} + \tilde{W} \subset \tilde{U}$.

Let $U \in \mathcal{B}$ and $x \in K^{\times}$. Then there exists $V \in \mathcal{B}$ with $xV \subset U$. We can suppose, moreover, that $V \subset U$, taking $V \cap U$ instead of V, if necessary. Then

$$x\tilde{V} = \frac{xV}{(1 + (V \setminus \{-1\})} \subset \frac{U}{1 + (U \setminus \{-1\})} = \tilde{U}$$

which gives (B_5). It remains to prove the continuity of inversion in $(K, \mathcal{S})$, i.e., a validity of (B_6). Let $U \in \mathcal{B}$ be arbitrary and let V satisfy $V + V \subset U$. Note that for arbitrary $a, b \in K$,

$$\left(1 + \frac{a}{1 + b}\right)^{-1} = \frac{1 + b}{1 + a + b} = 1 - \frac{a}{1 + a + b}$$

Taking $a \in V$ and $b \in V \setminus \{-1\}$, this implies that

$$(1 + \tilde{V})^{-1} \subset 1 + \frac{V}{1 + V + (V \setminus \{-1\})} \subset 1 + \frac{U}{1 + (U \setminus \{-1\})} = \tilde{U}$$

The last statement of Theorem 4 follows from Lemma 4(ii).

DEFINITION 5 A topological ring (or a field) $(K, \mathcal{T})$ is called *additively generated* if K^+ is the only open subgroup of the additive group K^+ of K.

LEMMA 5 Let $(K, \mathcal{T})$ be a minimal topological field. Then either it has a base of the neighborhoods of zero consisting of open subgroups of K^+ or $(K, \mathcal{T})$ is additively generated.

Proof: Let $gp(U)$ be the subgroup of K^+ generated by $U \subset K$. Suppose that $(K, \mathcal{T})$ is minimal and is not additively generated. Let $H < K^+$ be an open proper subgroup. Let $\mathcal{B}$ be a base of neighborhoods of zero for $(K, \mathcal{T})$. It is easy to see that $\mathcal{B}_1 = \{gp(U) : U \in \mathcal{B}\}$ defines a ring topology $\mathcal{T}_1$ on K, $\mathcal{T}_1 \le \mathcal{T}$. The topology $\mathcal{T}_1$ is nontrivial since there exists $U \subset H$. It implies that $gp(U) < H$. Moreover, there exists $\mathcal{T}_2 \in \mathcal{L}_f(K)$, satisfying $\mathbf{0}_K \neq \mathcal{T}_2 \leq \mathcal{T}_1$, by Theorem 4. Since $\mathcal{T}$ is a minimal topology, this implies that $\mathcal{T} = \mathcal{T}_1 = \mathcal{T}_2$. In particular, $\mathcal{B}_1$ is a base for $\mathcal{T}$ and consists of open additive subgroups of K^+.

4. EXISTENCE OF FIELD TOPOLOGIES ON INFINITE FIELDS

We are going to prove that every infinite field admits a proper field topology. We shall use the following general

THEOREM 5 (Arnautov [3, Cor. 1]) A countable, not necessarily commutative ring A admits a proper ring topology if and only if for every finite set of polynomials $f_1, \ldots, f_n \in A[X]$ satisfying $f_i(0) \neq 0$ for $i = 1, 2, \ldots, n$ there exists a nonzero $a \in A$ such that $f_i(a) \neq 0$ for $i = 1, 2, \ldots, n$.

Before giving a proof we shall introduce some notation. Assume that A is a ring, not necessarily commutative. Let V_i and S_{i-1} ($i \in \mathbb{N}$) be symmetric subsets of A. Let us put

$$f_1(S_1,V_1) = S_1V_1 + V_1S_1 + V_1V_1 + V_1 + V_1$$

Suppose that we have defined $f_k(S_1,\ldots,S_k,V_1,\ldots,V_k)$. Then define

$$f_{k+1}(S_1,\ldots,S_{k+1},V_1,\ldots,V_{k+1})$$
$$= f_1(S_1,V_1 \cup f_k(S_2,\ldots,S_{k+1},V_2,\ldots,V_{k+1}))$$

$$\varphi_k(S_0,\ldots,S_k,V_1,\ldots,V_k)$$
$$= \left\{x \in A : \bigcup_{i=0}^{k-1} f_{k-i}(S_{i+1},\ldots,S_k,V_{i+1},\ldots,V_k \cup \{\pm x\}) \cap S_i \neq \emptyset\right\}$$

It can be verified that the sets $f_k(\cdot)$ and $\varphi_k(\cdot)$ have the following properties:

LEMMA 6 (i) $f_k(X_1,\ldots,S_k,\{0\},\ldots,\{0\}) = \{0\}$.
(ii) f_k and φ_k are monotone functions of sets.
(iii) If all V_i and S_i are finite, then $f_k(S_1,\ldots,S_k,V_1,\ldots,V_k)$ is also finite.

(iv) If $0 \in V_i$ for every $i \in \mathbb{N}$, then $f_k(S_1,\dots,S_k,V_1,\dots,V_k) = f_{k+n}(S_1,\dots,S_{k+n},V_1,\dots,V_k,\{0\},\dots,\{0\})$ holds for every $n \in \mathbb{N}$.

(v) $f_k(S_1,\dots,S_k,V_1 \cup f_{k-1}(S_2,\dots,S_k,V_2,\dots,V_k),\dots,V_{k-1} \cup f_1(S_k,V_k),V_k) = f_k(S_1,\dots,S_k,V_1,\dots,V_k)$ holds for every $k \geq 2$.

Let $A[X]$ be the ring of noncommuting variable X over a ring A, i.e., the set of all linear combinations of monomials $a_1X^{i_1}a_2X^{i_2}\cdots a_NX^{i_N}$, where $a_j \in A$ and i_j is a nonnegative integer and $a = aX^0 = X^0a$ for $a \in A$.

LEMMA 7 Let S_i, V_i $(i = 1, \dots, k)$ and S be finite symmetric subsets of A such that $0 \in V_i$ and $f_k(S_1,\dots,S_k,V_1,\dots,V_k) \cap S = \emptyset$. Then there exists a finite set of polynomials $A_j \in A[X]$, $A_j(0) \neq 0$ for every j, such that every $a \in A$ satisfying

$$f_k(S_1,\dots,S_k,V_1,\dots,V_{k-1},V_k \cup \{\pm a\}) \cap S \neq \emptyset$$

is a zero of a suitable A_j.

Proof: Indeed, by Lemma 6(iii), $f_k(S_1,\dots,S_k,V_1,\dots,V_{k-1},V_k \cup \{\pm X\})$ is a finite subset of $A[X]$ consisting of, say, $Q_1, \dots, Q_n$. Let $b_j = Q_j(0)$. Then $b_j \notin S$, since

$$\begin{aligned} b_j &\in f_k(S_1,\dots,S_k,V_1,\dots,V_{k-1},V_k \cup \{0\}) \\ &= f_k(S_1,\dots,S_k,V_1,\dots,V_k) \end{aligned}$$

Let $S = \{c_i : i = 1, \dots, r\}$. Let us consider polynomials $P_{ji} = Q_j - c_i$. The set of all polynomials P_{ji} has the desired properties. Indeed, let $a \in A$ satisfy

$$f_k(S_1,\dots,S_k,V_1,\dots,V_{k-1},V_k \cup \{\pm a\}) \cap S \neq \emptyset$$

Then there exist i and j with $Q_j(a) = c_i$, i.e., $P_{ji}(a) = Q_j(a) - c_i = 0$.

Now let A be a countable ring. Let $A = \{0,a_1,a_2,\dots\}$ and $S_k = \{\pm a_1,\pm a_2,\dots,\pm a_k\}$. We define sets V_{ij} and S_{ij} $(i,j \geq 0)$ by induction. Put $V_{0j} = \{0\}$, $S_{0j} = S_j$ for every j. Suppose that V_{ij} and S_{ij} are defined for all j's and all $i \leq k$. If $k + 1$ is even, define

$S_{k+1,j} = S_{kj}$ for all j's and $V_{k+1,j} = V_{kj} \cup f_{k-1+1}(S_{j+1},\dots,S_{k+1}, V_{k,j+1},\dots,V_{k,k+1})$ if $j \le k$, $V_{k+1,j} = \{0\}$ if $j > k$. If $k + 1$ is odd and $a \in A$, then put $V_{k+1,0} = V_{k0} \cup \{\pm a\}$, $S_{k+1,0} = \emptyset$. Suppose that $S_{k+1,j}$ and $V_{k+1,j}$ are defined for $j \le m$. Then we define

$$S_{k+1,m+1} = S_{m+1} \cup \varphi_{m+1}(S_0,\dots,S_{m+1},V_{k+1,1},\dots,V_{k+1,m},V_{k,m+1})$$

Let $a \in A \setminus S_{k+1,m+1}$. Put

$$V_{k+1,m+1} = \begin{cases} V_{k,m+1} \cup \{\pm a\} & \text{if } m < k \\ \{0\} & \text{if } m \ge k \end{cases}$$

If $S_{k+1,m+1} = A$, then $V_{k+1,m+1} = V_{k,m+1}$. Finally, let $W_j = \bigcup_{i\ge 0} V_{ij}$.

Properties of the sets V_{ij} are formulated in the following.

LEMMA 8 (i) $f_m(S_{j+1},\dots,S_{j+m},V_{i,j+1},\dots,V_{i,j+m}) \subset f_{i-j-1}(S_{j+1},\dots, S_{i-1},V_{i,j+1},\dots,V_{i,i-1})$ holds for all m's and $j < i - 1$.

(ii) $f_m(S_{j+1},\dots,S_{j+m},V_{i,j+1},\dots,V_{i,j+m}) \subset V_{i+1,j}$.

(iii) $f_{k-\ell+1}(S_{\ell+1},\dots,S_{k+1},V_{k+1,\ell+1},\dots,V_{k+1,\ell+m},V_{k,\ell+m+1},\dots, V_{k,k+1}) \subset f_m(S_{\ell+1},\dots,S_{\ell+m},V_{k+1,\ell+1},\dots,V_{k+1,\ell+m})$ holds for every k, ℓ, m satisfying $m + \ell \le k$, $m \ge 1$.

(iv) V_{ij} are finite symmetric sets.

(v) $f_m(S_{j+1},\dots,S_{j+m},V_{i,j+1},\dots,V_{i,j+m}) \cap S_j = \emptyset$ holds for every i, j, m.

(vi) For every i, j there exists a finite set Σ_{ij} of polynomials f satisfying $f(0) \ne 0$ such that every element $s \in S_{ij}$ is a zero of a polynomial of Σ_{ij}.

Proof: We shall prove only (vi). Let $i = 0$. Since $S_{0j} = S_j$ does not contain 0 and is finite, the set $\Sigma_j = \{X - a : a \in S_j\}$ has the desired properties.

Let $i \ge 1$. Since $S_{ij} = S_{i-1,j}$ if i is even, we take an odd i. Then, by definition,

$$S_{ij} = S_j \cup \varphi_j(S_0,\dots,S_j,V_{i1},\dots,V_{i,j-1},V_{i-1,j})$$

Since $S_k \cap f_{j-k}(S_{k+1},\dots,S_j,V_{i,k+1},\dots,V_{i,j-1},V_{i-1,j)} = \emptyset$ by (v), then applying Lemma 7 we can find a finite subset Σ_{kij} of $A[X]$, consisting of f's with $f(0) \neq 0$, such that every $a \in A$ satisfying

$$S_k \cap f_{j-k}(S_{k+1},\ldots,S_j,V_{i,k+1},\ldots,V_{i,j-1},V_{i-1,j} \cup \{\pm a\}) = \emptyset$$

is a zero of a polynomial of Σ_{kij}. Taking $\Sigma_{ij} = \Sigma_j \cup (\bigcup_{k=0}^{j-1} \Sigma_{kij})$, the definition of

$$\varphi_j(S_0,\ldots,S_j,V_{i1},\ldots,V_{i,j-1},V_{i-1,j})$$

implies that every element of S_{ij} is a zero of a polynomial $g \in \Sigma_{ij}$.

LEMMA 9 If A is a countable ring, then the family of sets $W_i = \bigcup_{k\geq 0} V_{ki}$, $i \geq 0$, can be taken for a base of neighborhoods of zero defining a ring topology (perhaps discrete) of A.

We omit an easy proof.

Now we are ready to prove Theorem 5.

Proof of Theorem 5: Assume that for every finite set of polynomials $f_i \in A[X]$, $f_i(0) \neq 0$ $(i = 1, \ldots, n)$, there exists a nonzero element $a \in A$ satisfying $f_i(a) \neq 0$ $(i = 1, \ldots, n)$. Then by Lemma 8(vi) there exists $a_{ij} \notin S_{ij}$. Consequently, we can take $V_{ij} = V_{i-1,j} \cup \{\pm a_{ij}\}$ if i is odd. It means that $W_j = \bigcup_{i\geq 0} V_{ij} \neq \{0\}$; i.e., the topology defined as in Lemma 9 is nondiscrete.

Conversely, suppose that there exists a proper ring topology on A. Let $f_i(X) = P_i(X) + b_i$, $P_i(0) = 0$ $(i = 1, \ldots, n)$. Then there exists a neighborhood V of zero such that $-b_i \notin V$ for every $i = 1, \ldots, n$. Since algebraic operations of A are continuous, there exists a neighborhood U of zero satisfying $P_i(U) \subset V$ $(i = 1, \ldots, n)$. As the topology of A is nondiscrete, there exists a nonzero element $a \in U$. Then $P_i(a) + b_i \in V + b_i$. Since $-b_i \notin V$, we obtain finally that $0 \notin V + b_i$; i.e., $P_i(a) + b_i \neq 0$ holds for every $i = 1, \ldots, n$.

THEOREM 6 Every infinite field K admits a proper field topology. Furthermore, $|L_f(K)| = 2^{2^{|K|}}$.

Proof: We shall prove only the first statement of Theorem 6. The second will be proved in subsequent chapters. Let K be an infinite field. We can write $K = P(B)(A)$, where P is the prime field of K, B a transcendency basis of K over P, and A the family of all elements of K, algebraic over the field P(B).

Case 1. $B \neq \emptyset$. We shall prove that there exists an absolute value on K, defining a proper field topology. Let $b \in B$ be arbitrary and put $k = P(B \setminus \{b\})$. Then $K = k(b)(A)$. Defining the b-adic absolute value ω_b on k(b) (Example 3) we can extend it from k(b) to K, since ω_b is non-Archimedean and K is algebraic over k(b) (see Section 5).

Case 2. $B = \emptyset$. In the case when K has zero characteristic, i.e., if P is the rational number field $\mathbb{Q}$, there is nothing to do, since then $K = \mathbb{Q}(A)$ is a subfield of the complex number field $\mathbb{C}$ and the usual absolute value induces a nondiscrete and nontrivial topology on K.

In the case when K is an infinite algebraic extension of a finite field $\mathbf{F}_p$, where p is the characteristic of K, there exists no nondiscrete absolute value on K. Indeed, every element $a \in K^\times$ satisfies an equation of the form $a^q = a$, where q is a power of p. If w were an absolute value on K, we would have $w(a^q) = w(a)^q = w(a)$, or $w(a)^{q-1} = 1$, implying that $w(a) = 1$ for every $a \in K^\times$.

Thus let K be an infinite subfield of an algebraic closure of the finite field $\mathbf{F}_p$. Applying Theorem 5 to our case, we find that there is a proper ring topology $\mathcal{T}$ on K. Applying Theorem 4, we can find a proper field topology $\mathcal{T}_1 \leq \mathcal{T}$.

For a proof of the last statement of Theorem 6, see Kiltinen [4, Cor. 2.2].

Remark 1 One can, instead of Theorem 5, use the following result: Every countable commutative ring with 1 admits a proper ring topology (Arnautov [6]).

Kiltinen [1] proved directly that every infinite field admits a proper field topology.

Remark 2 Taĭmanov [2] gave an alternative proof of existence of proper field topologies on every countable field.

5. LOCALLY BOUNDED AND LOCALLY UNBOUNDED TOPOLOGIES

Let K be a field and $\mathcal{T}$ a ring topology on K. Let $\mathcal{B}$ be a base of zero neighborhoods for $\mathcal{T}$.

DEFINITION 6 Let K be a field and $\mathcal{T} \in L_r(K)$, $A \subset K$. we say that A is bounded if for every $U \in \mathcal{B}$ there exists $V \in \mathcal{B}$ such that $AV \subset U$.

LEMMA 10 In a topological field (ring) every compact set is bounded.

Proof: Let A be compact in $(K,\mathcal{T})$. The continuity of multiplication implies that for every $x \in A$ there exist $V_x \in \mathcal{B}$ and a neighborhood U_x of x, with $U_x V_x \subset U$. Since A is compact, there exists a finite set of elements $x_1, \ldots, x_n \subset A$ with $U_{x_1} \cup U_{x_2} \cup \cdots \cup U_{x_n} \supset A$. Taking

$$V = V_{x_1} \cap V_{x_2} \cap \cdots \cap V_{x_n}$$

we get $AV \subset U$.

We now list some properties of bounded sets.

LEMMA 11 Let $(K,\mathcal{T})$ be a topological field (or topological ring). Then:

(i) A subset of a bounded set is bounded.

(ii) The closure of a bounded set is bounded.

(iii) If A and B are bounded subsets of K, then $A \cup B$, AB, $A + B$ are also bounded.

(iv) Every convergent sequence is bounded.

(v) If $x_n \to 0$ and (y_n) is a bounded sequence, then $x_n y_n \to 0$.

Proof: Straightforward.

DEFINITION 7 A topological ring $(K,\mathcal{T})$ is called *locally bounded* if there exists a bounded and open subset A of $(K,\mathcal{T})$, $A \neq \emptyset$.

The notion of boundedness was introduced by Bourbaki (see also Šafarevič [1]) and studied first by Kaplansky [2] and Weiss [1]. For other properties of locally bounded topological rings, see Więsław [11]. All topological fields in Examples 1-7 are easily seen to be locally bounded. Moreover, the p-adic topology in Example 9 is also locally bounded.

DEFINITION 8 A topological ring $(K,\mathcal{T})$ is called *locally unbounded* if it is not locally bounded.

Topological fields in Examples 8 and 10-12 are locally unbounded. The fact that the topological field in Example 8 is locally unbounded can be proved either directly or can be deduced from the following result: The only locally bounded extensions of the reals (with the ordinary topology) are the real and the complex number fields (see Chapter 9, Section 4).

Local unboundedness of the topological field in Examples 10 and 11 was proved by Gould [1, p. 278] (resp. Mutylin [3]).

Now we consider Example 12. Let $U \in \mathcal{B}$ be arbitrary. The neighborhood U has the form

$$U = \{x \in \mathbb{Q} : v_{p_i}(x) \geq n_i \text{ for } i = 1, 2, \ldots, s\} \tag{26}$$

Let $p_0 \neq p_i$ for $i = 1, 2, \ldots, s$ be a prime number and let $V = \{x \in Q : v_{p_0}(x) \geq 1\}$. Take arbitrary W of the form (26), but with $q_1, \ldots, q_t$ instead of p_i's and $m_1, \ldots, m_t$ instead of n_i's. Then $x = a^{-1}q_1^{m_1}\cdots q_t^{m_t} \in W$, where $a \in \mathbb{Z}$ and $(a,p_i) = (a,q_j) = 1$ for $i = 1, \ldots, s$ and $j = 1, \ldots, t$. Taking $y = p_0^{-N}p_1^{n_1}\cdots p_s^{n_s}$ (N sufficiently large, $N \in \mathbb{N}$), we have

$$xy = a^{-1}p_0^{-N}p_1^{n_1}\cdots p_s^{n_s}q_1^{m_1}\cdots q_t^{m_t} \notin V$$

since $v_{p_0}(xy) = -N$ if $p_0 \neq q_j$ for all j's and $v_{p_0}(xy) = m_j - N$ otherwise. Thus $UW \not\subset V$.

Heine [2] proved that every field K which fails to be an algebraic extension of a finite field admits at least $2^{|K|}$ locally unbounded field topologies (for a proof, see Chapter 13, Theorem 4).

LEMMA 12 Let K be a field and $\mathcal{T} \in L_r(K)$, $\mathcal{T}$ be nondiscrete. Then the following conditions are equivalent:

(i) $A \subset K$ is bounded in $(K,\mathcal{T})$.

(ii) $\forall\, U \in \mathcal{B}\ \exists\, x \in K^\times$, $Ax \subset U$.

Proof: Since (i) ⇒ (ii) is trivial, assume that (ii) holds. Let $U \in \mathcal{B}$ be arbitrary. There exists $U_1 \in \mathcal{B}$ with $U_1U_1 \subset U$, by (B_4). By assumption (ii), there is an $x \in K^\times$ with $Ax \subset U_1$. Thus $U_1Ax \subset U_1U_1 \subset U$. Applying (B_5), we find that $V \in \mathcal{B}$ with $Vx^{-1} \subset U_1$. Then

$$AV \subset AU_1x \subset U_1U_1 \subset U$$

It will sometimes be useful to use condition (ii) instead of Definition 6.

DEFINITION 9 Let K be a field. A subset R of K is called an *almost-order* if the following conditions are satisfied:

(O_1) 0, 1, -1 $\in R$, $R \neq K$.

(O_2) $RR \subset R$.

(O_3) $K^\times = R^\times(R^\times)^{-1}$ where $R^\times = R \setminus \{0\}$.

(O_4) $\exists\, z \in R^\times$, $z(R \pm R) \subset R$ (such a z is called a *2-addiator*).

An almost-order R of K is called a *full almost-order* if, moreover,

(O_5) $\forall\, x \in K^\times$, $x \in R$, or $x^{-1} \in R$.

An almost-order R is an *order* if $z = 1$ is possible in (O_4), i.e., if R is a subring of K with identity and with quotient K.

DEFINITION 10 Let R and S be two almost-orders of a field K. We say that R and S are *equivalent* if there exist elements $a,b \in K^\times$ with $aR \subset S$ and $bS \subset R$.

It is not difficult to see that this relation is an equivalence relation on the set of almost-orders of K.

THEOREM 7 (Kowalsky and Dürbaum [1]) Let K be a field. Then there is a one-to-one correspondence between the proper locally bounded ring topologies on K and the classes of equivalent almost-orders of K.

Proof: We shall prove that for every locally bounded $\mathcal{T} \in \mathcal{L}_r(K)$ there exists an almost-order R defining the topology, i.e., R such that

$$\{aR : a \in K^\times\} \tag{27}$$

is a base of the neighborhoods of zero for $\mathcal{T}$, and conversely, every almost-order R defines a ring topology on K, for which (27) is a base of zero neighborhoods.

Let $(K,\mathcal{T})$ be locally bounded and $U \subset K$ be a bounded neighborhood of zero. Put

$$R = \{x \in K : xU \subset U\} \tag{28}$$

We prove that R is bounded, $R \in \mathcal{B}$, R induces $\mathcal{T}$ and is an almost-order. Since U is a bounded neighborhood of zero, there exists $V \in \mathcal{B}$ satisfying $UV \subset U$. Thus we have $V \subset R$. Moreover, the inclusion $RU \subset U$ implies that for every $u \in U$, $u \neq 0$, $Ru \subset U$, i.e., $R \subset u^{-1}U$, proving that R induces $\mathcal{T}$. It remains to prove that R is an almost-order. Condition (O_1) obviously holds. Let $x,y \in R$. Then $(xy)U = x(yU) \subset xU \subset U$, proving (O_2).

To prove (O_3), take $x \in K^\times$. Then there exists $V \in \mathcal{B}$ satisfying $xV^\times \subset R^\times$. If $u \in V^\times \cap R^\times$, then $x \in R^\times u^{-1} \subset R^\times(R^\times)^{-1}$.

To obtain (O_4), take $V \in \mathcal{B}$ with $V + V \subset R$. Since R is bounded, one can find $z \in K^\times$ with $zR \subset V$. Finally,

$$z(R \pm R) \subset zR + zR \subset V + V \subset R$$

Since $0,1 \in R$, then $z = (1 + 0)z \in z(R + R) \subset R$; hence $z \in R^\times$.

Now let R be an almost-order of a field K. We shall prove that the sets (27) form a base of zero neighborhoods for some ring topology $\mathcal{T}$ on K. It is obvious that if R and S are equivalent almost-orders, the topologies defined by them coincide.

Suppose that (B_1) does not hold: There exists $x \neq 0$, $x \in \bigcap_{a \in K^\times} aR$. In particular, for $a \in R^\times$, $x \in aR \subset R$ by (O_2). By assumption, for every $a \in K^\times$ there exists $r_a \in R^\times$ with $x = ar_a$. This implies that $a = xr_a^{-1} \in x(R^\times)^{-1}$, or $K^\times \subset x(R^\times)^{-1}$. Since $x^2 \in K^\times$, it has the form $x^2 = x/b$, $b \in R^\times$. Thus $x = b^{-1}$ and $b = x^{-1} \in R^\times$. Finally,

$$K^\times \subset (x^{-1})^{-1}(R^\times)^{-1} = (x^{-1}R^\times)^{-1} \subset (R^\times)^{-1} \qquad \text{by } (O_2)$$

giving $K^\times = (K^\times)^{-1} = (R^\times)^{-1}$ or $K^\times = R^\times$, $K = R$, a contradiction with (O_1). Thus (B_1) holds.

To prove (B_2), note that for $a,b \in R$, $abR \subset aR$, $abR \subset bR$, so $(ab)R \subset aR \cap bR$. Let $a = c/d$, $b = e/f$ with $c,d,e,f \in R^\times$. Then $c = ad$, $cR = adR \subset aR$. Similarly, $eR \subset bR$. Multiplying $cR \subset aR$ by e and $eR \subset bR$ by c, we obtain $ecR \subset aeR \subset aR$, $ecR \subset beR \subset bR$, i.e., $ecR \subset aR \cap bR$.

Let $U = aR \in \mathcal{B}$ be arbitrary. Multiplying (O_4) by a, we have

$$azR + azR \subset U$$

proving (B_3).

Since $RR \subset R$ by (O_2), then $aR \cdot R \subset aR$ for every $a \in K^\times$. Putting $V = aR \cap R$, we obtain $VV \subset aR$, hence (B_4). Let $x, a \in K^\times$; then $xx^{-1}R = R$ or $x(x^{-1}aR) = aR$. Putting $U = aR$, $V = x^{-1}aR$, we prove (B_5).

Now we can state the following.

THEOREM 8 Let K be a field equipped with a proper topology $\mathcal{T} \in \mathcal{L}_r(K)$. Then the following conditions are equivalent.

(i) $(K,\mathcal{T})$ is locally bounded.

(ii) There exists a base of neighborhoods of zero in $(K,\mathcal{T})$ consisting of bounded open sets.

(iii) For every bounded neighborhood U of zero, the sets aU, $a \in K^\times$, form a base of neighborhoods of zero in $(K,\mathcal{T})$.

(iv) There exists an almost-order defining topology $\mathcal{T}$ on K.

Proof: The equivalence (i) $\Leftrightarrow$ (iv) follows from Theorem 7, and the implications (iv) $\Rightarrow$ (iii) $\Rightarrow$ (ii) $\Rightarrow$ (i) can easily be deduced.

LEMMA 13 Let K be a field and $\mathcal{T},\mathcal{S} \in \mathcal{L}_r(K)$ locally bounded topologies, having T and S as the almost-order inducing them. Then the following conditions are equivalent:

(i) $\mathcal{T} \le \mathcal{S}$.

(ii) $\exists\, a \in K^\times$ such that $aS \subset T$.

Proof: (ii) $\Rightarrow$ (i) Recall that by Theorem 8, $\mathcal{B}(\mathcal{T}) = \{bT : b \in K^\times\}$ and similarly, $\mathcal{B}(\mathcal{S}) = \{cS : c \in K^\times\}$. Thus for every $U = bT \in \mathcal{B}(\mathcal{T})$ there exists $V = (ab)S \in \mathcal{B}(\mathcal{S})$ with $V \subset U$, thus $\mathcal{T} \le \mathcal{S}$.

(i) $\Rightarrow$ (ii) Since $T \in \mathcal{B}(\mathcal{T})$ and $\mathcal{T} \le \mathcal{S}$, there is $aS \in \mathcal{B}(\mathcal{T})$, $aS \subset T$.

THEOREM 9 Let $\mathcal{O}_i$ be an almost-order inducing a locally bounded topology $\mathcal{T}_i \in \mathcal{L}_r(K)$ on a field K, $i \in I$. Put

$$\mathcal{O}_I = \bigcap_{i \in I} \mathcal{O}_i \tag{29}$$

and assume that $\mathcal{O}_I$ is an almost-order of K. Let $T_I \in L_r(K)$ be the locally bounded topology having $\mathcal{O}_I$ as its almost-order. Then:

(i) T_I is a locally bounded ring topology on K, stronger than every T_i $(i \in I)$.

(ii) If $|I| < \infty$ and every $T_i \in L_t(K)$, then $T_I \in L_t(K)$ and $T_I = \sup_t\{T_i : i \in I\}$, t = r or t = f.

Remark It can happen that $\mathcal{O}_I$ is neither an almost-order nor is $\mathcal{O}_I^\times$ a multiplicative subgroup of $K^\times$ (see Dürbaum [1, pp. 88-89]).

LEMMA 14 Suppose that $\mathcal{O}_i$ $(i \in I)$ are almost-orders in a field K. If for every $a_i, \ldots, a_i \in K^\times$, $a_{i_1}\mathcal{O}_{i_1} \cap \cdots \cap a_{i_n}\mathcal{O}_{i_n} \neq \{0\}$ $(i_j \in I)$, then $\mathcal{O}_I$ defined by (29) is an almost-order in K.

Proof of Lemma 14: To prove this lemma we shall verify conditions (O_1)-(O_4) of Definition 9. Since $0,1 \in \mathcal{O}_i$ for every $i \in I$, then $0,1 \in \mathcal{O}_I$, and $\mathcal{O}_I \neq K$ follows from $\mathcal{O}_i \neq K$. Similarly, if $x,y \in \mathcal{O}_I$, then $x,y \in \mathcal{O}_i$ for each i, and $xy \in \mathcal{O}_i$ for all $i \in I$. Thus $xy \in \mathcal{O}_I$. To prove (O_3), let $x \in K^\times$ be arbitrary. Since the topology T_i induced in K by $\mathcal{O}_i$ is a ring topology, there exists $a_i \in K^\times$ such that $xa_i \in \mathcal{O}_i$. Let $a \in a_1\mathcal{O}_1 \cap \cdots \cap a_n\mathcal{O}_n \cap \mathcal{O}_1 \cap \cdots \cap \mathcal{O}_n$ be a nonzero element. Thus $a = a_ib_i$ with suitable $b_i \in \mathcal{O}_i$ leading to $xy = xa_ib_i \in \mathcal{O}_ib_i \subset \mathcal{O}_i\mathcal{O}_i \subset \mathcal{O}_i$ for every $i \in I$, and finally $xa \in \mathcal{O}_I$. Since $a \in \mathcal{O}_I = \mathcal{O}_1 \cap \cdots \cap \mathcal{O}_n$, by assumption, $x \in \mathcal{O}_I a^{-1} \subset \mathcal{O}_I(\mathcal{O}_I^\times)^{-1}$.

Let $z_i \in \mathcal{O}_i$, $z_i \neq 0$, be a 2-addiator for $\mathcal{O}_i$, i.e., $z_i(\mathcal{O}_i \pm \mathcal{O}_i) \subset \mathcal{O}_i$. Then every element $z \in z_1\mathcal{O}_1 \cap \cdots \cap z_n\mathcal{O}_n$, $z \neq 0$, is a 2-addiator for $\mathcal{O}_I$. Indeed, $z = z_ic_i$ $(i \in I)$, $c_i \in \mathcal{O}_i$. If $x,y \in \mathcal{O}_I$, then $z(x + y) = z_ic_i(x + y) = z_i(c_ix + c_iy) \in z_i(\mathcal{O}_i + \mathcal{O}_i) \subset \mathcal{O}_i$, i.e., $z(x + y) \in \mathcal{O}_I$. This proves (O_4).

Proof of Theorem 9: By assumption, $\mathcal{O}_I$ is an almost-order in K. Thus $T_I \in L_r(K)$ and T_I is locally bounded. Since $\mathcal{O}_I \subset \mathcal{O}_i$ for every $i \in I$, Lemma 13 implies that $T_i \leq T_I$ for every $i \in I$.

To prove (ii), suppose that $I = \{1,2,\ldots,n\}$ and $T_i \in L_r(K)$ for $i \in I$. Let T_I be the locally bounded ring topology on K having $\mathcal{O}_I$ as its almost-order (Lemma 14). Recall that every element of $\mathcal{B}(T_I)$

has a form $a\mathcal{O}_I = a(\mathcal{O}_1 \cap \cdots \cap \mathcal{O}_n) = a\mathcal{O}_1 \cap \cdots \cap a\mathcal{O}_n$, $a \in K^\times$, and every element of $\mathcal{B}(\sup_r(\mathcal{T}_1,\ldots,\mathcal{T}_n))$ has the form $d_1\mathcal{O}_1 \cap \cdots \cap d_n\mathcal{O}_n$, where $d_i \in K^\times$. Thus

$$\mathcal{T}_I \leq \sup_r(\mathcal{T}_1,\ldots,\mathcal{T}_n) \text{ in } \mathcal{L}_r(K)$$

But $\mathcal{T}_i \leq \mathcal{T}_I$ in $\mathcal{L}_r(K)$ $(i \in I)$, by Lemma 13. Finally, $\mathcal{T}_I = \sup_r(\mathcal{T}_1, \ldots,\mathcal{T}_n)$. In the case when $\mathcal{T}_i \in \mathcal{L}_f(K)$, we can apply Lemma 4. For the reader's convenience we also present a direct proof. We shall prove that division in $(K,\mathcal{T}_I)$ is continuous. Indeed, by assumption, there exists $d_i \in K^\times$ such that

$$(1 + d_i\mathcal{O}_i)^{-1} \subset 1 + \mathcal{O}_i \qquad i \in I$$

Let $d_I \in d_1\mathcal{O}_1 \cap \cdots \cap d_n\mathcal{O}_n$, $d_I \neq 0$. Taking an arbitrary $t \in (1 + d_I\mathcal{O}_I)^{-1}$, we obtain

$$t = (1 + d_I r_I)^{-1} = (1 + d_i r_i r_I)^{-1}$$

where $d_I = d_i r_i$, $r_i \in \mathcal{O}_i$ for $i \in I$ and $r_I \in \mathcal{O}_I$. This implies that $t \in (1 + d_i\mathcal{O}_i)^{-1} \subset 1 + \mathcal{O}_i$ for $i \in I$. Hence

$$(1 + d_I\mathcal{O}_I)^{-1} \subset (1 + \mathcal{O}_1) \cap \cdots \cap (1 + \mathcal{O}_n) \subset 1 + \mathcal{O}_I$$

Obviously, $\mathcal{T}_I$ is Hausdorff since

$$\bigcap_{a \in K^\times} a\mathcal{O}_I \subset \bigcap_{a \in K^\times} a\mathcal{O}_i = \{0\} \qquad (i \in I)$$

LEMMA 15 Let $(K,\mathcal{T})$ be a bounded topological field. Then $\mathcal{T} = \mathbf{1}_K$ or $\mathcal{T} = \mathbf{0}_K$.

Proof: Suppose to the contrary that $\mathcal{T} \neq \mathbf{1}_K$. Let $U \in \mathcal{B}(\mathcal{T})$ be such that $1 \notin U$. Since $(K,\mathcal{T})$ is bounded, there exists $a \in K^\times$ satisfying $aK \subset U$, i.e., $aa^{-1} = 1 \in U$, a contradiction.

COROLLARY Every compact, not antidiscrete topological field is finite, and thus discrete.

Proof: Let $(K,\mathcal{T})$ be a compact topological field. Then K is bounded (Lemma 10), hence discrete (Lemma 15). But every discrete compact topological space is finite.

Remark It follows from Lemma 15 that every bounded Hausdorff topological field is discrete. The converse is not true: There exists a simple transcendental field extension $F = K(x)$ and a proper locally bounded field topology $\mathcal{T}$ on F such that $\mathcal{T}|K = \mathbb{1}_K$ and K is not bounded with respect to $\mathcal{T}$ (see J.-A. Cohen [5]; Weber [4, (2.2)]).

LEMMA 16 (Weber [2]) Let K be a field and $\mathcal{T}, \mathcal{T}_1, \ldots, \mathcal{T}_n \in \mathcal{L}_r(K)$. If for every $k = 1, 2, \ldots, n$ and every $U \in \mathcal{B}(\mathcal{T})$, U is unbounded in $(K,\mathcal{T}_k)$, then for every M_k bounded in $(K,\mathcal{T}_k)$ $(k = 1, 2, \ldots, n)$ and every $U \in \mathcal{B}(\mathcal{T})$, $U \not\subset M_1 \cup \cdots \cup M_n$ holds.

Proof: We proceed by induction on n. $n = 1$: This is clear by the assumption on $\mathcal{T}$. Now let M_k be bounded in $(K,\mathcal{T}_k)$ and $U \in \mathcal{B}(\mathcal{T})$ $(k = 1, \ldots, n)$. Take $V \in \mathcal{B}(\mathcal{T})$ with $V + V \subset U$, and define $A_1 = M_1 - M_1$, $A_k = M_k$ $(k = 2, \ldots, n - 1)$. By the inductive assumption there exist elements $y \in V \setminus (A_1 \cup \cdots \cup A_{n-1})$ and $z \in V \setminus (B_2 \cup \cdots \cup B_n)$, where $B_j = M_j \cup (M_j - y)$, $j = 2, \ldots, n$. If $z \notin M_1$, then $z \in U \setminus (M_1 \cup \cdots \cup M_n)$. Let $z \in M_1$. Then $x = y + z \in U \setminus (M_1 \cup \cdots \cup M_n)$. Indeed, $x = y + z \in V + V \subset U$. If $x \in M_1$, then $y = x - z \in M_1 - M_1 = A_1$, a contradiction. If $x \in M_j$ for some j $(2 \le j \le n)$, then $z = x - y \in M_j - y \subset B_j$, again a contradiction. Thus $U \not\subset M_1 \cup \cdots \cup M_n$.

LEMMA 17 Let $\mathcal{T}, \mathcal{T}_1, \ldots, \mathcal{T}_n \in \mathcal{L}_r(K)$, $\mathcal{T}_i$ being locally bounded $(i = 1, \ldots, n)$. Then the following conditions are equivalent:

(i) $\mathcal{T}_j \nleq \mathcal{T}$ for every $j = 1, \ldots, n$.

(ii) For every $U \in \mathcal{B}(\mathcal{T})$ and for every $j = 1, \ldots, n$, U is unbounded in $(K,\mathcal{T}_j)$.

The proof is immediate.

Now we return to examples of locally bounded topological fields given in Section 2 and we list the almost-orders and their 2-addiators for all of them.

Example 1 $K = \mathbb{R}$, $R = [-1,1]$, $z = 1/2$; $K = \mathbb{C}$, $R = \{x \in C : |x| \leq 1\}$, $z = 1/2$.

Example 2 $g(R)$, $z = 1/2 = g(1/2)$.

Example 3 $R = \{x \in K : v_p(x) \geq 0\}$, $z = 1$.

Example 4 $R = \{f \in M : v_{x_0}(f) \geq 0\}$, $z = 1$.

Example 5 $R = \{a \in K : \varphi(a) \leq 1\}$, $z = 1$.

Example 6 $R = \{x \in \mathbb{Q} : |x| \leq 1, |x|_p \leq 1\}$, $z = 1/q$, where $q \neq p$ is a prime number.

Example 7 $R = \{x \in K : |x| \leq 1\}$, $z = 1/2$.

Example 8 $R = \{x \in K : v_p(x) \geq 0\}$, $z = 1$.

Other examples of locally bounded topological fields are presented in subsequent chapters. A general structure of locally bounded fields is given in Chapter 9, Section 2. These results are due to Nakano [1-3].

6. TOPOLOGICAL FIELDS AND GEOMETRY

DEFINITION 11 By a projective plane C we shall mean a set of points P and a set of lines $B \subset 2^P$ satisfying the following properties:

(P1) For every $a,b \in P$, $a \neq b$, there exists exactly one line $\ell \in B$ with $a,b \in \ell$.

(P2) For every $\ell,m \in B$, $\ell \neq m$, there exists exactly one $a \in P$ satisfying $\ell \cap m = \{a\}$.

(P3) There exist four different points such that no three of them lie on a line.

Let X be a set and let $(X \times X)^*$ denote the set of all pairs (x,y) $X \times X$ satisfying $x \neq y$.

DEFINITION 12 A projective plane C is said to be a topological projective plane if the sets P and B are provided with topologies such that the mappings

$$h: (P \times P)^* \to B, \ h((a,b)) = \overline{(a,b)} \text{ (the line through a and b)}$$

and

$$s: (B \times B)^* \to P, \quad s((m,n)) = m \cap n$$

are continuous. [Here $(P \times P)^*$ and $(B \times B)^*$ are equipped with relative topologies induced by the product topologies.]

THEOREM 10 (Salzmann [1]) Let C be a topological projective plane. Then every affine plane of C is homeomorphic in the induced topology with the product of two affine lines, both equipped with the induced topology.

THEOREM 11 (Misfeld [1, Satz (1.3)]) Let K be the coordinate field of the Desarguesean topological projective plane C. Then K is a topological field with respect to the induced topology.

DEFINITION 13 By a *projective space over a topological field* $(K,\mathcal{T})$ we shall mean the projective space $(K^{n+1})^\times/K^\times$ provided with the quotient topology of the product topology induced from K.

The following definition and theorem generalize Definition 12 and Theorem 10.

DEFINITION 14 A projective space S of finite projective dimension n is said to be a *topological projective space* if the point set P and the set H of hyperplanes are provided with topologies such that the joining of n independent points and the intersection of n independent hyperplanes are continuous mappings. [Here the set of n-tuples of independent points (resp. hyperplanes) is equipped with the relative topology induced by the product topology.]

THEOREM 12 (Sörensen [1]) Every Desarguesean topological projective space of finite projective dimension is isomorphic to a projective space over a topological field $(K,\mathcal{T})$. [The topology on the set of hyperplanes is also the topology of $(K^{n+1})^\times/K^\times$.]

Topological projective planes were studied by L. A. Skornjakov [1] and Salzmann [1]. For related results, see, e.g., Karzel [1], Doignon [1], Sörensen [1], Heise and Sörensen [1], Pieper and Biallas [1], Jauer [1], and Szambien [1]. Some fundamental results dealing with topological fields and topological geometry can be found in Salzmann [2].

NOTES

We shall not discuss here the general problem of existence of proper topologies on universal algebras, compatible with their algebraic structure. We simply mention here that there exist infinite groupoids, semigroups, quasi-groups, and nonassociative and noncommutative rings possessing only the trivial topologies compatible with their algebraic structure (see Hanson [1], Taĭmanov [1], Suvorov and Ufnarovskaja [1], and Arnautov [4], respectively). Every infinite module over a discrete ring admits a proper module topology over that ring (Arnautov [7]), but the one-dimensional vector space $M = \mathbb{Q}z$ over $\mathbb{Q}$ as a module over the ring $\mathbb{Q}[X]$ equipped with the X-adic topology, with the multiplication $(a_0 + a_1X + \cdots + a_NX^N) \cdot (rz) = (a_0 + a_1 + \cdots + a_N)r \cdot z$ (a_j, $r \in \mathbb{Q}$) admits no proper module topology over this topological ring (Arnautov [7]). On the other hand, every infinite vector space over a field K provided with a field topology $\mathcal{T}$ admits a proper vector space topology over $(K,\mathcal{T})$ (as can be deduced from Gould [1] after a slight modification of his proof). Every infinite associative and commutative ring possesses a proper ring topology (Arnautov [5,6]; Hochster [1]; Hochster and Kiltinen [1]). It is well known that every infinite abelian group admits a proper group topology (Kertesz and Szele [1]), but assuming the continuum hypothesis, there exists an infinite nonabelian group possessing no proper group topology (Shelah [1]). Kiltinen [5] proved that the number of group topologies on a commutative group G equals $2^{2^{|G|}}$, i.e., is the maximal one. Alas [1] showed that the additive group $\mathbb{R}^+$ admits 2^{c} metrizable group topologies. Zobel [2] proved that if countable noncommutative group G admits a proper group topology, then G possesses $2^{2^{\aleph_0}}$ group topologies. Moreover, if a countable group possesses a proper linear group topology, then it possesses a proper nonlinear group topology. Hesse [1] presents general results describing topologizations of universal algebras with some applications to groups. Some problems concerning existence of topologies on universal algebras are discussed in Banaschewski [1], Podewski [3], and Paljutin, Seese, and Taĭmanov [1].

Fundamental properties of topological rings and modules are presented in the book by Arnautov, Vodinčar, and Mihalev [1]. Taǐmanov [2] gave criteria for the topologizability of countable algebras, e.g., groups, rings, and fields. For example, every countable field admits a proper field topology. Applications of model theory in the topological algebra can be found in Stone [1] and Prestel and Ziegler [1,2]. A general theory of topological near-fields is contained in Wefelscheid [1,2].

Larson and Andima [1] considered the lattice $L(X)$ of all topologies on a set X. The lattice $L_r(K)$ of all ring topologies on a field K clearly is a subset of $L(K)$. Some historical remarks on the topological algebra can be found in Freudenthal [1] and van der Waerden [1].

2

Fundamental Properties of Absolute-Valued Fields

In this chapter we collect rather well-known facts from valuation theory, most of them without proof.

1. ARTIN-WHAPLES' APPROXIMATION THEOREM, HENSEL'S LEMMA

DEFINITION 1 A field K is said to be *absolute-valued* if there exists a function $|\cdot|: K \to \mathbb{R}_+$, called an *absolute value*, such that for all $x, y \in K$,

(AV_1) $|x| \geq 0$, $|x| = 0 \Leftrightarrow x = 0$

(AV_2) $|xy| = |x||y|$

(AV_3) $|x + y| \leq |x| + |y|$

An absolute value is called *non-Archimedean* if it satisfies

(AV_3') $|x + y| \leq \max(|x|, |y|)$

An absolute value is called *Archimedean* if there exists $n \in \mathbb{N}$ such that $|n| > 1$. The system of balls

$$U_\varepsilon = \{a \in K : |a| < \varepsilon\} \qquad \varepsilon \in \mathbb{R}_+^\times$$

defines a field topology in K.

The absolute value φ_0 defined as $\varphi_0(a) = 1$ for $a \in K^\times$, $\varphi_0(0) = 0$, is called trivial. It induces the discrete topology on K. In Chapter 1, Examples 1-5 and 9, the topology is induced by a suitable absolute value. In the sequel, let T_φ denote the field topology induced by an absolute value φ of a field K.

DEFINITION 2 Two absolute values φ, ψ of a field K are called *equivalent* if $T_\varphi = T_\psi$.

THEOREM 1 Let φ, ψ be nontrivial absolute values of a field K. Then the following conditions are equivalent:

(i) φ is equivalent to ψ.

(ii) $\varphi = \psi^{\alpha}$ for some $\alpha \in \mathbb{R}_{+}^{\times}$.

(iii) $T_{\varphi} \le T_{\psi}$.

(iv) $\forall\, x \in K,\ \varphi(x) < 1 \Rightarrow \psi(x) < 1$.

(v) $\forall\, x \in K,\ \varphi(x) \le 1 \Leftrightarrow (x) \le 1$.

Proof: See Endler [7, pp. 2-3].

THEOREM 2 Let $\varphi\colon K \to \mathbb{R}_{+}$ be a function satisfying (AV_1) and (AV_2). Then the following conditions are equivalent:

(i) φ is a non-Archimedean absolute value.

(ii) φ satisfies (AV_3) and $\varphi(\mathbb{N})$ is bounded.

(iii) $\forall\, x \in K,\ \varphi(x) \le 1 \Rightarrow \varphi(1 + x) \le 1$.

(iv) φ satisfies (AV_3) and $\varphi(n \cdot 1) \le 1$ for all $n \in \mathbb{N}$.

(v) $\forall\, \alpha \in \mathbb{R}_{+}^{\times}$, φ^{α} is an absolute value of K.

Proof: See Endler [7, pp. 5-6].

THEOREM 3 (Artin-Whaples) If there are given any n nontrivial, pairwise nonequivalent absolute values φ_s of K, an element a_s for each $s = 1, 2, \ldots, n$ and an $\varepsilon \in \mathbb{R}_{+}^{\times}$, then there exists an element $a \in K$ such that

$$\varphi_s(a - a_s) < \varepsilon \qquad \text{for every } s = 1, 2, \ldots, n$$

Proof: See Endler [7, pp. 3-4].

This theorem (called the approximation theorem) implies, in particular, that if T is the supremum of a finite family of topologies induced by pairwise inequivalent absolute values of K, the completion $\hat{K}$ of K at T is the direct sum of the completions of K at these absolute values.

The approximation theorem implies also that inequivalent absolute values are multiplicatively independent:

THEOREM 4 If $\varphi_1, \varphi_2, \ldots, \varphi_n$ are nontrivial and inequivalent, the relation

$$\varphi_1(x)^{k_1}\varphi_2(x)^{k_2}\cdots\varphi_n(x)^{k_n} = 1$$

is true for all nonzero $x \in K$ if and only if all $k_i = 0$.

Proof: See Artin and Whaples [1].

In subsequent chapters we shall need the following

THEOREM 5 (Strong Approximation Theorem) Let K be the quotient field of a Dedekind domain A and $\mathbb{P}$ the set of all nonzero prime ideals of A. Let T be a finite subset of $\mathbb{P}$. Then for every $a_p \in K$ and every $\varepsilon > 0$ there exists an $a \in K$ such that $|a - a_p|_p < \varepsilon$ if $p \in T$ and $|a|_p \le 1$ if $p \in \mathbb{P} \setminus T$.

Proof: See Endler [7, Th. (12.15)]; also O'Meara [1, 21:2].

DEFINITION 3 Let K be a non-Archimedean valued field and φ its absolute value. The set $V_\varphi = \{x \in K : \varphi(x) \le 1\}$ is called the *valuation ring*, and $P_\varphi = \{x \in K : \varphi(x) < 1\}$ is the *valuation ideal*. The field $\bar{k}_\varphi = V_\varphi/P_\varphi$ is called the *residue class field* of φ. By the *value group* of φ we understand the multiplicative subgroup $\varphi(K^\times)$ of the positive reals $\mathbb{R}_+^\times$. Finally, an absolute value φ is called *discrete* if its value group is a discrete subgroup of $\mathbb{R}_+^\times$

Let K be a field with a non-Archimedean absolute value φ. The canonical homomorphism $V_\varphi \to \bar{k}_\varphi$ can be extended in a natural way to the homomorphism $V_\varphi[X] \to \bar{k}_\varphi[X]$. We shall denote the image of $f \in V_\varphi[X]$ by $\bar{f}$.

THEOREM 6 (Hensel's Lemma) Let K be a field complete with respect to a nontrivial and non-Archimedean absolute value φ. Then for any $F \in V_\varphi[X]$, $a \in V_\varphi$ and $g, h \in \bar{k}_\varphi[X]$ such that $(g,h) = 1$, with $\bar{F} = gh$ and the leading coefficient of g equal to $\bar{a}$, there exist polynomials $G, H \in V_\varphi[X]$ such that $F = GH$, $\deg G = \deg g$, $\bar{G} = g$, $\bar{H} = h$, and a is the leading coefficient of G.

Proof: See, e.g., Endler [7, pp. 120-121]. For equivalent forms of Hensel's lemma, see Ribenboim [11].

Note that if φ is a nontrivial absolute value of a field K, then for every $\varepsilon > 0$, $\varphi(K^\times) \cap (0,\varepsilon) \neq \emptyset$. Indeed, there exists, by assumption, a $c \in K^\times$ such that $\varphi(c) \neq 1$. We may assume that $\varphi(c) < 1$, since otherwise we could take c^{-1} instead of c. Consequently, $\varphi(c^n) < \varepsilon$ for large $n \in \mathbb{N}$, since $\varphi(c^n) = \varphi(c)^n$ tends to zero.

THEOREM 7 (Kürschak [1, pp. 250-253]) Let K be an algebraically closed field with an absolute value φ. Then the completion $\hat{K}$ of K at φ is also algebraically closed.

Proof: (Staum [1]) We may assume that φ is nontrivial, since otherwise $\hat{K} = K$. Let $p(X) = X^n + a_{n-1}X^{n-1} + \cdots + a_0 \in \hat{K}[X]$. Let φ denote the absolute value of K and also its extension to $\hat{K}$. We will consider two cases.

Case 1. $\varphi(a_k) < 1/2n$ for $k = 0, 1, \ldots, n - 1$.

(1) For every $\varepsilon > 0$ there exists $x \in K$ with $\varphi(x) \leq 1$ and $\varphi(p(x)) < \varepsilon$.

To prove this, let $0 < \varepsilon < 1$. Since K is dense in $\hat{K}$, there are elements $b_0, b_1, \ldots, b_{n-1} \in K$ such that $\varphi(b_k - a_k) < \varepsilon/2n$ for every $k = 0, 1, \ldots, n - 1$. Then $\varphi(b_k) < 1/n$ for $k = 0, 1, \ldots, n - 1$. Let $q(X) = X^n + b_{n-1}X^{n-1} + \cdots + b_0 \in K[X]$. Since K is algebraically closed, there is an $x \in K$ with $q(x) = 0$. The element x must satisfy $\varphi(x) \leq 1$, since otherwise $\varphi(x^n) > \varphi(b_{n-1}x^{n-1} + \cdots + b_1x + b_0)$, i.e., $q(x) \neq 0$, a contradiction. Thus

$$\varphi(p(x)) = \varphi(p(x) - q(x)) = \varphi\left(\sum_{k=0}^{n-1} (a_k - b_k)x^k\right) < \frac{\varepsilon}{2} < \varepsilon$$

proving (1). Iterating, we obtain the following assertion:

(2) There exists a sequence $x_n \in K$ such that $\varphi(x_n) \leq 1$ for every $n \in \mathbb{N}$ and $\varphi(p(x_n)) \to 0$ for $n \to \infty$.

Now we prove

(3) There is a subsequence (x_{1n}) of (x_n) with $\varphi(x_{1n} - x_{1m}) < 1/2$ for each pair $m,n \in \mathbb{N}$.

To prove (3), take $\varepsilon = 1/4$. Then there are, by (1), $b_0, b_1, \ldots, b_{n-1} \in K$, satisfying $\varphi(a_k - b_k) < \varepsilon^n/2n$ for $k = 0, 1, \ldots, n - 1$. Let $q(X) = X^n + b_{n-1}X^{n-1} + \cdots + b_0 \in K[X]$. Since K is an algebraically closed field, one can write $q(X) = (X - c_1)(X - c_2)\cdots(X - c_n)$

with suitable $c_1, \ldots, c_n \in K$. Hence

$$p(X) = q(X) + (p(X) - q(X)) = (X - c_1)\cdots(X - c_n) + \sum_{k=0}^{n-1} (a_k - b_k)X^k$$

Let $S(a,\varepsilon) = \{b \in K : \varphi(a - b) < \varepsilon\}$ be a ball with center a and radius ε. Suppose that there is $x \notin S(c_1,\varepsilon) \cup \cdots \cup S(c_n,\varepsilon)$. Then

$$\varphi((x - c_1)\cdots(x - c_n)) \geq \varepsilon^n \quad \text{and} \quad \varphi\left(\sum_{k=0}^{n-1} (a_k - b_k)x^k\right) < \frac{\varepsilon^n}{2}$$

giving

$$\varphi(p(x)) = \varphi\left((x - c_1)\cdots(x - c_n) + \sum_{k=0}^{n-1} (a_k - b_k)x^k\right) > \frac{1}{2}\varepsilon^n$$

Since all elements of the sequence (x_m) except a finite number of them satisfy

$$\varphi(p(x_m)) < \frac{1}{2}\varepsilon^n$$

there exists $m_0 \in \mathbb{N}$ such that for every $m \geq m_0$, $x_m \in S(c_1,\varepsilon) \cup \cdots \cup S(c_n,\varepsilon)$. This means that there is a subsequence (x_{1m}) of (x_m) lying in $S(c_k,\varepsilon)$; hence $\varphi(x_{1s} - x_{1t}) < 2\varepsilon = 1/2$ holds for every $s,t \in \mathbb{N}$.

In the same way as in (3), taking (x_{1t}) instead of (x_t), we find a subsequence (x_{2t}) of (x_{1t}) satisfying $\varphi(x_{2s} - x_{2t}) < 1/4$ for every $s,t \in \mathbb{N}$. By induction, we find subsequences $(x_{st})_{t\in\mathbb{N}}$ ($s = 1, 2, \ldots$) such that $(x_{si} - x_{sj}) < 2^{-s}$ for every $i,j,s \in \mathbb{N}$. The diagonal sequence $(x_{ii})_{i\in\mathbb{N}}$ is a Cauchy sequence with respect to φ and, moreover, $\varphi(p(x_{ii})) \to 0$ for $i \to \infty$. Thus there exists $x_0 \in \hat{K}$, $x_0 = \lim x_{ii}$, $p(x_0) = 0$.

Case 2. There exists k, $0 \leq k \leq n - 1$, with $\varphi(a_k) \geq 1/2n$. Since φ is nontrivial, one can find $c \in K$ satisfying

$$0 < \varphi(c) < \frac{1}{2n\varphi(a_k)} \qquad \text{for every } k = 0, 1, \ldots, n - 1$$

Let $r(X) = X^n + ca_{n-1}X^{n-1} + (c^2a_{n-2})X^{n-2} + \cdots + (c^{n-1}a_1)X + c^na_0$.

Since $r \in \hat{K}[X]$ and satisfies the assumptions of case 1, there exists $z \in \hat{K}$ with $r(z) = 0$. Then ac^{-1} is a root of $p(X) = 0$.

Remark It is not difficult to see that if φ is a nontrivial absolute value of an algebraically closed field K, then the value group $\varphi(K^\times)$ is dense in $\mathbf{R}_+^\times$.

2. DESCRIPTION OF ABSOLUTE VALUES OF SOME FIELDS

The following theorem follows easily (see the proof of Theorem 6 of Chapter 1).

THEOREM 8 Let K be a field. Then the following conditions are equivalent.

(i) K admits a nontrivial absolute value.

(ii) K is not an algebraic extension of a finite field.

Modifying the proof of Theorem 6 of Chapter 1, we obtain

THEOREM 9 Let K be a field, P its prime subfield, and B a transcendence basis of K over P. Let B be nonempty. Then:

(i) K admits at least $2^{|B|}$ pairwise inequivalent absolute values.

(ii) In particular, if $|K| = |B|$, then K admits exactly $2^{|K|}$ pairwise inequivalent absolute values.

For example, the fields R, $\mathbb{C}$, and $\mathbb{Q}_p$ admit 2^c inequivalent absolute values.

Proof: Let $K = P(B)(A)$, where elements of A are algebraic over $P(B)$. Take arbitrary subset $\emptyset \neq B_0 \subsetneq B$. Let $k_0 = P(B \setminus B_0)$, $k = k_0(B_0)$. Let $\deg_0(f)$ be degree of $f \in k_0[B_0]$ with respect to variables from B_0, and put

$$\varphi(f) = e^{-\deg_0(f)} \tag{4}$$

if $f \in k_0[B_0]$ is nonzero and $\varphi(0) = 0$. Since k is the quotient field of $k_0[B_0]$ and

$$\deg_0(fg) = \deg_0(f) + \deg_0(g) \tag{5}$$

$$\deg_0(f + g) \leq \max(\deg_0(f), \deg_0(g)) \tag{6}$$

we can extend φ from $k_0[B_0]$ to k, obtaining an absolute value, which again can be extended to an algebraic extension $K = k(A)$ of k (see Section 4). In this way, with every subset B_0 of B we associate an absolute value. Moreover, if ψ is associated with B_1 and $b \in B_1 \setminus B_0$, then $\psi(b) = e^{-1}$ but $\varphi(b) = 1$. This shows (Theorem 1) that φ is not equivalent to ψ. Hence there is one-to-one correspondence between subsets of the transcendence basis of K over its prime subfield and constructed absolute values of K. Since every absolute value is a function φ: $K \to \mathbb{R}$, the set of all absolute values of K has cardinality not greater than $|\mathbb{R}^K| = |\mathbb{R}|^{|K|} = 2^{\aleph_0 |K|} = 2^{|B|}$ in case (ii). Thus K has exactly $2^{|K|}$ inequivalent absolute values.

DEFINITION 4 By a *global field* we shall mean a finite field extension either of $\mathbb{Q}$ or the rational function field $\mathbb{F}_q(t)$ of one variable t over a finite field $\mathbb{F}_q$.

The following theorem describes the structure of nontrivial absolute values on global fields.

THEOREM 10 (Ostrowski [4]) Let K be a finite extension of the rational number field $\mathbb{Q}$. Then the only nontrivial absolute values of K (up to equivalence) are the *p*-adic absolute values, where *p* is a prime ideal in the Dedekind ring $\mathbb{Z}_K$ of algebraic integers of K (see Chapter 1, Example 9), and absolute values of the form

$$\varphi(x) = |g(x)| \qquad x \in K$$

where g is an embedding of K into the complex number field, and $|\cdot|$ denotes the ordinary absolute value. If φ and ψ are absolute values defined by different embeddings, say g and h, then φ and ψ are equivalent if and only if g and h are complex conjugates [i.e., $g(x) = \overline{h(x)}$ holds for every $x \in K$]. If $K = \mathbb{Q}(a)$, where a is an algebraic integer and f is the minimal polynomial of a, $\deg(f) = n$, then there are exactly n embeddings of K into $\mathbb{C}$ (see Borewicz and Šafarevič [1, Th. 1, p. 317]; or Narkiewicz [1, Th. 3.2, p. 88]).

If K is an algebraic function field of one variable over a finite field $\mathbb{F}_q$, i.e., a finite extension of the rational function field $\mathbf{F}_q(t)$, then every nontrivial absolute value of K is a p-adic absolute value, where p is a prime ideal of the integral closure either of $\mathbf{F}_q[t]$ or $\mathbf{F}_q[t^{-1}]$ in K.

For the reader's convenience we shall give the proof. Let us remark, first, that if K is the quotient field of a domain R, then any absolute value v of K is uniquely determined by its values on R. Indeed, if $a \in K$, $a = b/c$ $(b,c \in R)$, then $v(a) = v(b)/v(c)$. Moreover, for any other fraction b'/c' representing a, $b/c = b'/c'$, i.e., $bc' = b'c$, and $v(b)v(c') = v(b')v(c)$, implying that $v(a) = v(b)/v(c) = v(b')/v(c')$. Thus v(a) does not depend on the representation of a as a quotient of elements of R.

Proof of Theorem 10: Let us consider first the case $K = \mathbb{Q}$. Let w be a nontrivial absolute value of $\mathbb{Q}$. The remark above allows us to consider w only on $\mathbb{Z}$.

Case 1. w is Archimedean. Let n and m be arbitrary integers greater than 1. Write

$$m = a_r n^r + \cdots + a_1 n + a_0 \qquad 0 \le a_j < n,\ a_j \in \mathbb{Z}$$

Since $n^r \le m$, it follows that $r \le \log m/\log n$. The triangle inequality implies that $w(k) \le k$ if $k \in \mathbb{N}$: thus $w(a_i) < n$ for every i. Consequently,

$$\begin{aligned} w(m) &\le \sum_{i=0}^{r} w(a_i)w(n)^i < n \sum_{i=0}^{r} w(n)^i \le n(r+1)\max(1,w(n)^r) \\ &\le n\left(\frac{\log m}{\log n} + 1\right)\max(1,w(n)^{\log m/\log n}) \\ &= n\left(\frac{\log m}{\log n} + 1\right)(\max(1,w(n)))^{\log m/\log n} \end{aligned}$$

Substituting for m the positive power m^s, it follows that

$$w(m) \le \left(n\left(s\,\frac{\log m}{\log n} + 1\right)\right)^{1/s}(\max(1,w(n)))^{\log m/\log n}$$

If $s \to \infty$, then

$$w(m) \le (\max(1,w(n)))^{\log m/\log n} \tag{7}$$

holds for every $n,m > 1$. The inequality (7) implies that $w(n) > 1$ is satisfied for every $n \in \mathbb{N}$, $n > 1$. Otherwise, $w(n) \leq 1$ for some $n > 1$ would imply [by (7)] that $w(m) \leq 1$ for every $m \in \mathbb{N}$, a contradiction, as w is Archimedean, by the assumption. Thus

$$w(m) \leq w(n)^{\log m/\log n} \qquad \text{i.e.,} \quad w(m)^{1/\log m} \leq w(n)^{1/\log n}$$

Exchanging m and n gives the following equality:

$$w(n)^{1/\log n} = w(m)^{1/\log m} = c \qquad c > 0 \text{ being a constant}$$

Consequently, $w(n) = c^{\log n} = n^r$, $r = \log c > 0$; i.e., w is equivalent to the ordinary absolute value of $\mathbb{Q}$.

Case 2. w is non-Archimedean. Then $P = P_w \cap \mathbb{Z}$ is a prime ideal of $\mathbb{Z}$, i.e., $P = p\mathbb{Z}$, with a suitable prime number p. Hence $w(p) = c$, $0 < c < 1$. If $a \in \mathbb{Z} \setminus P$, then $w(a) = 1$, and $|a|_p = 1$. Finally, if $n \in \mathbb{Z}$ is arbitrary, $n \neq 0$, then

$$n = p^{v_p(n)} a \qquad (p,a) = 1$$

and

$$w(n) = w(p)^{v_p(n)} w(a) = c^{v_p(n)}$$

(v_p denotes the p-adic exponential valuation). Consequently, w is equivalent to the p-adic absolute value $|\cdot|_p$.

Now let $K \neq \mathbb{Q}$ be a finite extension of $\mathbb{Q}$. Let w be a nontrivial absolute value of K. We shall prove first that

$$w|\mathbb{Q} \text{ is nontrivial} \tag{8}$$

Indeed, suppose that w is trivial on $\mathbb{Q}$, i.e., $w(x) = 1$ holds for every $x \in Q^{\times}$. Let $b_1, \ldots, b_n$ be a base of K over $\mathbb{Q}$ and take an arbitrary $x \in K$. If $x = x_1 b_1 + \cdots + x_n b_n$ $(x_i \in \mathbb{Q})$, then

$$\begin{aligned} w(x) &\leq w(x_1)w(b_1) + \cdots + w(x_n)w(b_n) \\ &\leq w(b_1) + \cdots + w(b_n) \end{aligned}$$

This implies that K is bounded in the topology $\mathcal{T}_w$ induced by w;

i.e., either $T_w = \mathbb{1}_K$ or $T_w = \mathbb{0}_K$ (by Chapter 1, Lemma 15), contradicting the hypothesis that (K, T_w) is a proper topological field (or, equivalently, that w is a nontrivial absolute value).

Recall that the integral closure $\mathbb{Z}_K$ of $\mathbb{Z}$ in K, the ring of integers of the field K, is a Dedekind domain (see any textbook on algebraic number theory, e.g., Borewicz and Šafarevič [1] or Narkiewicz [1]).

Case 2. w is non-Archimedean. It was proved [see (8)] that $w|\mathbb{Q}$ is nontrivial. As w is non-Archimedean, $w|\mathbb{Z}$ is also nontrivial. Now since w is non-Archimedean, and nontrivial on $\mathbb{Z}$, $p = P_w \cap \mathbb{Z}_K$ is a prime ideal of the Dedekind domain $\mathbb{Z}_K$. We are going to prove that $\mathbb{Z}_K \subset V_w$. Indeed, let $z \in \mathbb{Z}_K$ satisfy an equation $z^n + a_{n-1}z^{n-1} + \cdots + a_0 = 0$, $a_j \in \mathbb{Z}$. Since w is non-Archimedean, $w(k) \le 1$ if $k \in \mathbb{Z}$. Thus $w(z^n) = w(z)^n \le \max\{w(a_i z^i) : i = 0, 1, \ldots, n-1\} \le \max\{w(z)^i : i = 0, 1, \ldots, n-1\} = w(z)^k$ with $0 \le k \le n-1$. Consequently, $w(z)^n \le w(z)^k$, i.e., $w(z)^{n-k} \le 1$, which gives that $w(z) \le 1$ or $z \in V_w$.

Let $x \in K$ satisfy $v_p(x) = 0$, i.e., $|x|_p = 1$. Hence the fractional ideal $x\mathbb{Z}_K = ab^{-1}$, where a and b are ideals of $\mathbb{Z}_K$ such that $p \nmid ab$. Let $b \in b \setminus p$. Then $xb \in a \setminus p$. Thus $w(b) = w(bx)$ by the definition of p, i.e., $w(x) = 1$.

We have proved that

$$\forall\, x \in K,\ |x|_p = 1 \Rightarrow w(x) = 1 \tag{9}$$

Let us fix now $p \in p \setminus p^2$. Take an arbitrary $x \in K^\times$ with $n = v_p(x)$. If $y = xp^{-n}$, then $|y|_p = |x|_p |p|_p^{-n} = 1$, since $v_p(p) = 1$. Thus, applying (9), we see that $w(y) = w(x)w(p)^{-n} = 1$, i.e., $w(x) = (w(p)^{-1})^n = (w(p)^{-1})^{v_p(x)}$. This means, however, that $w \sim |\cdot|_p$.

It remains to consider

Case 1. w is Archimedean. We shall first prove that

The ordinary absolute value of $\mathbb{R}$ has unique extension to $\mathbb{C}$. (10)

It is sufficient to prove that if f is any extension of the absolute value of $\mathbb{R}$, then

$$\forall\ z \in \mathbb{C},\ |z| = 1 \Rightarrow f(z) = 1 \tag{11}$$

Indeed, if (11) is satisfied, then take any $w \in \mathbb{C}^{\times}$. As $|w/|w|| = 1$. so $f(w/|w|) = 1$, by (11), i.e., $f(w/|w|) = f(w)|w|^{-1} = 1$, and consequently, $f(w) = |w|$.

To prove that (10) is fulfilled, suppose to the contrary that there exists $z \in \mathbb{C}$ such that $|z| = 1$ and $f(z) > 1$. Note that $f(i) = 1$ for any absolute value f of $\mathbb{C}$, extending the ordinary absolute value, since $i^2 = -1$, i.e., $f(i)^2 = 1$. Now take an integer $n > [\log 2/\log f(z)]$. Then $f(z)^n > 2$. Write $z^n = a + ib$. Since $|z| = 1$, then $\max(|a|, |b|) \leq 1$. Consequently,

$$\begin{aligned} 2 < f(z)^n &= f(z^n) = f(a + ib) \leq f(a) + f(i)f(b) \\ &= |a| + |b| \leq 2 \end{aligned}$$

a contradiction!

Let K be any field that can be embedded in the field $\mathbb{C}$. For any embedding g of K into $\mathbb{C}$, the function f_g defined by $f_g(x) = |g(x)|$, $x \in K$, is an Archimedean absolute value of K. We shall prove that if K is a finite extension of $\mathbb{Q}$, then every Archimedean absolute value of K has this form.

Now let f be any absolute value of K extending the ordinary absolute value of $\mathbb{Q}$. We may suppose that $K = \mathbb{Q}(a)$, by Abel's theorem on primitive elements. Let $F \in \mathbb{Q}[X]$ be a minimal polynomial of a over $\mathbb{Q}$, $n = \deg(F)$. Suppose that $a_1, \ldots, a_s \in \mathbb{R}$ and $a_j \notin \mathbb{R}$ ($j = s + 1, \ldots, n$), $n = s + 2t$, are roots of F. Let

$$g_j : K \to \mathbb{C} \qquad j = 1, \ldots, n$$

be the embedding of K into $\mathbb{C}$, defined by

$$g_j(b_0 + b_1 a + \cdots + b_{n-1}a^{n-1}) = b_0 + b_1 a_j + \cdots + b_{n-1}a_j^{n-1}$$

($b_i \in \mathbb{Q}$). Consider the following diagram of absolute-valued fields:

$$\begin{array}{ccc} (\mathbb{Q}, T_\infty) & \longrightarrow & (K, T_f) \\ \downarrow & & \downarrow \\ (\mathbb{Q}, T_\infty) & \longrightarrow & (\hat{K}, T_{\hat{f}}) \end{array}$$

where $\hat{K}$ denotes the completion of $(K, T_{\bar{f}})$ and $\bar{\mathbb{Q}}$ is the closure of $\mathbb{Q}$ in K with respect to the ordinary topology T_∞. Since $T_f | \mathbb{Q} = T_\infty$, $\bar{\mathbb{Q}}$ is topological isomorphic with the reals $\mathbb{R}$. Let g be the (unique) topological isomorphism of $\bar{\mathbb{Q}}$ onto $\mathbb{R}$. This means, however, that

$$f(x) = |g(x)| \qquad x \in K \tag{12}$$

Assume first that $a = a_j$ $(1 \le j \le s)$. Then $a \in \bar{\mathbb{Q}}$, and g induces a real embedding $g|K: K \to \mathbb{C}$ such that $f(x) = \bar{f}(x) = |g(x)|$, $g = g_j$.

Let now $a = a_j$ $(s + 1 \le j \le n)$. Then $[\bar{\mathbb{Q}}(a):\bar{\mathbb{Q}}] = 2$ and the embedding $g: \bar{\mathbb{Q}} \to \mathbb{R}$ can be extended exactly in two ways to the embedding $\bar{g}: \bar{\mathbb{Q}}(a) \to \mathbb{C}$, by putting either $\bar{g}(t + ua) = g(t) + g(u)a$, or $\bar{g}(t + ua) = g(t) + g(u)\bar{a}$, where $\bar{a}$ denotes the conjugate number to a. This implies by (10) that $\bar{f}(x) = |\bar{g}(x)|$ for every $x \in \bar{\mathbb{Q}}(a)$. Consequently, $f(x) = |g(x)|$ $(x \in K)$, as desired.

If g and h are embeddings of K into $\mathbb{C}$, and $g = \bar{h}$ [i.e., $g(x) = \overline{h(x)}$ for every $x \in K$], then the absolute values $f_g(x) = |g(x)|$ and $f_h(x) = |h(x)|$ $(x \in K)$ are equivalent:

$$f_g(x) = |g(x)| = |\overline{h(x)}| = |h(x)| = f_h(x)$$

Conversely, let $f_g \sim f_h$. We may assume that $f_g = f_h$, taking a suitable power of one of them, if necessary. Let $a \in K$ and $b = g(a)$, $c = h(a)$. Then, by assumption, $f_g(a) = f_h(a)$ and $f_g(1 + a) = f_h(1 + a)$, implying that

$$\begin{aligned} f_g(1 + a)^2 &= g(1 + a)\overline{g(1 + a)} \\ &= 1 + b + \bar{b} + b\bar{b} = 1 + c + \bar{c} + c\bar{c} \\ &= h(1 + a)\overline{h(1 + a)} = f_h(1 + a)^2 \end{aligned}$$

Thus either $b = c$ or $b = \bar{c}$.

Let $K_1 = \{a \in K : g(a) = h(a)\}$ and $K_2 = \{a \in K : g(a) = \overline{h(a)}\}$. Obviously, K_1 and K_2 are subfields of K, and $K = K_1 \cup K_2$ by the remark above. Assume to the contrary that $K_1 \not\subset K_2$ and $K_2 \not\subset K_1$ and take $a_1 \in K_1 \setminus K_2$ and $a_2 \in K_2 \setminus K_1$. Put $a_3 = a_1 + a_2$. We have either $a_3 \in K_1$ or $a_3 \in K_2$, since $K = K_1 \cup K_2$. If $a_3 \in K_1$, then $a_2 = a_3 - a_1 \in K_1$, a contradiction! Consequently, $a_3 \in K_2$. But then $a_1 = a_3 - a_2 \in K_2$, again a contradiction!

Remark One could apply here a much more general result due to Soundararajan [1]: If fields $F_1, \ldots, F_n$ have a common overfield and K is a subring of $F_1 \cup \cdots \cup F_n$, with unity, then $K \subset F_i$ for some i $(1 \le i \le n)$. In particular, a field cannot be union of finitely many of its proper subfields.

To finish the proof of Theorem 10, we have to consider the case of nonzero characteristic. We shall consider only $K = \mathbf{F}_q(t)$. The case we consider is contained in Theorem 11.

THEOREM 11 Let F(X) be a simple transcendental extension of a field F and φ a nontrivial absolute value of F(X), trivial on F. Then φ is equivalent either to a p-adic absolute value ($p \in F[X]$ a prime element) or $|a|_\infty$ [where for $f/g \in F(X)$, $f,g \in F[X]$, $|f/g|_\infty = e^{\deg f - \deg g}$].

Proof: Case 1. $\varphi(X) > 1$. Then $\varphi(cX^k) = \varphi(X)^k > \varphi(X)^\ell = \varphi(dX^\ell)$ for every $k > \ell \ge 0$, and $c,d \in F^\times$, since $\varphi(X) > 1$ implies that $\varphi(X)^N = \varphi(X^N) > 1$ for every $N \in \mathbb{N}$. This absolute value φ is non-Archimedean since it extends the trivial absolute value. As for $\varphi(a) \ne \varphi(b)$, we have

$$\varphi(a + b) = \max(\varphi(a), \varphi(b)) \qquad a,b \in F(X)$$

it follows that $\varphi(f(X)) = \varphi(a_M X^M + \cdots + a_0) = \varphi(X^M) = \varphi(X)^{\deg f}$ for every $f \in F[X]$. Thus due to (AV_2),

$$\varphi\left(\frac{f}{g}\right) = \varphi(X)^{\deg f - \deg g}$$

Case 2. $0 < \varphi(X) \le 1$. This implies that $\varphi(h) \le 1$ for every $h \in F[X]$. Let

$$R_\varphi = \{f \in F[X] : \varphi(f) < 1\}$$

Observe that we must have $\varphi(f/g) = \varphi(p)^N$, where $f/g = p^N(f_1/g_1)$, $(f_1,p) = (g_1,p) = 1$, f, g, f_1, g_1, $p \in F[X]$, and p is a prime element of F[X]. Indeed, since R_φ is a prime ideal in F[X], it is generated by a prime element $p \in F[X]$. So $\varphi(k) = 1$ if and only if $(k,p) = 1$ for $k \in F[X]$ and

$$\varphi\left(\frac{f}{g}\right) = \varphi\left(p^n \frac{f_1}{g_1}\right) = \varphi(p)^N = c^N \qquad c = \varphi(p) \le 1$$

Since φ is nontrivial on $F(X)$, then $0 < c < 1$ and φ is equivalent to the p-adic absolute value.

We formulate without proof the so-called "product formula for valuations," proved for the first time by Artin and Whaples [1,2].

THEOREM 12 (product formula) Let K be a global field. Then one can select a set V of nontrivial and pairwise nonequivalent absolute values of K (called *normalized absolute values*) such that:

(i) Every nontrivial absolute value of K is equivalent to exactly one element of V.

(ii) For every $a \in K^\times$ the set $\{v \in V : v(a) \neq 1\}$ is finite.

(iii) $\prod_{v \in V} v(a) = 1$ holds for every $a \in K^\times$.

For a proof, see, e.g., Borewicz and Šafarevič [1, Th. 2, pp. 318-319]; also Narkiewicz [1, Th. 3.3, p. 91].

3. TOPOLOGICAL CHARACTERIZATIONS OF ABSOLUTE-VALUED FIELDS

We shall start with the following.

LEMMA 1 Let K be a field and let $\varphi: K \to \mathbb{R}$ satisfy conditions (AV_1)-(AV_2). Then the following conditions are equivalent.

(i) $\exists\, C > 0\ \forall\, x,y \in K,\ \varphi(x + y) \le C \max(\varphi(x),\varphi(y))$.

(ii) $\exists\, \alpha > 0,\ \varphi^\alpha$ is an absolute value of K.

(iii) $\exists\, C > 0\ \forall\, x \in K,\ \varphi(x) \le 1 \Rightarrow \varphi(1 + x) \le C$.

Proof: (i) $\Rightarrow$ (ii) Let $\alpha > 0$ satisfy $C \le 2$ and write $\tilde{\varphi} = \varphi^\alpha$. Thus

$$\tilde{\varphi}(a + b) \le 2 \max(\tilde{\varphi}(a),\tilde{\varphi}(b)) \tag{13}$$

iterating (13), we get for $n = 2^k$: $\tilde{\varphi}(a_1 + \cdots + a_n) \le n \max_{1 \le i \le n} \tilde{\varphi}(a_i)$; in particular, $\tilde{\varphi}(n) \le n$. Now we shall prove that

$$\tilde{\varphi}(x) \le 2x \tag{14}$$

holds for every $x \in \mathbb{N}$. Let k be chosen so that $2^k \le x < 2^{k+1}$. We have $x = (x - 2^k) + 2^k$. If $\tilde{\varphi}(2^k) \le \tilde{\varphi}(x - 2^k)$, then

$$\tilde{\varphi}(x) \le 2 \max(\tilde{\varphi}(2^k), \tilde{\varphi}(x - 2^k)) \le 2 \cdot 2^k \le 2x$$

Otherwise,

$$\tilde{\varphi}(x) \le 2\tilde{\varphi}(x - 2^k) \le 2 \cdot 2(x - 2^k) = 4x - 2^{k+2} \le 2x$$

follows by recurrence. Now we apply the inequality $\tilde{\varphi}(a_1 + \cdots + a_n) \le n \max(\tilde{\varphi}(a_1), \ldots, \tilde{\varphi}(a_n))$ to $\tilde{\varphi}(1 + x)^{n-1}$, where n is a power of 2. Thus

$$\tilde{\varphi}(1 + x)^{n-1} = \tilde{\varphi}((1 + x)^{n-1})) \le n \max_{0 \le i \le n-1} \tilde{\varphi}\left(\binom{n - 1}{i}\right)\tilde{\varphi}(x^i)$$
$$\le 2n(1 + \tilde{\varphi}(x))^{n-1}$$

or $\tilde{\varphi}(1 + x) \le (2n)^{1/(n-1)}(1 + \tilde{\varphi}(x))$. If $n \to \infty$, it follows that $\tilde{\varphi}(1 + x) \le 1 + \tilde{\varphi}(x)$. Since $\tilde{\varphi}(xy) = \tilde{\varphi}(x)\tilde{\varphi}(y)$, then $\tilde{\varphi}(x + y) = \tilde{\varphi}(x(1 + y/x)) = \tilde{\varphi}(x)\tilde{\varphi}(1 + y/x) \le \tilde{\varphi}(x)(1 + \tilde{\varphi}(y/x)) = \tilde{\varphi}(x) + \tilde{\varphi}(y)$.

(ii) ⇒ (iii) If $\psi = \varphi^\alpha$ is an absolute value, then $\psi(x) \le 1$ implies that $\psi(1 + x) \le \psi(1) + \psi(x) \le 2$, i.e., $\varphi(1 + x) \le 2^{1/\alpha} = C$.

(iii) ⇒ (i) Let $x,y \in K^\times$ and $\varphi(y) \le \varphi(x)$. Since φ satisfies (AV_2), we have

$$\varphi(x + y) = \varphi(x)\varphi(1 + z) \le \varphi(x)C$$
$$= C \max(\varphi(x), \varphi(y)) \qquad \text{with } z = yx^{-1}$$

LEMMA 2 (E. Artin) Let K be a field and let $\varphi: K \to \mathbb{R}$ satisfy (AV_1)-(AV_2). Then the following conditions are equivalent.

(i) If T is the weakest topology on K such that φ is continuous, then $(x,y) \mapsto x + y$ is continuous at (0,0).

(ii) There exists an $\alpha > 0$ such that φ^α is an absolute value on K.

Proof: (i) ⇒ (ii) In view of Lemma 1, it is sufficient to prove that the function $f(x) = \varphi(1 + x)$ is bounded on $A = \{a \in K : \varphi(a) \le 1\}$. If f is not bounded, there would be a sequence $a_n \in A$ with $\varphi(1 + a_n)^{-1} \to 0$, $\varphi(a_n)\varphi(1 + a_n)^{-1} = \varphi(a_n(1 + a_n)^{-1}) \to 0$, i.e., $(1 + a_n)^{-1} \to 0$ and $a_n(1 + a_n)^{-1} \to 0$ in the topology T, contradicting the continuity of addition at (0,0) in the topology T since $(1 + a_n)^{-1} + a_n(1 + a_n)^{-1} = 1$ holds for every $n \in \mathbb{N}$.

(ii) ⇒ (i) Let $\psi = \varphi^\alpha$. Then

$$U_\varepsilon = \varphi^{-1}([0,\varepsilon)) = \{x \in K : \varphi(x) < \varepsilon\} = \psi^{-1}([0,\varepsilon^\alpha)) \tag{15}$$

is an open set in the topology T for every $\varepsilon > 0$. Since ψ is a valuation, then for every $x,y \in U_\delta$ with $\delta = (1/2)^{1/\alpha}\varepsilon$, we have

$$\psi(x + y) \leq \psi(x) + \psi(y) < \frac{1}{2}\varepsilon^\alpha + \frac{1}{2}\varepsilon^\alpha = \varepsilon^\alpha \qquad \text{i.e., } \varphi(x + y) < \varepsilon$$

Hence the addition $(x,y) \mapsto x + y$ is continuous at $(0,0)$ in the topology T.

THEOREM 13 Let φ be a nontrivial absolute value on a field K. Then the field topology T_φ induced by φ is a minimal element of $L_r(K)$. Moreover, T_φ is a locally bounded topology.

Proof: Suppose, to the contrary, that there is a ring topology T on K satisfying

$$\mathbf{0}_K \neq T < T_\varphi \tag{16}$$

Let $\mathcal{B}$ be a base of the neighborhoods of zero for T. Since T is properly weaker than T_φ, then for every $U \in \mathcal{B}$ there exists $\varepsilon > 0$ such that

$$\{x \in K : \varphi(x) < \varepsilon\} \subset U$$

As $T \neq T_\varphi$, there exists $a > 0$ with the property that every $U \in \mathcal{B}$ contains an x satisfying $\varphi(x) \geq a$.

Let $U \in \mathcal{B}$ be arbitrary and let $V \in \mathcal{B}$ satisfy $VV \subset U$. Choose $b > 0$ so that

$$\{x \in K : \varphi(x) < b\} \subset V$$

Since φ is nondiscrete, there is $y \in K^\times$ with $\varphi(y) \geq (ab)^{-1}$. Let $W \in \mathcal{B}$ satisfy $yW \subset V$. We can find a $z \in W$ with $\varphi(z) \geq a$. Then

$$\varphi(yz) = \varphi(y)\varphi(z) > a^{-1}b^{-1}a = b^{-1} \qquad \text{or} \qquad \varphi((yz)^{-1}) \leq b$$

In other words, $(yz)^{-1} \in V$. This implies that

$$1 = (yz)(yz)^{-1} \in VV \subset U$$

a contradiction, since $U \in \mathcal{B}$ was arbitrarily chosen and $\mathcal{T}$ was supposed to be a nontrivial ring topology.

DEFINITION 5 An element x of a topological field (ring) $(K,\mathcal{T})$ is called *topologically nilpotent* if $x^n \to 0$ for n tending to infinity, *inversely nilpotent* if x^{-1} exists and is topologically nilpotent, and *neutral* if it is neither nilpotent nor inversely nilpotent.

Let $(K,\mathcal{T})$ be a topological field. Put

$$N_0 = \{a \in K : a^n \to 0 \text{ at } \mathcal{T}\} \tag{17}$$

$$N = K \setminus (N_0 \cup (N_0^\times)^{-1}) \qquad \text{where } N_0^\times = N_0 \setminus \{0\} \tag{18}$$

$$\tilde{N}_0 = (K \setminus N_0)^{-1} = \{x \in K : x^{-1} \notin N_0\} \tag{19}$$

In other words, N_0 is the set of topological nilpotents and N the set of neutral elements of the field $(K,\mathcal{T})$.

Now we can formulate

THEOREM 14 (Šafarevič [1]; Kaplansky [2]) Let $(K,\mathcal{T})$ be a proper topological field. The following conditions are equivalent.

(i) The topology $\mathcal{T}$ is induced by an absolute value.

(ii) N_0 is open and $\tilde{N}_0$ is bounded.

(iii) N_0 is open and bounded and $N_0(N_0 \cup N) \subset N_0$.

(iv) N_0 is open and $N \cup N_0$ is bounded.

(v) N_0 is open and for every A, disjoint with some neighborhood of zero, A^{-1} is bounded.

Proof: (i) ⇒ (ii) Let $\mathcal{T} = \mathcal{T}_\varphi$, where φ is an absolute value. Then $x \in N_0 \Leftrightarrow x^n \to 0$ at $\mathcal{T}_\varphi \Leftrightarrow \varphi(x^n) = \varphi(x)^n \to 0$ in $\mathbb{R} \Leftrightarrow \varphi(x) < 1$. Thus $N_0 = \{x \in K : \varphi(x) < 1\}$ and N_0 is open in $(K,\mathcal{T}_\varphi)$. Let $U_\varepsilon = \{x \in K : \varphi(x) < \varepsilon\}$. We have to find $\delta > 0$ with $\tilde{N}_0 U_\delta \subset U_\varepsilon$. Let $x \in \tilde{N}_0$, $y \in U_\delta$. We must have $\varphi(xy) = \varphi(x)\varphi(y) < \varepsilon$. But $x \in \tilde{N}_0 \Leftrightarrow x^{-1} \notin N_0 \Leftrightarrow \varphi(x^{-1}) = \varphi(x)^{-1} \geq 1 \Leftrightarrow \varphi(x) \leq 1$. Thus it is sufficient to take $\delta = \varepsilon$ to have $\tilde{N}_0 U_\varepsilon \subset U_\varepsilon$.

(i) ⇒ (iii) One proves this similarly to (i) ⇒ (ii).

(iii) ⇒ (i) Note first that $N^{-1} \subset N$ and $NN \subset N$. Indeed, since the first statement is obvious, let $a,b \in N$. If $ab \notin N$, then $ab \in N_0$

or $ab \in N_0^{-1}$. Let $ab \in N_0$. Since $a^{-1} \in N$, then (iii) gives $(ab)a^{-1} = b \in N_0$, contrary to our assumption. If $ab \in N_0^{-1}$, then $(ab)^{-1} \in N_0$ and, again by (iii), we get $(ab)^{-1}b = a^{-1} \in N_0$, a contradiction, as $a^{-1} \in N$. Thus N is a subgroup of $K^\times$.

Define an order in $K^\times/N$, putting

$$aN \geq bN \Leftrightarrow ab^{-1} \in N \cup N_0 \tag{20}$$

We shall prove that the order in $K^\times/N$ is linear, i.e.,

(1°) $aN \geq aN$.

(2°) $aN \geq bN,\ bN \geq aN \Rightarrow aN = bN$.

(3°) $aN \geq bN,\ bN \geq cN \Rightarrow aN \geq cN$.

(4°) For every $a,b \in K^\times$, $aN \geq bN$ or $bN \geq aN$.

Let $ab^{-1} \in N \cup N_0$, $ba^{-1} \in N \cup N_0$. If $ba^{-1} \in N_0$, then $1 = (ab^{-1})(ab^{-1})^{-1} \in N_0$, which is impossible, since $1 \notin N_0$. This proves (2°). Thus $ba^{-1} \in N$.

To prove (3°), let $ab^{-1}, bc^{-1} \in N \cup N_0$. Then

$$(ab^{-1})(bc^{-1}) = ac^{-1} \in (N \cup N_0)(N \cup N_0) \subset N \cup N_0 \quad \text{by (iii)}$$

To prove (4°), let $a,b \in K^\times$. We must show that ab^{-1} or $(ab^{-1})^{-1}$ lies in $N \cup N_0$, or that for every $x \in K^\times$, $x \in N \cup N_0$ or $x^{-1} \in N \cup N_0$. Suppose that for some $x \in K^\times$, $x \notin N \cup N_0$ and $x^{-1} \notin N \cup N_0$. This gives a contradiction, since $x \notin N_0$ and $x^{-1} \notin N_0$ imply that $x \in N$.

(5°) The order in $K^\times/N$ is Archimedean.

Note at first that

$$aN > bN \Leftrightarrow ab^{-1} \in N_0$$

Indeed, if $ab^{-1} \in N$, then $aN = bN$. Thus $cN > 1 \cdot N = N \Rightarrow c \in N_0$. Let $a,b \in K^\times$ be arbitrary and $bN > N$. Thus $b \in N_0$, i.e., $b^n \to 0$ at T. Since the multiplication in (K,T) is continuous, hence $b^n a^{-1} \to 0$, then $b^n a^{-1} \in N_0$ for sufficiently large n. Now

$$(b^n a^{-1})N > N \Leftrightarrow b^n N > aN \Leftrightarrow (bN)^n > aN$$

hence (5°).

Thus there exists an order isomorphism $f: K^\times/N \to \mathbb{R}_+^\times$ (see the Appendix). Now we are able to construct an absolute value for K, which induces $\mathcal{T}$. Let us consider the following diagram:

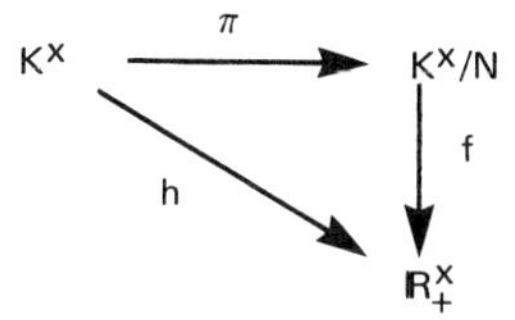

$h = f \circ \pi$

where π is the canonical epimorphism of $K^\times$ onto $K^\times/N$. Put

$$\varphi(x) = h(x^{-1}) \qquad \text{for } x \in K^\times \qquad \varphi(0) = 0$$

Note that $\varphi: K \to \mathbb{R}_+^\times$, $\varphi(x) > 0$ for $x \neq 0$, $\varphi(xy) = h(x^{-1}y^{-1}) = h(x^{-1})h(y^{-1}) = \varphi(x)\varphi(y)$. Moreover,

$$x \in N_0 \Leftrightarrow \pi(x) > \pi(1) \text{ in } K^\times/N \Leftrightarrow f(xN) > f(N) \text{ in } \mathbb{R}_+^\times$$
$$\Leftrightarrow h(x) > h(1) = 1 \Leftrightarrow \varphi(x) = h(x^{-1}) < 1$$

This means that

$$N_0 = \{x \in K : \varphi(x) < 1\} \tag{21}$$

Since N_0 is open and bounded, the sets aN_0, $a \in K^\times$ (Chapter 1, Theorem 8) form a base of zero neighborhoods for $(K,\mathcal{T})$. It follows from (21) that $\mathcal{T} = \mathcal{T}_\varphi$. Thus addition is continuous with respect to $\mathcal{T}_\varphi$. Lemma 2 shows that by taking φ^α with a suitable $\alpha > 0$ instead of φ, we obtain an absolute value. This proves that $\mathcal{T}$ is induced by an absolute value.

(iii) $\Rightarrow$ (iv) Since N_0 is open and bounded, then (Chapter 1, Theorem 8) let $\mathcal{B} = \{aN_0 : a \in K^\times\}$ be a base of the neighborhoods of zero for $\mathcal{T}$. Our assumption $N_0(N_0 \cup N) \subset N_0$ implies that $aN_0(N_0 \cup N) \subset aN$ for $a \in K^\times$, which means that $N_0 \cup N$ is bounded.

(iv) $\Rightarrow$ (v) Since N_0 is open and bounded, $\mathcal{T}$ is locally bounded and $\mathcal{B} = \{aN_0 : a \in K^\times\}$ is a base of the neighborhoods of zero for $\mathcal{T}$ (Chapter 1, Theorem 8). Let $A \cap aN_0 = \emptyset$ for some $a \in K^\times$. Then $A \subset K \setminus aN_0 = a(K \setminus N_0)$, and $A^{-1} \subset (K \setminus aN_0)^{-1} = a^{-1}(K \setminus N_0)^{-1}$.

Since $(K \setminus N_0)^{-1} \subset N \cup N_0$, thus $A^{-1} \subset a^{-1}(N \cup N_0)$ and A^{-1} is bounded as a subset of a bounded set $a^{-1}(N \cup N_0)$.

(v) ⇒ (ii) Let $A = K \setminus N_0$. A is disjoint from some neighborhood of zero, since N_0 is open and contains zero. Hence $\tilde{N}_0 = A^{-1}$ is bounded.

(ii) ⇒ (iii) If $x,y \in N_0$, then $(xy)^n = x^n y^n \to 0$ as $n \to \infty$. Hence $N_0 N_0 \subset N_0$. Moreover, since $N = \tilde{N}_0 \cap \tilde{N}_0^{-1}$, N is bounded by (ii). Note first that $K^\times = \tilde{N}_0 \cup \tilde{N}_0^{-1}$. Suppose to the contrary that there is $x \in K^\times \setminus (\tilde{N}_0 \cup \tilde{N}_0^{-1})$. Thus $x \notin (K \setminus N_0)^{-1}$, i.e., $x^{-1} \in N_0$. Similarly, $x \notin \tilde{N}_0^{-1}$ gives $x \in N_0$. Hence $1 = xx^{-1} \in N_0$, a contradiction! Now $K^\times = \tilde{N}_0 \cup \tilde{N}_0^{-1}$ implies that $N_0^\times = K^\times \setminus (K \setminus N_0) \subset (K \setminus N_0)^{-1} = \tilde{N}_0$, i.e., $N_0 = N_0^\times \cup \{0\}$ is bounded. This proves that N_0 is open and bounded, hence $\mathcal{T}$ is locally bounded. Applying Theorem 8 of Chapter 1, one can take the sets

$$\mathcal{B} = \{aN_0 : a \in K^\times\} \tag{22}$$

for a base of the neighborhoods of zero. If $y \in N$, then $y^m \in N$ for every $m \in \mathbb{N}$. Indeed, if $y^m \notin N$, then either $y^m \in N_0$ or $y^{-m} \in N_0$, leading to either $y \in N_0$ or $y^{-1} \in N_0$, a contradiction. Hence $D_y = \{y^m : m \in \mathbb{N}\} \subset N$ is bounded. As N and N_0 are bounded, so is NN_0; i.e., there is $d \in K^\times$ with $dNN_0 \subset N_0$. Take an arbitrary $x \in N_0$. Then $x^m \in dN_0$ for large m. Thus $(xy)^m = x^m y^m \in dN_0 D_y \subset dN_0 N \subset N_0$. This implies that $xy \in N_0$, and finally, $N_0 N \subset N_0$.

4. ALGEBRAIC CHARACTERIZATIONS OF ABSOLUTE VALUES

Let K be a field and R an almost-order of K. Let Z_R be the set of all 2-addiators of the almost-order R, i.e.,

$$Z_R = \{z \in R : z(R \pm R) \subset R\}$$

DEFINITION 6 Let Z be a subset of K. An almost-order R of K is called Z-*almost-order* if $Z_R = Z$.

DEFINITION 7 Suppose that R is an almost-order, $a \in K$ is an arbitrary element, and $Z \subset Z_R$. Let $R_Z[a]$ be the intersection of all Z-almost-orders of K containing R and a.

It is clear from this definition that $R_Z[a]$ is an Z-almost-order containing R and a. $R_Z[a]$ is called the *Z-almost-order generated by a over R*.

DEFINITION 8 A subset M of K is called an *R,Z-module* if $RM \subset R$ and $Z(M + M) \subset M$.

Now take a subset A of K. Let $[A]_{R,Z}$ be the intersection of all R,Z-modules of K containing A. The R,Z-module $[A]_{R,Z}$ is called the *R,Z-module generated by A*.

PROPOSITION 1 For a subset A of K, set $A_0 = RA$, and inductively $A_{n+1} = A_n \cup Z(A_n + A_n)$. Then $[A]_{R,Z} = \bigcup_{n\geq 0} A_n$. If, moreover, $AA \subset A$ (i.e., A is a semigroup) and $-1, 1 \in A$, then $[A]_{R,Z}$ is an Z-almost-order containing R.

Proof: Since $A_n \subset [A]_{R,Z}$, it is sufficient to prove that $\bigcup_{n\geq 0} A_n$ is an R,Z-module. Indeed, we obtain immediately from the definition of $[A]_{R,Z}$ that $[A]_{R,Z} \subset \bigcup_{n\geq 0} A_n$. By induction, $RA_n \subset A_n$ and $Z(A_{n-1} + A_{n-1}) \subset A_n$. Since $A_{n-1} \subset A_n$ holds for every $n \in \mathbb{N}$, $\bigcup_{n\geq 0} A_n$ also is an R,Z-module.

Now let A be a semigroup containing -1 and 1. Then $A_nA_n \subset [A]_{R,Z}$. Indeed, if n = 0, then $A_0A_0 = RARA \subset R(AA) \subset RA$. Let now $A_nA_n \subset [A]_{R,Z}$. Then

$$\begin{aligned} A_{n+1}A_{n+1} &= (A_n \cup Z(A_n + A_n))(A_n \cup Z(A_n + A_n)) \\ &\subset A_nA_n \cup A_nZ(A_n + A_n) \cup Z(A_n + A_n)Z(A_n + A_n) \\ &\subset A_n \cup Z(A_n + A_n) \cup Z(Z(A_nA_n + A_nA_n) + Z(A_nA_n + A_nA_n)) \\ &\subset A_n \cup A_{n+1} \cup Z(A_{n+1} + A_{n+1}) \subset [A]_{R,Z} \end{aligned}$$

COROLLARY 1 For every $b \in K$, $b[A]_{R,Z} = [bA]_{R,Z}$.

COROLLARY 2 Let $A = \{a^n : n \geq 0\} \cup \{-a^n : n \geq 0\}$, $a \in K$. Then

$$[A]_{R,Z} = R_Z[a]$$

Put $a^{(n)} = \{\pm a^m : 0 \leq m \leq n\}$.

COROLLARY 3 If $a^{m+1} \in [a^{(m)}]_{R,Z}$, then $R_Z[a] = [a^{(m)}]_{R,Z}$.

PROPOSITION 2 Let $b \in R_Z[a]$, $c \in R$. If $b \in [a^{(n)}]_{R,Z}$ and $ca^k \in R$ holds for every $1 \leq k \leq n$, then $cb \in R$.

Proof: Indeed, $cb \in cR_Z[a] = c[a^{(n)}]_{R,Z} = [ca^{(n)}]_{R,Z} \subset R$.

LEMMA 3 Let R be an almost-order of K, $Z = Z_R$. Assume that R is a maximal Z-almost-order of K; i.e., there exists no Z-almost-order S satisfying $R \subsetneq S \subsetneq K$. Let $G = \{x \in R : x^{-1} \in R\}$ be the group of units of R. Then

(i) $\forall\, d \in R \setminus G\ \forall\, b \in K\ \exists\, n \in \mathbb{N},\ bd^n \in R$.

(ii) $K^\times = R^\times \cup (R^\times)^{-1}$, i.e., R is a full Z-almost-order of K.

Proof: (i) Since $d^{-1} \notin R$, then $R_Z[d^{-1}] = K$, by the maximality of R. Moreover, if $b \in K$ is any element, then $b \in R_Z[d^{-1}]$, i.e., $b \in [(d^{-1})^{(n)}]_{R,Z}$ with a suitable $n \in \mathbb{N}$, by Proposition 1. Thus $bd^n \in [(d^{-1})^{(n)}]_{R,Z} = [(d^{-1})^{(n)}d^n]_{R,Z} \subset R$.

(ii) Suppose the contrary: There exists $a \in K^\times$ such that neither a nor a^{-1} lies in R. Then $R_Z[a] = R_Z[a^{-1}] = K$. Since $R_Z[a^{-1}] = K$, there is $m \in \mathbb{N}$ such that $a \in [(a^{-1})^{(m)}]_{R,Z}$. It follows from Corollary 1 that $a^{m+1} \in [a^{(m)}]_{R,Z}$, i.e., $R_Z[a] = [a^{(m)}]_{R,Z}$ by Corollary 3.

Let $d \in R \setminus G$ be arbitrary. Then there exists, by (i), an $n \in \mathbb{N}$ such that $ad^n \in R$. Generally, for every $1 \leq k \leq m$ there exists $n_k \in \mathbb{N}$ with $a^k d^{n_k} \in R$. Let $N = \max(n_1, \ldots, n_m)$. Then multiplying $a^k d^{n_k}$ by d^{N-n_k}, one obtains

$$a^k d^N \in RR \subset R \qquad \text{for } 1 \leq k \leq m$$

Since $K = [a^{(m)}]_{R,Z}$, the last statement implies that $cd^N \in R$ holds for every $c \in K$, by Proposition 2, a contradiction, since for $c = d^{-N-1}$, $cd^N = d^{-1} \notin R$, by the assumption.

DEFINITION 9 We say that an almost-order R is *maximal* if it is maximal as Z_R-almost-order.

Now we can prove the following.

THEOREM 15 (Dürbaum [1, Satz 1]) Every maximal almost-order R of a field K is an almost-order of an absolute value of K; i.e., there exists an absolute value φ of K such that $R = \{a \in K : \varphi(a) \leq 1\}$. The correspondence between the maximal almost-orders of K and the classes of equivalent absolute values of K is one-to-one.

Proof: Keeping notations as in Lemma 3, put $H = K^{\times}/G$. The quotient group H becomes a linearly ordered group with respect to the ordering $aG \leq bG \Leftrightarrow ab^{-1} \in R$. This order is linear, by Lemma 3(ii). Let

$$\varphi(a) = aG \text{ if } a \in K^{\times} \qquad \varphi(0) = 0$$

The group H is an Archimedean ordered group. Indeed, if $\varphi(d) < 1$ and $\varphi(b) < 1$, then $d \notin G$ and $b \notin G$. Applying Lemma 3(i) to the pairs (d,b) and (b,d), we can find $n,m \in \mathbb{N}$ satisfying $d^n b^{-1} \in R$ and $b^m d^{-1} \in R$, i.e., $\varphi(d)^n \leq \varphi(b)$ and $\varphi(b)^m \leq \varphi(d)$, implying that the order of H is Archimedean (see the Appendix). Thus H can be identified with the subgroup of the ordered group $\mathbb{R}_+^{\times}$ of positive real numbers. It is easy to see that the addition and the multiplication of K (equipped with the topology induced by R) are continuous with respect to $\varphi: K \to \mathbb{R}_+$ and $R = \{a \in K : \varphi(a) \leq 1\}$. We shall not give the details here; a similar construction was presented with a detailed proof in Section 3.

5. EXTENSIONS OF ABSOLUTE-VALUED FIELDS

DEFINITION 10 Let $L > K$ be a field extension. If ψ is an absolute value on L and φ an absolute value on K, we say that ψ *extends* φ if $\psi|K = \varphi$; i.e., if for every $x \in K$, $\psi(x) = \varphi(x)$.

In this section we shall summarize fundamental properties of this notion.

THEOREM 16 Let $L > K$ be a finite field extension of an absolute-valued field K, provided with an absolute value φ. Then there is an absolute value on K, extending φ.

To prove it, we shall need some more general facts.

DEFINITION 11 Let $(L,\mathcal{T}_L)$ and $(K,\mathcal{T}_K)$ be topological fields (rings), where $L > K$ is a field (ring) extension. We say that $(L,\mathcal{T}_L)$ is a *topological extension* of $(K,\mathcal{T}_K)$ if, moreover, $\mathcal{T}_L|_K = \mathcal{T}_K$.

THEOREM 17 Let $(K,\mathcal{T})$ be a topological field and $L = K(a)$ be a simple algebraic extension. Then the Cartesian topology on L, as a finite-dimensional vector space over K, is a field topology. Moreover, if $(K,\mathcal{T})$ is locally bounded, then the Cartesian topology on L is locally bounded.

Proof: It is standard that as a vector space over K, L is, with the Cartesian topology $\mathcal{T}_L = \mathcal{T} \times \cdots \times \mathcal{T}$ (n times, where $n = [L:K]$), a topological vector space over $(K,\mathcal{T})$. Moreover, we can define multiplication in K in terms of the basis $1, a, \ldots, a^{n-1}$. Explicitly, there exist constants $m_{ijk} \in K$, $0 \le i,j,k \le n - 1$, such that $a^i a^j = \Sigma_{k=0}^{n-1} m_{ijk} a^k$ and hence

$$\left(\sum_{i=0}^{n-1} a_i a^i\right)\left(\sum_{j=0}^{n-1} b_j a^j\right) = \sum_{k=0}^{n-1}\left(\sum_{i,j=0}^{n-1} a_i b_j m_{ijk}\right) a^k$$

Since L is a topological vector space over the topological field $(K,\mathcal{T})$, the mapping

$$\left(\sum_{i=0}^{n-1} a_i a^i, \sum_{j=0}^{n-1} b_j a^j\right) \mapsto \sum_{k=0}^{n-1}\left(\sum_{i,j=0}^{n-1} a_i b_j m_{ijk}\right) a^k$$

is a continuous form $(L,\mathcal{T}_L) \times (L,\mathcal{T}_L) \to (L,\mathcal{T}_L)$. Moreover, the inversion is continuous since if $0 \ne b = \Sigma_{i=0}^{n-1} a_i a^i$, then

$$b^{-1} = \sum_{i=0}^{n-1} \frac{f_i(a_0,a_1,\ldots,a_{n-1})}{g(z_0,a_1,\ldots,a_{n-1})} a^i$$

where f_i, $g \in K[X_1,\ldots,X_n]$ and g is not identically zero.

Finally, to show that the topology $\mathcal{T}_L$ on L is locally bounded, we shall take as a basis of the neighborhoods of zero all sets of the form

$$U_L = U + aU + \cdots + a^{n-1}U \tag{23}$$

where $U \in \mathcal{B} = \mathcal{B}(\mathcal{T})$ is bounded.

(We can assume, without restricting the generality, that all elements of $\mathcal{B}$ are bounded.) Let $\mathcal{B}_L = \{U_L : U \in \mathcal{B}\}$. Let $U_L, V_L \in \mathcal{B}_L$ be arbitrary. We shall prove that U_L is bounded in $(L, \mathcal{T}_L)$. Indeed, since U is bounded in $(K, \mathcal{T})$, there is an element $x \in K^\times$ satisfying $xU \subset V$. Then

$$xU_L = xU + a(xU) + \cdots + a^{n-1}(xU) \subset V + aV + \cdots + a^{n-1}V = V_L$$

Now we are able to prove Theorem 16.

Proof of Theorem 16: It suffices to consider the case of a simple algebraic extension only. Let $L = K(a)$. Suppose that $\mathcal{T} = \mathcal{T}_\varphi \in \mathcal{L}_f(K)$, where φ is a nontrivial absolute value. (The case of trivial φ is trivial!) We know (Theorem 13) that the topology $\mathcal{T}_\varphi$ is minimal and locally bounded. Let $(\hat{K}, \hat{\mathcal{T}})$ be the completion of $(K, \mathcal{T})$ and $\tilde{L} = \hat{K}(a)$.

Suppose that $[\tilde{L}:\hat{K}(a)] = s \leq n$. Then the Cartesian product topology $\mathcal{T}_{\tilde{L}} = \hat{\mathcal{T}} \times \cdots \times \hat{\mathcal{T}}$ (s times), which is (by Theorem 17) a locally bounded field topology, is also minimal (Banaschewski [1]; Nachbin [1]). Theorem 8 of Chapter 5 implies that $\mathcal{T}_{\tilde{L}}$ is induced by an absolute value, say ψ, since $(\hat{K}, \hat{\mathcal{T}})$, and thus $(\tilde{L}, \mathcal{T}_{\tilde{L}})$, has nonzero topological nilpotents. Since $\mathcal{T}_{\tilde{L}}|K = \mathcal{T}$, so $\psi|K$ and φ are equivalent. Thus for a suitably chosen real number $r > 0$, $\psi^r = \varphi$ on K. Since ψ^r is again an absolute value, the theorem is proved.

Remark For standard direct proofs of Theorem 16, see Artin [2], Bachman [1], Endler [7], or Zariski and Samuel [1, Vol. II].

LEMMA 4 (Kaplansky [2, Th. 13]) Let $(K, \mathcal{T})$ be a topological field and L a finite field extension of K. If $(K, \mathcal{T})$ has a bounded open neighborhood U of zero consisting of topological nilpotents, then L endowed with the Cartesian product topology has a bounded open neighborhood of zero consisting of topological nilpotents.

Proof: Let $b_1, \ldots, b_n$ be a basis for L over K, where $b_i b_j = \Sigma_{k=1}^n c_{ijk} b_k$ and $c_{ijk} \in K$. Since U is bounded, U^2 and $\Sigma_{i,j=1}^n U^2 c_{ijk}$ are also bounded. Hence we can find $t \in U$, $t \neq 0$, such that $tU \subset U$ and $t\, \Sigma_{i,j=1}^n U^2 c_{ijk} \subset U$ for every $k = 1, 2, \ldots, n$.

Take arbitrary $x = x_1b_1 + \cdots + x_nb_n \in L$ with $x_i \in t^2U$. Then for every $k \geq 2$, $x^{k-1} \in t^k(Ub_1 + \cdots + Ub_n)$. Indeed, if $k = 2$, then

$$x = x_1b_1 + \cdots + x_nb_n \in t(Ub_1 + \cdots + Ub_n)$$

Suppose that $x^{k-1} \in t^k(Ub_1 + \cdots + Ub_n)$. Then

$$\begin{aligned}
x^k = x^{k-1}x \in xt^k(Ub_1 + \cdots + Ub_n) &= \sum_{i=1}^{n} x_ib_it^k \sum_{j=1}^{n} Ub_j \\
&= \sum_{i,j=1}^{n} x_it^kb_ib_jU \subset t^k \sum_{i,j=1}^{n} t^2U^2b_ib_j \\
&= t^{k+1} \sum_{k=1}^{n} b_kt \sum_{i,j=1}^{n} U^2c_{ijk} \subset t^{k+1} \sum_{k=1}^{n} b_kU \\
&\subset t^{k+1}(Ub_1 + \cdots + Ub_n)
\end{aligned}$$

Thus $t^2(Ub_1 + \cdots + Ub_n)$ has the desired properties.

Remark Lemma 4 can be applied to obtain another proof of Theorem 16 by showing that conditions of Theorem 14 are satisfied.

Below we describe how one can effectively extend an absolute value from a field to its finite extension, and the number of such extensions. Omitting a proof, we shall point out suitable references.

Let K be an arbitrary field and let L be its finite extension of degree $[L:K] = n$. Let φ be an absolute value on K and $\varphi_1, \varphi_2, \ldots, \varphi_g$ be its inequivalent prolongations to L. Denote by G the value group of φ and by G_i the value groups of φ_i, $i = 1, 2, \ldots, g$. Similarly, let k and k_i denote the residue-class fields of φ and φ_i, respectively. Then g is finite, $[k_i:k] < \infty$ and G has finite index in each G_i, $(G_i:G) < \infty$. If, moreover, $f_i = [k_i:k]$, i.e., if the *inertial degree of* φ *relative to* φ_i is f_i and $e_i = (G_i:G)$ is the *ramification index* of φ, then

THEOREM 18 $\Sigma_{i=1}^{g} e_if_i \leq n$. If, moreover, L is a Galois extension of K, then we have

$$n = p^{\delta} \sum_{i=1}^{g} e_i f_i \qquad \text{if } \mathrm{char}(K) = p \neq 0\ (\delta \geq 0)$$

and

$$n = \sum_{i=1}^{g} e_i f_i \qquad \text{if } \mathrm{char}(K) = 0$$

Proof: See, e.g., Artin [2], Bachman [1], Cohen and Zariski [1], and Endler [7].

Applying the Kuratowski-Zorn lemma and using Theorem 16, one can prove the following:

THEOREM 19 Let K be a field and φ a nontrivial absolute value on K. Then φ can be extended to the algebraic closure $\bar{K}$ of K. In particular, every absolute value can be extended from a field to an arbitrary algebraic extension.

An extension of an absolute value, mentioned in Theorem 19, need not be unique. But

THEOREM 20 Let φ be a nontrivial absolute value on K and L be a finite extension of K, of degree $[L:K] = n$. If (K,T_φ) is complete, the extension ψ of φ to L is unique and is given by the formula

$$\psi(x) = [\varphi(N_{L/K}(x))]^{1/n} \tag{24}$$

Proof: See, e.g., Artin [2], Bachman [1], or Endler [7]. It can be shown (Dock [1]; Viswanathan [3]) that φ admits a unique prolongation to L if and only if (24) defines an absolute value of L.

DEFINITION 12 An absolute-valued field (K,φ) is called *Henselian* if for every finite field extension L of (K,φ) admits a unique prolongation to L.

THEOREM 21 Let K be a field and φ its non-Archimedean and nontrivial absolute value. The following conditions are equivalent.

(i) (K,φ) is Henselian.

(ii) Hensel's lemma holds for (K,φ).

Proof: See, e.g., Endler [7].

THEOREM 22 Let (K,φ) be a non-Archimedean absolute-valued field and L its arbitrary extension. Then there exists an absolute value ψ on L, extending φ.

Proof: Let $L = K(B)(A)$, where B is a transcendence basis of L over K and A the set of elements of L, algebraic over K(B). If B is empty, there is nothing to do (Theorem 19). So let B be nonempty and let $b \in B$. We can extend φ to φ from K to K(b) by putting

$$\varphi_b\left(\frac{a_N b^N + \cdots + a_1 b + a_0}{c_M b^M + \cdots + c_1 b + c_0}\right) = \frac{\max(\varphi(a_N),\ldots,\varphi(a_1),\varphi(a_0))}{\max(\varphi(c_m),\ldots,\varphi(c_1),\varphi(c_0))} \qquad (25)$$

where $a_i, c_j \in K$ for $i = 0, 1, \ldots, N$; $j = 0, 1, \ldots, M$; $a_N, c_M \in K^{\times}$. With the help of the Kuratowski-Zorn lemma we are able to extend φ to K(B), and later, applying Theorem 19, to L.

In general, there are many ways to effect prolongation of φ from K to its arbitrary extensions. In the case of Archimedean absolute values it is, generally, not always possible (see Chapter 3, Section 2).

DEFINITION 13 Let K be a non-Archimedean valued field and φ its absolute value. We say that (K,φ) is *maximally complete* if φ cannot be extended without changing the value group $\varphi(K^{\times})$ or the residue-class field $\bar{k}_{\varphi}$.

DEFINITION 14 A non-Archimedean valued field (K,φ) is called *spherically complete* if every decreasing sequence of balls has a common point.

It can be shown (see Schilling [6, pp. 36-39]) that a non-Archimedean valued field is spherically complete if and only if it is maximally complete. This notion plays an important role in non-Archimedean analysis (see Monna [1] or Narici, Beckenstein, and Bachman [1]).

DEFINITION 15 An absolute-valued field (K,φ) is said to have the *Hahn-Banach property* if for every non-Archimedean normed vector space V over (K,φ), every linear functional defined on a subspace of V possesses an extension of the same norm defined on the whole of V.

"Good" non-Archimedean normed vector spaces are characterized by the following

THEOREM 23 (Ingleton [1, Th. 3]) A non-Archimedean absolute-valued field has the Hahn-Banach property if and only if it is spherically complete.

For a detailed proof, see Narici, Beckenstein, and Bachman [1, Chap. 3.3, especially Th. 1, p. 78].

NOTES

An absolute value on a division ring is defined in the same way as for fields. We retain the notation of Chapter 2.

It is natural to ask when a division ring topology can be induced by an absolute value. Let $(K,\mathcal{T})$ be a topological division ring, let N_0 be the set of all topological nilpotents of $(K,\mathcal{T})$, and let N be the set of all neutral elements. Kaplansky [2] proved the following

THEOREM For a topological division ring $(K,\mathcal{T})$, the following conditions are equivalent.

(i) $\mathcal{T}$ is induced by an absolute value.

(ii) The set N_0 is open, $N \cup N_0$ is right bounded, and the commutator group of the multiplicative group $K^\times$ of K is right bounded.

There are examples of topological division rings such that N_0 is open, $N \cup N_0$ is right bounded, the commutator group of K is not right bounded, and the topology is not induced by an absolute value (see Kowalsky and Dürbaum [1]).

It should be interesting to describe all absolute values on some classes of division rings, as in the case of fields. Here rather little is known. We shall mention some results of Satyanarayanamoorthy [1,2], stated without proof.

THEOREM Let K be a composition division algebra over the rational numbers $\mathbb{Q}$. An Archimedean absolute value v on K is defined as in the case of fields. We further impose the condition $v(x) = v(\bar{x})$, where the bar denotes the conjugation of K.

(i) If K admits an Archimedean absolute value v, then $v = v_0^t$ $(0 < t \le 1)$, where $(v_0(x))^2 = |N(x)|$, $x \in K$, $N(x) = x\bar{x}$ being the norm of K. In particular, v is unique up to equivalence.

(ii) If K is octonian, it admits an Archimedean absolute value.

(iii) If K is a quaternion algebra with the norm form $N(x) = x_0^2 + ax_1^2 + bx_2^2 + abx_3^2$, then K admits an Archimedean absolute value if and only if a and b are positive.

(iv) If K is quartic with the norm form $N(x) = x_0^2 + ax_1^2$, K admits an Archimedean absolute value if and only if a is positive.

A counterpart to the preceding theorem is the following

THEOREM Let K be a composition division algebra over the rational numbers $\mathbb{Q}$. We consider a non-Archimedean absolute value v on K satisfying $v(x) = v(\bar{x})$. If K admits a non-Archimedean absolute value v, then $v = v_0^t$ $(0 < t \le 1)$, where $(v_0(x))^2 = |N(x)|_p$, $N(x) = x\bar{x}$ being the norm of K and $|\cdot|_p$ the p-adic absolute value of $\mathbb{Q}$. Moreover,

(i) If K is octonian, it does not admit any nontrivial non-Archimedean absolute value.

(ii) If $K = A(m,n)$ (m, n odd positive integers) is a quaternion division algebra with the norm form $N(x) = x_0^2 + mx_1^2 + nx_2^2 + mnx_3^2$, then v_0 [where $(v_0(x))^2 = |N(x)|_2$] is a non-Archimedean absolute value if and only if $m \equiv n \equiv 1 \pmod 4$.

(iii) v [where $(v(x))^2 = |N(x)|_p$, p an odd prime number] is a non-Archimedean absolute value of $K = A(pm,pn)$ [here $(m,p) = (n,p) = 1$] if and only if $(mn/p)(-1/p) = -1$.

Some related results dealing with extensions of absolute values can be found in Mothon [1].

3

Normed Fields and Their Properties

In this chapter we consider a generalization of the notion of an absolute value—the notion of a norm. The norms of fields and rings were introduced and examined in a series of papers by Mahler [1-4].

1. EXAMPLES AND PROPERTIES OF NORMED FIELDS

DEFINITION 1 A *norm on a field* K is defined as a real-valued function $p: K \to \mathbf{R}_+$ such that

(N_1) $p(0) = 0$; there exists $x \in K^\times$ such that $p(x) > 0$.

(N_2) $p(xy) \le p(x)p(y)$.

(N_3) $p(x - y) \le p(x) + p(y)$.

A *norm* p is *non-Archimedean* if it satisfies the *ultrametric inequality*

(N_3') $p(x - y) \le \max(p(x), p(y))$.

It is not difficult to see that $x \in K^\times$ implies that $p(x) > 0$. In fact, the set $I = p^{-1}(0)$ is an ideal in K, and since p is not zero identically, we have $I = (0)$; thus $p(x) = 0$ implies that $x = 0$.

Note that if p is any norm on a field K, then

(N_2') $p(ab^{-1}) \ge p(a)p(b)^{-1}$ holds for every $a,b \in K$, $b \ne 0$.

Indeed, write $a = (ab^{-1})b$ and apply (N_2).

Example 1 (Compare Chapter 1, Example 6) Let $K = \mathbb{Q}$. Define

$$p(x) = \max(|x|, |x|_q) \qquad q \text{ a prime number}$$

It is immediate that p is a nontrivial norm on $\mathbb{Q}$.

Example 2 Let $\varphi_1, \varphi_2, \ldots, \varphi_n$ be absolute values on a field K. Put

$$p(x) = \max(\varphi_1(x), \varphi_2(x), \ldots, \varphi_n(x)) \tag{1}$$

The formula (1) again defines a norm on K. We shall see later that for many fields this formula gives all norms on K, up to equivalence.

Other examples of norms will be given in Section 5. Let $\mathcal{T}_p$ be the topology defined by the base $\mathcal{B}$ of the neighborhoods of zero, consisting of balls

$$U_\varepsilon = \{a \in K : p(a) < \varepsilon\} \qquad \varepsilon \in \mathbf{R}_+^\times \tag{2}$$

DEFINITION 2 Let p and q be two *norms* on a field K. We say that p and q are *equivalent*, $p \sim q$, if they define the same topology, i.e., if $\mathcal{T}_p = \mathcal{T}_q$.

There is no equivalent condition for $p \sim q$, similar to that in the case of absolute values (Chapter 2, Theorem 1). For example, if p is a non-Archimedean norm and $f: \mathbf{R} \to \mathbf{R}$ is a strictly increasing function such that $f(0) = 0$, $f(xy) \le f(x)f(y)$ and f is continuous near 0, then $f \circ p$ is a norm; moreover, $f \circ p \sim p$.

THEOREM 1 Let p be a norm on a field K. Then the topology $\mathcal{T}_p$ having the sets (2) for a base of the neighborhoods of zero is a field topology.

Proof: If p is trivial, i.e., $p(x) \ge c > 0$ (c a constant) for $x \in K^\times$, then $\mathcal{T}_p = \mathbf{1}_K$ and there is nothing to do.

Suppose that p is nontrivial. The continuity of multiplication follows from (N_2) and (N_3). Indeed, let $p(x - x_0) < \varepsilon$ and $p(y - y_0) < \varepsilon$. Then, using $xy - x_0y_0 = x(y - y_0) + y_0(x - x_0)$, we have

$$p(xy - x_0y_0) \le p(x)p(y - y_0) + p(y_0)p(x - x_0)$$
$$< \varepsilon(p(x_0) + p(y_0) + \varepsilon)$$

The continuity of addition and subtraction follows from (N_3), since $p(-y) = p(y)$.

Finally, to prove the continuity of the inverse, let $p(x - x) < \varepsilon$, $x, x_0 \in K^\times$. We have

$$|p(x^{-1}) - p(x_0^{-1})| \leq p(x^{-1} - x_0^{-1}) = p(x^{-1}(x_0 - x)x_0^{-1})$$
$$\leq p(x^{-1})p(x_0 - x)p(x_0^{-1}) \quad (3)$$

since (N_3) implies that

$$|p(a) - p(b)| \leq p(a - b) \text{ holds for all } a,b \in K \quad (4)$$

Let

$$U = \left\{x \in K : p(x_0 - x)p(x_0^{-1}) < \frac{1}{2}\right\} \quad (5)$$

For arbitrary $x \in U$, applying (3), we find that

$$p(x^{-1}) - p(x_0^{-1}) \leq |p(x^{-1}) - p(x_0^{-1})| < \frac{1}{2}\, p(x^{-1}) \quad \text{or}$$
$$p(x^{-1}) < 2p(x_0^{-1}) \quad \text{for } x \in U \quad (6)$$

It implies, again by (3), that $|p(x^{-1}) - p(x_0^{-1})| < 2(p(x_0^{-1}))^2 p(x - x_0)$ holds for all $x \in U$. Since U is a neighborhood of x_0, this proves the continuity of division in (K,T_p).

LEMMA 1 Let p be a nontrivial norm on a field K and A an arbitrary subset of K. The following conditions are equivalent.

(i) A is a bounded subset of (K,T_p).

(ii) There exists $c \in \mathbb{R}_+^\times$ with $A \subset U_c$.

Proof: By definition, A is a bounded subset of (K,T_p) if and only if for every $\varepsilon > 0$ there exists $\delta > 0$ such that $p(x) < \delta$ implies that $p(ax) < \varepsilon$ for every $a \in A$. Suppose that p is nondiscrete.

(i) $\Rightarrow$ (ii) Putting $\varepsilon = 1$, we find $\delta > 0$ such that $p(x) < \delta$ implies $p(ax) < 1$ for all $a \in A$. Since T_p is nondiscrete, there is an element $x \in K^\times$ such that $p(x) < \delta$. Then $p(ax) < 1$ for $a \in A$ and $p(a) = p(axx^{-1}) \leq p(ax)p(x^{-1}) < p(x^{-1})$; thus $A \subset U_c$, $c = p(x^{-1})$.

(ii) $\Rightarrow$ (i) If $p(a) < c$ for every $a \in A$ and $p(x) < \delta$, then $p(ax) \leq p(a)p(x) < c\delta$. Thus the definition is satisfied with $\delta = \varepsilon c^{-1}$.

To finish this section, note that if p is a nontrivial norm on a field K, then the set N_0 of topological nilpotents of (K, T_p) is nonzero. Indeed, since $T_p \neq \mathbb{1}_K$, there is an $a \in K^\times$ satisfying $p(a) < 1$. But then $p(a^n) \le p(a)^n \to 0$, i.e., $a \in N_0$.

2. THE GELFAND-MAZUR THEOREM

In this section we present an elementary proof of the Gelfand-Mazur theorem, due to Tornheim [1].

DEFINITION 3 Let $L > K$ be a field extension and let q, p be norms on L and K, respectively. We say that q is an extension of p if $q = p$ on K.

In particular, if q extends p, then $T_q|_K = T_p$, but not conversely.

LEMMA 2 If a norm p on a field L extends an absolute value φ of its subfield K, then

$$p(xy) = \varphi(x)p(y) \tag{7}$$

holds for every $x \in K$, $y \in L$. Conversely, if (7) is satisfied for all $x \in K$, $y \in L$, then $T_p|_K = T_\varphi$.

Proof: Let $x \in K^\times$, $y \in L$. Then

$$p(y) = p((xy)x^{-1}) \le p(xy)p(x^{-1}) = p(xy)\varphi(x^{-1}) = p(xy)\varphi(x)^{-1}$$

or $\varphi(x)p(y) \le p(xy)$. But, by definition (N_2), $p(xy) \le p(x)p(y) = \varphi(x)p(y)$, which gives the equality (7).

If (7) holds, we have $p(x) = \varphi(x)p(1)$. From Definition 1 it follows only that $p(1) \ge 1$. It means that $T_p|_K = T_\varphi$.

THEOREM 2 (Gelfand [1]; Mazur [1]) Every normed field over the reals $\mathbb{R}$ is either the real number field $\mathbb{R}$ or the complex number field $\mathbb{C}$.

THEOREM 3 The complex number field $\mathbb{C}$ is the only normed extension of itself.

At first we shall prove Theorem 3, and then we shall deduce Theorem 2 from it. In Chapter 9, Section 4, we present a more general result.

Proof of Theorem 3: Let K be a normed extension of $\mathbb{C}$ and p its norm. Putting in (5) and (6) $x_0 = 1$, $x = 1 - y$, one observes that

$$p(y) < (2p(1))^{-1} \quad \text{implies} \quad p((1 - y)^{-1}) < 2p(1) \tag{8}$$

We shall prove that for every $y \in K$ there exists $c \in \mathbb{C}$ such that $y - c$ does not have a reciprocal. Then since K is a field, $y - c = 0$ and $y = c \in \mathbb{C}$. Note first that

For every $y \in K^{\times}$ there exists $k > 0$ such that if $y \neq c$ and $|c| > k$, then $p((y - c)^{-1}) < p(y^{-1})$. (9)

In fact, $(y - c)^{-1} = -c^{-1}(1 - y/c)^{-1}$; if $p(y/c) < 1/2$, i.e., if $|c| > 2p(y)$, then applying (8), $p((1 - y/c)^{-1}) < 2p(1)$. Thus $p((y - c)^{-1}) < 2|c|^{-1}p(1)$. Since the right side tends to zero as $|c| \to \infty$, we have (9).

Suppose that $y - c$ has a reciprocal for every $c \in \mathbb{C}$. This implies that $p((y - c)^{-1})$ is a continuous function of c. Therefore, $p((y - c)^{-1})$ attains a maximum M on the compact ball $\{c \in \mathbb{C} : |c| \leq k\}$, where k is given by (9). This value is not attained for $|c| > k$, since then $p((y - c)^{-1}) < p(y^{-1}) \leq M$. Let $C_0 = \{c \in \mathbb{C} : p((y - c)^{-1}) = M\}$. The set C_0 is nonempty and compact. Let $c_0 \in C_0$ be a boundary point of it. Moreover, let $d \notin C_0$ be such that $r = |d - c_0| < 1/M$. Let $S = \{c \in \mathbb{C} : |c - c_0| = r\}$ and consider the vertices $c_k = c_0 + r\varepsilon^k \in S$ $(k = 1, 2, \ldots, n)$, where $\varepsilon = \exp(2\pi i/n)$, of a regular n-gon. Let

$$S(n) = \frac{1}{n} \sum_{k=1}^{n} \frac{1}{y - c_k} \tag{10}$$

If $f(X) = X^n - r^n = \Pi_{k=1}^{n} (X - r\varepsilon^k)$, the logarithmic derivative of f equals

$$\frac{nX^{n-1}}{X^n - r^n} = \sum_{k=1}^{n} \frac{1}{X - r\varepsilon^k} \tag{11}$$

Putting $x = y - c_0$ in (11) and dividing by n, we obtain

$$S(n) = \frac{x^{n-1}}{x^n - r^n} = x^{-1} \frac{1}{1 - (r/x)^n} \tag{12}$$

It follows from the definition of r and from (8) that

$$\lim_{n\to\infty} p(S(n)) = p(x^{-1}) = M \tag{13}$$

since $p(r/x) = rp((y - c_0)^{-1}) = rM < 1$. Indeed,

$$p(S(n)) \le \frac{1}{n} \sum_{k=1}^{n} p((y - c_k)^{-1}) \le \frac{1}{n} nM = M$$

by (9) and from the definition of M. On the other hand, applying (N_2') to (12), by putting $a = x^{-1}$ and $b = 1 - (r/x)^n$ in (12), one obtains, finally

$$\begin{aligned} M \ge p(S(n)) &\ge p(x^{-1})p\left(1 - \left(\frac{r}{x}\right)^n\right) \ge p(x^{-1})\left(p(1) - p\left(\left(\frac{r}{x}\right)^n\right)\right) \\ &\ge p(x^{-1})(1 - p(rx^{-1})^n) = p(x^{-1})(1 - (rp(x^{-1}))^n) \\ &\ge M(1 - (rM)^n) \end{aligned}$$

Letting $n \to \infty$, (13) follows.

Now d lies on the circle S and not in the closed set $S \cap C_0$. Therefore, d lies on a nonempty open arc of $S \setminus C_0$ and hence on a closed arc S_1 in $S \setminus C_0$ of length $2\pi r/m$, for $m \in \mathbb{N}$ sufficiently large. Hence if mq points are taken on S equally spaced, q of them will lie on S_1. Because S_1 is closed and bounded, $p((y - c)^{-1})$ attains a maximum M_1 for $c \in S_1$ and $M_1 < M$, since $S_1 \cap C_0 = \emptyset$. Then, since q values of c_k lie in S_1,

$$p(S(mq)) \le \frac{1}{mq} (qM_1 + (mq - q)M)$$

by a simple calculation. This implies that $p(S(mq)) \le M - [(M - M_1)/m]$, so that $p(S(mq))$ is bounded outside M for all values of q, but this contradicts (13). Hence the theorem.

Proof of Theorem 2: Let K be a normed field over $\mathbb{R}$, and define $L = K(i)$. L becomes a normed field over $\mathbb{C}$ by putting $q(a + ib) = p(a) + p(b)$, where p is the norm of K. Applying Theorem 3, we have

$L = \mathbb{C}$, hence $\mathbb{R} \subset K \subset \mathbb{C}$; since K is a field, we must have $K = \mathbb{R}$ or $K = \mathbb{C}$.

COROLLARY 1 The complex number field is algebraically closed.

Proof: (Iseki [1]) Let $A = \mathbb{C}[X]$ and put $p(f) = |a_0| + \cdots + |a_m|$ if $f(X) = a_m X^m + \cdots + a_0$. It is immediate that A is a normed algebra over $\mathbb{C}$ and that $p(a) = |a|$ for $a \in \mathbb{C}$. Let $g \in \mathbb{C}[X]$ be a fixed unitary polynomial of degree $n \geq 1$. Define an operator F: $A \to A$, putting

$$F(f) = \begin{cases} f & f = 0 \text{ or } \deg(f) < n \\ f - a_m X^{m-n} g & \text{if } f(X) = a_m X^m + \cdots + a_0,\ a_m \neq 0,\ m \geq n \end{cases} \tag{14}$$

We have

$$p(F(f)) \leq p(f) + p(f)p(g) = Mp(f) \qquad [M = p(g) + 1] \tag{15}$$

and

$$p(F^n(f)) \leq M^n p(f) \tag{16}$$

Let g be irreducible over $\mathbb{C}$. We shall prove that $n = \deg(g) = 1$. Obviously, $K = \mathbb{C}[X]/(g)$ is a field extension of $\mathbb{C}$. Let $y = X + (g) \in K$, i.e., $g(y) = 0$ in K. For every $a \in K$ there exists exactly one $f_a \in A$ with $a = f_a(y)$, $\deg(f_a) < n$. Define

$$p_1(a) = p(f_a) \tag{17}$$

We have

$$p_1(a) \geq 0 \qquad p_1(a) = 0 \Leftrightarrow a = 0 \tag{18}$$

$$p_1(a + b) \leq p_1(a) + p_1(b) \qquad \text{since } f_{a+b} = f_a + f_b \tag{19}$$

$$p_1(ab) \leq M^n p_1(a) p_1(b) \tag{20}$$

Indeed, $f_{ab} \equiv f_a f_b \pmod g$, or $f_{ab} = F^n(f_a f_b)$, since $\deg(f_a f_b) \leq 2n - 2$. Thus

$$\begin{aligned} p_1(ab) &= p(F^n(f_a f_b)) \leq M^n p(f_a f_b) \leq M^n p(f_a) p(f_b) \\ &= M^n p_1(a) p_1(b) \end{aligned}$$

By putting

$$q(a) = M^n p_1(a) \qquad \text{for } a \in K$$

K becomes a normed field over $\mathbb{C}$. Theorem 3 gives that $K = \mathbb{C}$, thus $n = \deg(g) = 1$.

COROLLARY 2 (Ostrowski [3, p. 284]) Let $(K, \mathcal{T}_v)$ be a field with an Archimedean absolute value v. Then there exists an isomorphism g from K into $\mathbb{C}$ and a real number $t > 0$ such that $v(x) = |g(x)|^t$ holds for every $x \in K$. In particular, every complete Archimedean absolute-valued field is topologically isomorphic either to $\mathbb{R}$ or $\mathbb{C}$.

Proof: Indeed, it is sufficient to consider the completion of $\mathbb{Q}$ [char(k) = 0, since, by assumption, v is Archimedean] with respect to $\mathcal{T}_v$ and next apply Theorem 2.

Remark There exist metrizable field topologies on the rational function field $\mathbb{C}(X)$ inducing the ordinary topology on $\mathbb{C}$ (Boeme [1]; Williamson [1]).

Every such topology is locally unbounded (by Theorem 13 of Chapter 9). For a more general result in this direction, see Arnautov [9].

3. TOPOLOGICAL CHARACTERIZATIONS OF NORMED FIELDS

The question arises: What are the necessary and sufficient conditions for the field topology $\mathcal{T}$ of K to be induced by a norm? The answer is given by the following theorem of Cohn [1].

THEOREM 4 Let $(K, \mathcal{T})$ be a field with a proper ring topology. The following conditions are equivalent.

(i) $\mathcal{T}$ is induced by a norm.

(ii) K contains a nonempty open bounded set and there exists a nonzero topological nilpotent in $(K, \mathcal{T})$.

(iii) $(K, \mathcal{T})$ has a nonempty bounded open set consisting of topological nilpotents.

Moreover, a topology $\mathcal{T}$ can be induced by a non-Archimedean norm if and only if $(K, \mathcal{T})$ has a nonempty open and bounded subgroup H of K^+ satisfying $H \subset N_0$.

The original proof of Cohn's theorem is based on the notion of a gauge set.

DEFINITION 4 A subset A of a field K will be called a *gauge set* of K if $-1 \in A$, $AA \subset A$, there exists a $c \in K^{\times}$ such that $A + A \subset cA$, $A \neq K$, and there exists a $d \in K^{\times}$ with

$$\bigcup_{n \in \mathbb{N}} d^{-n}A = K \tag{21}$$

In other words, a gauge set A is an almost-order for which there is an element $d \in K^{\times}$, satisfying (21). It can be shown (Cohn [1]) that there is one-to one correspondence between the sets of equivalent classes of gauge sets and norms on K.

A subset B of K is said to be linear if given $x \in K^{\times}$, either $x \in B$ or $x^{-1} \in B$. It turns out that a nontrivial norm p on a field is equivalent to an absolute value if and only if it is associated with a linear gauge set (Cohn [1]).

To prove Theorem 4, we shall present a general result of Warner [3]. Cohn's theorem will follow from it as a corollary.

THEOREM 5 Let $(A,\mathcal{T})$ be a proper topological ring whose center contains a cancellable topological nilpotent a [i.e., $xa = ya$ (resp. $ax = ay$) implies that $x = y$] such that $x \mapsto ax$ is an open mapping. If $(A,\mathcal{T})$ has a bounded neighborhood of zero (an open and bounded additive subgroup), then the topology $\mathcal{T}$ of A is defined by a norm (a non-Archimedean norm).

Proof: Let $S = \{a, a^2, a^3, \ldots\}$ be the multiplicative semigroup generated by a. Let $A_1 = S^{-1}A$ be the quotient ring of A with respect to the multiplicative semigroup S. We topologize A_1 by declaring a filter base $\mathcal{B}$ of the neighborhoods of zero in $(A,\mathcal{T})$ as a base in A_1. Clearly, A_1 is a topological ring and A is its open subring, since every mapping $x \mapsto a^n x$ is open. Consequently, replacing A by A_1, if necessary, we may assume that a is invertible in A.

Let $\mathcal{B}$ be a basis of symmetric neighborhoods of zero in $(A,\mathcal{T})$. Let $V \in \mathcal{B}$ be a bounded set. Put

$$U = \{x \in A : xV \subset V\} \tag{22}$$

Since $V = -V$ and $V \in \mathcal{B}$ is bounded, there exists $V_1 \in \mathcal{B}$ such that $V_1 V \subset V$. Thus $V_1 \subset U$ and $U \in \mathcal{B}$. Now we show that U is bounded. Let $W \in \mathcal{B}$. Then there is $T \in \mathcal{B}$ satisfying $VT \subset W$, $TV \subset W$. Since $a^n \to 0$, we have $a^k \in V$ for some k. As $x \mapsto a^k x$ is open, $a^k T \in \mathcal{B}$ and $Ua^k T \subset VT \subset W$, and similarly, $a^k TU = TUa^k \subset W$, since a (and consequently a^k) lies in the center of A. Thus U is bounded and satisfies $UU \subset U$.

Let $U_1 \in \mathcal{B}$ be such that

$$U_1 \subset U \qquad \text{and} \qquad U_1 U + U_1 U + U_1 U \subset U \tag{23}$$

Replacing a by a power of a, if necessary, we may assume that $a \in U_1$. This gives $aU + aU + aU \subset U$. Clearly, $a^k U$, $k \in \mathbb{Z}$, is a decreasing sequence of neighborhoods of zero. It is a fundamental system of such neighborhoods. Indeed, let $V_2 \in \mathcal{B}$ be arbitrary. Then, as U is bounded, there exists $V_3 \in \mathcal{B}$ satisfying $UV_3 \subset V_2$ and $V_3 U \subset V_2$. Since $a^n \to 0$, one can find $s \in \mathbb{N}$ such that $a^s \in V_3$. It implies that $a^s U \subset V_3 U \subset V_2$. This implies that $\mathcal{T}$ is the topology induced by $a^k U$, $k \in \mathbb{Z}$. In particular, we have

$$\bigcap_{k \in \mathbb{N}} a^k U = \{0\} \tag{24}$$

Put

$$g(0) = 0, \qquad g(x) = 2^{-k} \text{ for } x \in A^\times \tag{25}$$

where $k = \max\{n \in \mathbb{Z} : x \in a^n U\}$. Let us write $k = v(x)$ if $x \neq 0$. Thus $g(x) = 2^{-v(x)}$. Such an integer exists, since $\lim a^n x = 0$, hence $x \in a^{-n} U$ for a suitable $n \geq 0$. If V is an additive subgroup of A, then $U \subset A$ is a subring and g is a non-Archimedean norm defining the topology $\mathcal{T}$.

Let

$$f(x) = \inf\left\{ \sum_{i=1}^{N} g(x_i) : x_1 + \cdots + x_N = x \right\} \tag{26}$$

Since $aU + aU + aU \subset U$, we have $a^{n+1}U + a^{n+1}U + a^{n+1}U \subset a^n U$ for every $n \in \mathbb{Z}$. The definition of v implies that $v(xy) \geq v(x) + v(y)$ and $v(x + y + z) \geq \min\{v(x),v(y),v(z)\} - 1$. Indeed, let $k = v(x)$, $\ell = v(y)$, $m = v(z)$, i.e., $x \in a^kU$, $y \in a^\ell U$, $z \in a^mU$. Then $a^kUa^\ell U \subset a^{k+\ell}U$, since a lies in the center of A and $UU \subset U$. Thus $v(xy) \geq k + \ell = v(x) + v(y)$. Now suppose that $m \leq \ell \leq k$. Then

$$x + y + z \in a^kU + a^\ell U + a^mU \subset a^mU + a^mU + a^mU \subset a^{m-1}U$$

This implies that $v(x + y + z) \geq m - 1 = \min\{v(x),v(y),v(z)\} - 1$. Thus, applying the inequalities above, we have

$$g(xy) \leq g(x)g(y) \qquad \text{and} \qquad g(x + y + z) \leq 2 \max\{g(x),g(y),g(z)\}$$

We shall prove a more general inequality

$$\frac{1}{2}\, g(x_1 + \cdots + x_N) \leq g(x_1) + \cdots + g(x_N) \tag{27}$$

For $N = 1, 2, 3$ it has just been proved. We shall proceed by induction on N. Let $x = x_1 + \cdots + x_N$. Suppose that $x \in U$, multiplying x, if necessary, by a suitable power a^n of a ($n \geq 0$). Put $d = \Sigma_{i=1}^N g(x_i)$. If $d \geq 1/2$, there is nothing to prove since $g(x) \leq 1$. Hence let $d < 1/2$. Let p be the maximal index such that $\Sigma_{i<p} g(x_i) \leq d/2$. This implies that $\Sigma_{i<p+1} g(x_i) > d/2$. Thus $\Sigma_{i>p} g(x_i) \leq d/2$. We have $g(x_1 + \cdots + x_{p-1}) \leq d$ and $g(x_{p+1} + \cdots + x_N) \leq d$, by the inductive assumption (27). Let $k > 0$ be the least integer satisfying $2^{-k} \leq d$, i.e., $k \geq 2$. We have $x_1 + \cdots + x_{p-1} \in a^kU$, $x_{p+1} + \cdots + x_N \in a^kU$. Moreover, $x_p \in a^kU$, since $g(x_p) \leq d$. Thus finally,

$$\begin{aligned} x = (x_1 + \cdots + x_{p-1}) + x_p + (x_{p+1} + \cdots + x_N) &\in a^kU \\ + a^kU + a^kU &\subset a^{k-1}U \end{aligned}$$

hence $v(x) \geq k - 1$. This implies that $g(x) \leq (1/2)^{k-1} = 2^{-k} \cdot 2 \leq 2d$. We obtain immediately from (27) that $(1/2)g(x) \leq f(x) \leq g(x)$ holds for every $x \in A$. We shall prove that f is a norm on A. Indeed, $f(x) = 0$ implies that $g(x) = 0$, i.e., $x = 0$ by definition of g. The triangle inequality (N_3) follows directly from the definition of f:

$$f(x + y) = \inf\left\{\sum_{i=1}^{k} g(z_i) : x + y = z_1 + \cdots + z_k\right\}$$

$$\leq \inf\left\{\sum_{i=1}^{n} g(x_i) + \sum_{j=1}^{m} g(y_j) : x = \sum_{i=1}^{n} x_i,\ y = \sum_{j=1}^{m} y_j\right\}$$

$$= \inf\left\{\sum_{i=1}^{n} g(x_i) : x = \sum_{i=1}^{n} x_i\right\}$$

$$+ \inf\left\{\sum_{j=1}^{m} g(y_j) : y = \sum_{j=1}^{m} y_j\right\}$$

$$= f(x) + f(y) \qquad (x_i, y_j, z_s \in A)$$

It remains to show that f satisfies (N_2), i.e., $f(xy) \leq f(x)f(y)$ holds for every $x,y \in A$. Indeed, let $x = \Sigma_{i=1}^{n} x_i$, $y = \Sigma_{j=1}^{m} y_j$. Then

$$f(xy) \leq \inf\left\{\sum_{i=1}^{n}\sum_{j=1}^{m} g(x_i y_j) : xy = \sum_{i=1}^{n}\sum_{j=1}^{m} x_i y_j\right\}$$

$$\leq \inf\left\{\sum_{i=1}^{n}\sum_{j=1}^{m} g(x_i)g(y_j) : x = \sum_{i=1}^{n} x_i,\ y = \sum_{j=1}^{m} y_j\right\}$$

since $g(ab) \leq g(a)g(b)$ holds for all $a,b \in A$. Consequently,

$$f(xy) \leq \inf\left\{\left(\sum_{i=1}^{n} g(x_i)\right)\left(\sum_{j=1}^{m} g(y_j)\right) : x = \sum_{i=1}^{n} x_i,\ y = \sum_{j=1}^{m} y_j\right\}$$

$$= \inf\left\{\sum_{i=1}^{n} g(x_i) : x = \sum_{i=1}^{n} x_i\right\} \cdot \inf\left\{\sum_{j=1}^{m} g(y_j) : y = \sum_{j=1}^{m} y_j\right\}$$

$$= f(x)f(y)$$

Thus f is a norm on A and $\mathcal{T}_f = \mathcal{T}$.

To prove Theorem 4, we shall apply the following.

LEMMA 3 Let $(K,\mathcal{T})$ be a proper locally bounded topological field. If the set N_0 of all topological nilpotents of $(K,\mathcal{T})$ contains a nonzero element, then it is open.

Proof: Let $0 \neq x \in N_0$ and let W denote a bounded neighborhood of zero in $(K,\mathcal{T})$ containing 1. Let $\mathcal{B}$ be a base of symmetric neighborhoods of zero. First we will prove that

There is $U \in \mathcal{B}$ such that for every $V \in \mathcal{B}$ there is an $n_0 \in \mathbb{N}$ such that $U^n \subset V$ for every $n \geq n_0$. (28)

Since $W \in \mathcal{B}$ is bounded, there exists $U \in \mathcal{B}$ satisfying $WU \subset xW$. Let $V \in \mathcal{B}$ be arbitrary. Then $x^n W \subset V$ for all $n \geq n_0$. Moreover,

$$V \supset x^n W = x^{n-1}(xW) \supset x^{n-1}WU = x^{n-2}(xW)U \supset x^{n-2}U^2W \supset \cdots \supset U^n$$
for all $n \geq n_0$

Thus we have found an open set $U \in \mathcal{B}$ such that

$$U \subset N_0 \text{ and, if } x \in N_0\text{, then for some } n \in \mathbb{N},\ x^n \in U \tag{29}$$

Observe that if for some x, $x^n \in U$, then $x \in N_0$. Indeed, in view of (29), the sequence $x^n, x^{2n}, x^{3n}, \ldots$ tends to zero, and by the continuity of multiplication, each of the sequences $k^{kn+j} = x^{kn}x^j$ $(k \in \mathbb{N},\ j = 1, 2, \ldots, n - 1)$ also tends to zero. Let $N \in \mathbb{N}$ be arbitrary. Applying the division algorithm, we can find $k \in \mathbb{N}$ and $1 \leq j \leq n - 1$, satisfying $N = kn + j$. Thus $x^N \to 0$ and $x \in N_0$. Concluding, we get

$$N_0 = \bigcup_{n \in \mathbb{N}} \{x \in K : x^n \in U\} \tag{30}$$

and since $x \mapsto x^n$ is continuous, the openness of N_0 follows.

Proof of Theorem 4: (i) ⇒ (iii) Let p be a nontrivial norm inducing $\mathcal{T}$. Then $U = \{x \in K : p(x) < 1\}$ is bounded and $U \subset N_0$, since $p(x^n) \leq p(x)^n \to 0$ for $x \in U$.

(iii) ⇒ (ii) Obvious.

(ii) ⇒ (i) This follows from Theorem 5. Indeed, since $a \in N_0$, $a \neq 0$, and $(K,\mathcal{T})$ is a topological field, it implies that the mapping $x \mapsto ax$ is open.

Lemma 3 shows, moreover, that N_0 is open. However, N_0 must not be bounded even if $(K,\mathcal{T})$ is locally bounded and $N_0 \neq \{0\}$, as we shall see later.

Theorem 4 can easily be extended to topological vector spaces (Singh [1]). Let $(V,\mathcal{T}_2)$ be a topological vector space over a proper topological field $(K,\mathcal{T}_1)$. Then $\mathcal{T}_1$ is induced by a nontrivial norm p and $\mathcal{T}_2$ is induced by a nontrivial pseudonorm $\|\cdot\|$ [i.e., $\|0\| = 0$; $\|x\| > 0$ for $x \in V$, $x \neq 0$; $\|x + y\| \leq \|x\| + \|y\|$ for $x, y \in V$; $\|ax\| \leq p(a)\|x\|$ for $a \in K$, $x \in V$] if and only if the following three conditions are satisfied.

(i) $(K,\mathcal{T}_1)$ has a nonempty bounded open set.
(ii) There exists a nonzero topological nilpotent in $(K,\mathcal{T}_1)$.
(iii) $(V,\mathcal{T}_2)$ has a nonempty bounded open set.

[Recall that a subset B of a topological vector space $(V,\mathcal{T}_2)$ over a topological field $(K,\mathcal{T}_1)$ is bounded if for every neighborhood U of zero in V there exists a neighborhood W of zero in K such that $BW \subset U$.]

THEOREM 6 Let K be a normed field and L a finite field extension of K. Then the topology of K can be extended to a topology of L, induced by a norm.

Proof: Let $\mathcal{T}$ be a topology of K induced by a norm, say p. Suppose that p is a nontrivial norm. Then the Cartesian topology $\mathcal{T}_L$ on L, as a finite-dimensional vector space over K, is a locally bounded field topology of L (Chapter 2, Theorem 17). Since $\mathcal{T} = \mathcal{T}_p \neq \mathbb{1}_K$, then $(K,\mathcal{T})$ has a nonzero locally bounded and open set consisting of topological nilpotents. Lemma 4 of Chapter 2 implies that $(L,\mathcal{T}_L)$ also has a nonzero open and bounded set consisting of topological nilpotents. Theorem 4 of this chapter shows that $\mathcal{T}_L$ is induced by a norm.

We shall give an alternative proof due to Vizitiu [4]. Before giving a proof, we state the following.

LEMMA 4 Let p be a norm on a field K, and $f \in K[X]$ be a monic polynomial. Then the norm p can be extended to a norm p_f on K[X], satisfying the following condition. Let $g \in K[X]$ be arbitrary. If q, $r \in K[X]$ are obtained by dividing g by f, i.e., $g = fq + r$, $\deg(r) < \deg(f)$, then $p_f(r) \leq p_f(g)$.

Proof: Let $f(X) = X^n + \Sigma_{i=0}^{n-1} a_i X^i$. There exists a positive real number r satisfying $r^n > \Sigma_{i=0}^{n-1} p(-a_i)r^i$. Let $p_f(g) = r$ if $g(X) = X$ and $p_f(g) = \Sigma_{j=0}^{N} p(c_j)r^j$ if $g(X) = \Sigma_{j=0}^{N} c_j X^j$, $c_N \neq 0$. We shall prove that p_f satisfics conditions (N_1)-(N_3) of Definition 1. Indeed, $p_f(a) = p(a)$ for every $a \in K$; i.e., p_f does not vanish identically. This implies (N_1). Moreover, p_f extends p. To prove (N_2) and (N_3), let

$$g_1(X) = \sum_{i=0}^{x} b_i X^i \qquad g_2(X) = \sum_{i=0}^{t} c_i X^i \qquad g_1, g_2 \in K[X]$$

We can assume that $s = t$, taking, if necessary, zero coefficients. Then

$$\begin{aligned} p_f(g_1 + g_2) &= \sum_{i=0}^{s} p(b_i + c_i)r^i \le \sum_{i=0}^{s} (p(b_i) + p(c_i))r^i \\ &= p_f(g_1) + p_f(g_2) \end{aligned}$$

This proves (N_3). Moreover,

$$\begin{aligned} p_f(g_1 g_2) &= p_f\left(\sum_{k=0}^{2s}\left(\sum_{i+j=k} b_i c_j\right)X^k\right) \\ &= \sum_{k=0}^{2s} p\left(\sum_{i+j=k} b_i c_j\right)r^k \le \sum_{k=0}^{2s}\left(\sum_{i+j=k} p(b_i)p(c_j)\right)r^k \\ &= \left(\sum_{i=0}^{s} p(b_i)r^i\right)\left(\sum_{j=0}^{s} p(c_j)r^j\right) \\ &= p_f(g_1)p_f(g_2) \end{aligned}$$

proving (N_2).

Now let $g \in K[X]$ be arbitrary, and let $g = qf + r$, where either $r = 0$ or $\deg(r) < \deg(f) = n$. We shall prove by induction with respect to $N = \deg(g)$ that $p_f(r) \le p_f(g)$. If N is less than n, there is nothing to do: $g = 0 \cdot f + r$, $r = g$, i.e., $p_f(r) = p_f(g)$. Suppose that it was proved for all g's such that $n \le N \le m$. If $g(X) = \Sigma_{i=0}^{m+1} b_i X^i$, then $g(X) = (b_{m+1}X^{m+1} + g_1(X))f(X) + r(X)$, $\deg(r) < n$,

$\deg(g_1) \le m - n$. Thus $g(X) - b_{m+1}X^{m+1-n}f(X) = g_1(X)f(X) + r(X)$. Since $\deg(g_1 f + r) \le m$, then by the inductive assumption, we have $p_f(r) \le p_f(g_1 f + r)$. We shall prove that $p_f(g_1 f + r) \ge p_f(g)$. Indeed,

$$g_1(X)f(X) + r(X) = \sum_{i=0}^{m} b_i X^i + \sum_{i=0}^{n-1} b_{m+1}(-a_i)X^{m+1-n+i}$$

This yields

$$\begin{aligned} p_f\left(\sum_{i=0}^{n-1} b_{m+1}(-a_i)X^{m+1-n+i}\right) &= \sum_{i=0}^{n-1} p(-b_{m+1}a_i)r^{m+1-n+i} \\ &\le \sum_{i=0}^{n-1} p(b_{m+1})p(-a_i)r^{m+1-n+i} \\ &= p(b_{m+1})r^{m+1-n}\sum_{i=0}^{n-1} p(-a_i)r^i \\ &< p(b_{m+1})r^{m+1-n}r^n = p(b_{m+1})r^{m+1} \end{aligned}$$

and

$$\begin{aligned} p_f(g_1 f + r) &= p_f\left(\sum_{i=0}^{m} b_i X^i + \sum_{i=0}^{n-1} b_{m+1}(-a_i)X^{m+1-n+i}\right) \\ &\le p_f\left(\sum_{i=0}^{m} b_i X^i\right) + p_f\left(\sum_{i=0}^{n-1} b_{m+1}(-a_i)X^{m+1-n+i}\right) \\ &\le \sum_{i=0}^{m} p(b_i)r^i + p(b_{m+1})r^{m+1} = \sum_{i=0}^{m+1} p(b_i)r^i = p_f(g) \end{aligned}$$

by a previous inequality. Consequently, $p_f(r) \le p_f(g)$.

Proof of Theorem 6: It suffices to consider the case of a simple algebraic extension only. Let $L = K(a)$ and let $f \in K[X]$ be the minimal polynomial of a, $\deg(f) = n$. Every element $b \in L$ has a form $b = g(a)$, $g \in K[X]$, $\deg(g) < n$. Put

$$\bar{p}(b) = p_f(g)$$

We shall prove that $\bar{p}$ is the norm of L extending p. Indeed,

$\bar{p}(b) = p_f(b) = p(b)$ holds for every $b \in K$, i.e., $\bar{p}|K = p$. Let $b_i = g_i(a)$, $\deg(g_i) < n$, be elements of L ($i = 1, 2$). Then $b_1 + b_2 = (g_1 + g_2)(a)$, $\deg(g_1 + g_2) < n$, and $\bar{p}(b_1 + b_2) = p_f(g_1 + g_2) \leq p_f(g_1) + p_f(g_2) = \bar{p}(b_1) + \bar{p}(b_2)$. Similarly, writing $b_1b_2 = g_1g_2(a) = r(a)$, where $g_1g_2 = fq + r$, $\deg(r) < n$, we obtain from Lemma 4 that

$$\bar{p}(b_1b_2) = p_f(r) \leq p_f(g_1g_2) \leq p_f(g_1)p_f(g_2) = \bar{p}(b_1)\bar{p}(b_2)$$

As a corollary from Theorem 6, by applying transfinite induction, we obtain

THEOREM 7 Every norm can be extended from a field to its algebraic closure. In particular, a norm can be extended from a field to an arbitrary algebraic extension.

4. REGULAR AND NONREGULAR NORMS ON FIELDS

DEFINITION 5 A norm p on a field K is called *regular* if the set N_0 of topological nilpotents in $(K, \mathcal{T}_p)$ is bounded. Otherwise, a norm p is called *nonregular*.

DEFINITION 6 Let A be a subset of the field K. Any element $x \in K$ is said to be *integrally dependent* on A if there is an element $a \in A$, $a \neq 0$, such that $ax^n \in A$ for every $n \in \mathbb{N}$. The set of all elements integrally dependent on A is called the *integral closure* of A and is denoted by A*. A is said to be *integrally closed* if A = A* (Cohn [1]; Singh [2]). A gauge set is called *regular* if it is equivalent to its integral closure.

Remark This notion is not equivalent to the classical notion of integral closure (see Zariski and Samuel [1, Vol. I, p. 256]).

Example 3 (A nonregular norm, Cohn [1]) Let $K = \mathbb{Q}(X_1, X_2, \ldots)$ be the rational function field in a countable set of indeterminates X_1, X_2, ... over $\mathbb{Q}$. Fix a prime number p. We shall define a gauge set A on K such that A is not regular. Then the norm associated with A provides an example of a nonregular norm.

A rational number $x \in \mathbb{Q}^\times$ is integral for p if $v_p(x) \geq 0$, where v_p is the p-adic exponential valuation of $\mathbb{Q}$. A polynomial with rational coefficients is said to be primitive if all its coefficients are integral for p and for at least one of them, say a, $v_p(a) = 0$.

We define A as the subring of K generated by $\{x \in \mathbb{Q} : v_p(x) \geq 0\}$, the elements f^{-1}, where f is a primitive element of $\mathbb{Q}[X_1, X_2, \ldots]$ and the elements $pX_1^n, p^2X_2^n, p^3X_3^n, \ldots$ $(n \in \mathbb{N})$. It can be shown (Cohn [1, pp. 176-177]) that A is not equivalent to A*.

DEFINITION 7 A norm p on a ring K is called *power multiplicative* if it satisfies $p(x^N) = p(x)^N$ for every $x \in K$ and every $N \in \mathbb{N}$.

The norms in Examples 1 and 2 are power multiplicative.

DEFINITION 8 Let p be a norm on a field K. The function $p_{sp} : K \to \mathbb{R}_+$ defined as $p_{sp}(x) = \lim_{n\to\infty}(p(x^n))^{1/n}$ is called the *spectral norm associated with* p (or the homogenization of p).

LEMMA 5 Let p be a norm on a field K. Then the homogenization p_{sp} of p is a power-multiplicative norm satisfying $p_{sp}(x) \leq p(x)$ for every $x \in K$.

Proof: We show first that the limit

$$\lim_{n\to\infty}(p(x^n))^{1/n} \tag{31}$$

exists for every $x \in K$. Indeed, let $a = \inf\{(p(x^n))^{1/n} : n \in \mathbb{N}\}$. Obviously, $0 \leq a \leq p(x)$. For every $\varepsilon > 0$ there exists $m \in \mathbb{N}$ satisfying

$$(p(x^m))^{1/m} \leq a + \varepsilon \tag{32}$$

Let n be arbitrary. Then $n = qm + r$, where $q, r \in \mathbb{N}$ and $0 \leq r < m$. It follows that

$$(p(x^n))^{1/n} \leq (p(x^m)^q p(x^r))^{1/n} = (p(x^m))^{q/n}(p(x^r))^{1/n}$$

Since

$$\frac{q}{n} = \frac{q}{qm + r} = \frac{1}{m}\left(\frac{mq + r}{mq + r} - \frac{r}{n}\right) = \frac{1}{m}\left(1 - \frac{r}{n}\right)$$

we have

$$(p(x^m))^{q/n} = [(p(x^m))^{1/m}]^{1-r/n} \leq (a + \varepsilon)^{1-r/n} \qquad \text{by (32)}$$

Since r/n tends to zero as n tends to infinity, we find that

$$(p(x^n))^{1/n} \leq (a+\varepsilon)^{1-r/n}(p(x^r))^{1/n} \leq (a+\varepsilon)^{1-r/n}p(x)^{r/n}$$
$$\leq a + 2\varepsilon$$

for sufficiently large n. Thus p_{sp} is well defined.

Now we will check Definition 1. Since $(p(x^n)^{1/n} \leq p(x)$ for every $n \in \mathbb{N}$ and $x \in K$, so $p_{sp}(x) \leq p(x)$, implying that (N_1) holds. The condition (N_2) for p_{sp} follows from the inequality $p((xy)^n) \leq p(x^n)p(y^n)$. The proof of the triangle inequality (N_3) for p_{sp} runs as follows.

Let us note first that (N_2) and (N_3) imply that

$$p((x+y)^n) = p\left(\sum_{k=0}^{n} \binom{n}{k} x^k y^{n-k}\right) \leq \sum_{k=0}^{n} \binom{n}{k} p(x^k)p(y^{n-k}) \tag{33}$$

Let $a > p_{sp}(x)$ and $b > p_{sp}(y)$ be arbitrary. Definition 8 implies that there exists $m \in \mathbb{N}$ such that

$$p(x^n)^{1/n} < a \qquad \text{and} \qquad p(y^n)^{1/n} < b$$

hold for every $n \geq m$. Let $c = p(x)$, $d = p(y)$. Then the inequalities $p(x^n)^{1/n} \leq c$ and $p(y^n)^{1/n} \leq d$ are satisfied for all $n \in \mathbb{N}$. Taking $n > 2m$ and applying (33), we obtain

$$\begin{aligned}
p((x+y)^n) &\leq \sum_{k=0}^{m-1} \binom{n}{k} c^k b^{n-k} + \sum_{k=m}^{n-m} \binom{n}{k} a^k b^{n-k} + \sum_{k=n-m+1}^{m} \binom{n}{k} a^k d^{n-k} \\
&= \sum_{k=0}^{m-1} \binom{n}{k} a^k b^{n-k} \left(\frac{c}{a}\right)^k + \sum_{k=m}^{n-m} \binom{n}{k} a^k b^{n-k} \\
&\quad + \sum_{k=n-m+1}^{m} \binom{n}{k} a^k b^{n-k} \left(\frac{d}{b}\right)^{n-k} \\
&\leq \left(\sum_{k=0}^{n} \binom{n}{k} a^k b^{n-k}\right)\left(\max_{0\leq k\leq m-1} \left(\frac{c}{k}\right)^k + \max_{0\leq k\leq m-1} \left(\frac{d}{b}\right)^k + 1\right) \\
&= (a+b)^n C
\end{aligned}$$

where the constant $C > 1$ is independent of n. This implies that

$$p_{sp}(x + y) = \lim_{n\to\infty} p((x + y)^n)^{1/n} \le \lim_{n\to\infty}(a + b)C^{1/n} = a + b \quad (34)$$

Let $\varepsilon > 0$ be arbitrary. Let us put $a = p_{sp}(x) + \varepsilon/2$, $b = p_{sp}(y) + \varepsilon/2$. Thus

$$p_{sp}(x + y) \le p_{sp}(x) + p_{sp}(y) + \varepsilon \quad (35)$$

Since $\varepsilon > 0$ is arbitrary, (35) implies (N_3) for p_{sp}.

Finally, note that the spectral norm p_{sp} is power multiplicative. Indeed, for every $N \in \mathbb{N}$ the equality

$$p_{sp}(x^N) = \lim_{n\to\infty}(p(x^{nN}))^{1/n} = \lim_{n\to\infty}(p(x^{nN})^{1/nN})^N = p_{sp}(x^N)$$

holds.

LEMMA 6 A norm p on a field K is power multiplicative if and only if $p(x^2) = p(x)^2$ holds for every $x \in K$.

Proof: Let $p(x^2) = p(x)^2$ hold for every $x \in K$. Then, by induction on k, $p(x^{2^k}) = p(x)^{2^k}$ is satisfied for every $k \in \mathbb{N}$ and $x \in K$. The definition of p_{sp} implies that

$$p_{sp}(x) = \lim_{k\to\infty} p(x^{2^k})^{1/2^k} = p(x)$$

Since p_{sp} is power multiplicative, p also is power multiplicative.

The converse implication is obvious.

We shall use the following notation: If f and g are two real-valued functions defined on the same set X, we write $f \le g$ if and only if $f(x) \le g(x)$ holds for every $x \in X$. Then we say that f is subordinate to g.

The following criteria were given by Cohn [1] and Aurora [2]:

THEOREM 8 Let $(K,\mathcal{T})$ be a normed field, $\mathcal{T} = \mathcal{T}_p$. The following conditions are equivalent.

(i) p is a regular norm.

(ii) $p \sim \sup_{i\in I} \varphi_i$, where φ_i $(i \in I)$ are absolute values of K and I is a finite set.

(iii) $p \sim q$, where q is a power-multiplicative norm of K.

(iv) p is associated with a regular gauge set.

Proof: For a proof of the equivalence (i) $\Leftrightarrow$ (iv), see Cohn [1, Th. 11.1].

(i) $\Leftrightarrow$ (iii) The original proof can be found in Cohn [1, Th. 13.3]. Assume first that $p \sim q$, q being power multiplicative. Then $\mathcal{T}_p = \mathcal{T}_q$, and $N_0(\mathcal{T}_p) = N_0(\mathcal{T}_q) = \{x \in K : q(x) < 1\}$, since $q(x^n) = q(x)^n \to 0 \Leftrightarrow q(x) < 1$. Hence $N_0(\mathcal{T}_p)$ is open and bounded, i.e., p is a regular norm.

Conversely, let $N_0(\mathcal{T}_p)$ be a bounded set in $(K,\mathcal{T}_p)$, p being nontrivial (i.e., $\mathbb{0}_K \neq \mathcal{T}_p \neq \mathbb{1}_K$). Note first that $N_0(\mathcal{T}_p) = N_0(\mathcal{T}_q) - \{x \in K : q(x) < 1\}$, where $q = p_{sp}$ is the homogenization of the norm p. It is sufficient to show that $N_0(\mathcal{T}_p) = \{x \in K : q(x) < 1\}$, since the last equality $N_0(\mathcal{T}_q) = \{x \in K : q(x) < 1\}$ is evident, as $q = p_{sp}$ is power multiplicative. Let $x \in N_0(\mathcal{T}_p)$. Since $q \le p$ and $p(x^n) \to 0$, we have (by Lemma 5) $q(x^n) = q(x)^n \to 0$, i.e., $q(x) < 1$. This gives the inclusion $\subset$. Conversely, let $a = p_{sp}(x) < 1$. Then for every $\varepsilon > 0$ there exists n_0 such that for every $n \ge n_0$, $p(x^n)^{1/n} - a < \varepsilon$. Since $a < 1$, we can take an $\varepsilon > 0$ with $a + \varepsilon < 1$. Then $p(x^n) < (a + \varepsilon)^n < 1$ implies that $p(x^n) \to 0$, i.e., $x \in N_0(\mathcal{T}_p)$. This gives the reverse inclusion. Thus $N_0(\mathcal{T}_q) = N_0(\mathcal{T}_p)$, and $N_0(\mathcal{T}_q)$ is open and bounded in $(K,\mathcal{T}_p)$, since $N_0(\mathcal{T}_p)$ is open and bounded in $(K,\mathcal{T}_p)$, by (i). This implies that $\mathcal{T}_p = \mathcal{T}_q$, since $\{aN_0(\mathcal{T}_q) : a \in K^\times\}$ can be taken for a base of the neighborhoods of zero for $\mathcal{T}_p$ (and $\mathcal{T}_q$) (see Chapter 1, Theorem 8). Thus $p \sim q = p_{sp}$, q being the power-multiplicative norm of K.

Now we shall prove (following Aurora [2]) that (ii) $\Leftrightarrow$ (iii). The implication (ii) $\Rightarrow$ (iii) is obvious. To obtain the converse implication (iii) $\Rightarrow$ (ii), it is sufficient to show that $p \sim \sup\{\varphi_i : i \in I\}$. This implies that the set I is finite. Indeed, otherwise we can assume that I is infinite and $\varphi_i \not\sim \varphi_j$ for every pair $i \neq j$. Then the topology

$$\mathcal{T} = \mathcal{T}_p = \sup_f\{\mathcal{T}_{\varphi_i} : i \in I\}$$

could not be locally bounded (see a proof that the topology in Chapter 1, Example 12, is locally unbounded).

We shall need some lemmas before presenting a proof of (iii) ⇒ (ii). Let p be a power-multiplicative norm of a field K. Take an arbitrary $a \in K^{\times}$. Then we have the inequalities

$$p(x) \geq \frac{p(xa)}{p(a)} \geq \frac{p(xa^2)}{p(a)^2} \geq \cdots \geq \frac{p(xa^n)}{p(a)^n} \geq \cdots \geq 0$$

Thus for any $a \in K^{\times}$,

$$p_a(x) = \lim_{n\to\infty} p(xa^n)p(a)^{-n} \tag{36}$$

is well defined.

LEMMA 7 Let p be a power-multiplicative norm of a field K. Then p_a is a power-multiplicative norm of K such that:

(i) p_a is subordinate to p: $p_a \leq p$.

(ii) $p_a(a) = p(a)$.

(iii) $p_a(ax) = p_a(a)p_a(x)$ holds for every $x \in K^{\times}$.

Proof of Lemma 7: It is easy to see that p_a is a norm of K. Indeed:

$$\begin{aligned} p_a(x+y) &= \lim_{n\to\infty} p((x+y)a^n)p(a)^{-n} \\ &\leq \lim_{n\to\infty}(p(xa^n) + p(ya^n))p(a)^{-n} = p_a(x) + p_a(y) \end{aligned}$$

$$\begin{aligned} p_a(xy) &= \lim_{n\to\infty} p(xya^n)p(a)^{-n} = \lim_{m\to\infty} p(xya^{2m})p(a)^{-2m} \\ &= \lim_{m\to\infty} p((xa^m)(ya^m))p(a)^{-2m} \\ &\leq \lim_{m\to\infty} P(xa^m)p(a)^{-m}p(ya^m)p(a)^{-m} = p_a(x)p_a(y) \end{aligned}$$

moreover, $p_a(a) = \lim_{n\to\infty} p(aa^n)p(a)^{-n} = p(a)$, since p is power multiplicative. Thus (ii) holds. Clearly, p_a is subordinate to p:

$$p_a(x) = \lim_{n\to\infty} p(xa^n)p(a)^{-n} \leq \lim_{n\to\infty} p(x)p(a)^n p(a)^{-n} = p(x)$$

To show that p_a is power multiplicative, note that

$$p_a(x)^2 = (\lim_{n\to\infty} p(xa^n)p(a)^{-n})^2 = \lim_{n\to\infty} p(xa^n)^2 p(a)^{-2n}$$

$$= \lim_{n\to\infty} p(x^2a^{2n})p(a)^{-2n} = p_a(x^2)$$

and apply Lemma 6. One proves (iii) similarly:

$$p_a(ax) = \lim_{n\to\infty} p(xaa^n)p(a)^{-n} = p(a) \lim_{n\to\infty} p(xa^{n+1})p(a)^{-(n+1)}$$

$$= p(a)p_a(x) = p_a(a)p_s(x)$$

DEFINITION 9 A nonempty set P of power-multiplicative norms on a field K is called a *hereditary system* if for every $p \in P$ and every $a \in K^\times$, $p_a \in P$.

LEMMA 8 A minimal element of a hereditary system (if it exists) is an absolute value.

Proof of Lemma 8: If p is a minimal element of a hereditary system, then $p_a = p$ for every $a \in K^\times$. Thus for an arbitrary $b \in K$ we have

$$p(ab) = p_a(a)p_a(b) = p(a)p(b) \qquad \text{by Lemma 7}$$

Let p be a power-multiplicative norm of a field K, $a \in K^\times$. Let [K;p;a] be the set of all power-multiplicative norms q of K, satisfying conditions of Lemma 7, i.e., such that

(i) $q \le p$.

(ii) $q(a) = p(a)$.

(iii) $q(ax) = q(a)q(x)$ for every $x \in K$.

(iv) $q(x^2) = q(x)^2$ for every $x \in K$.

LEMMA 9 [K;p;a] is a hereditary system and it contains minimal elements, i.e., absolute values of K.

Proof of Lemma 9: Note first that $[K;p;a] \neq \emptyset$, since $p_a \in [K;p;a]$, by Lemma 7. To show that [K;p;a] is a hereditary system, take an arbitrary $q \in [K;p;a]$ and $d \in K^\times$. Then we have $q_d \le q \le p$, hence (i) follows;

$$q_d(a) = \lim_{n\to\infty} q(ad^n)q(d)^{-n} = \lim_{n\to\infty} q(a)q(d^n)q(d)^{-n} = q(a) = p(a)$$

since q is power multiplicative; thus (ii) follows; moreover

$$q_d(ax) = \lim_{n\to\infty} q(axd^n)q(d)^{-n} = \lim_{n\to\infty} q(a)q(xd^n)q(d)^{-n}$$
$$= q(a)q_d(x) = q_d(a)q_d(x)$$

holds for every $x \in K$. Lemma 7 implies that q_d is power multiplicative.

Now we shall prove that [K;p;a] contains minimal elements. Let $\emptyset \neq C \subset [K;p;a]$ be a chain in a partially ordered set [K;p;a]. It is easy to see that the formula

$$p'(x) = \inf\{q(x) : q \in C\}$$

defines a norm of K, satisfying conditions (i)-(iii) of the definition of [K;p;a]. Since $p'(a) = p(a) > 0$ $(a \neq 0)$, $p'(x) > 0$ holds for every $x \in K^\times$ (see Definition 1), p' is a norm of K. The Kuratowski-Zorn lemma implies now that [K;p;a] contains minimal elements. Every minimal element of [K;p;a] is, by Lemma 9, an absolute value of K.

We are ready to finish the proof of Theorem 8.

Proof of Theorem 8: (iii) $\Rightarrow$ (ii) Let p be a power-multiplicative norm of a field K. Let M be the set of all minimal elements of the sets [K;p;a] as a runs over all nonzero elements of K. Then the elements q of M are absolute values of K subordinate to p, and consequently, p_0 defined as $p_0 = \sup\{q : q \in M\}$ satisfies $p_0 \leq p$. Let $a \in K^\times$ and take a minimal element $q \in [K;p;a]$. Then $q(a) = p(a)$, by Lemmas 7 and 9. Since $q \in M$, we obtain

$$q \leq p_0 \leq p$$

and finally, $p(a) = q(a) \leq p_0(a) \leq p(a)$. Thus for every $a \in K^\times$, $p(a) = p_0(a)$, implying that $p = \sup\{q : q \in M\}$, where M is a set of absolute values of K.

Theorem 8 is proved.

5. NORMS ON THE QUOTIENT FIELDS OF DEDEKIND DOMAINS

Let R be a Dedekind domain with the quotient field K, $\mathbb{P} = \mathbb{P}(K)$ be the set of all nonzero prime ideals of R. Let v_p denote the p-adic exponential valuation of R. Let us put for $x \in K^\times$

$$n(x) = \prod_{v_p(x)>0} p^{v_p(x)} \qquad \text{the numerator of } x$$

$$d(x) = \prod_{v_p(x)<0} p^{-v_p(x)} \qquad \text{the denominator of } x$$

$$R_{(x)} = \{y \in K : \forall\, p \mid n(x),\ v_p(y) \geq 0\} \tag{37}$$

Note that if $n(x) = R$, then $R_{(x)} = K$. If $A \subset \mathbb{P}$ is a set of prime ideals, we define

$$\mathcal{O}(A) = \{x \in K : \forall\, p \in A,\ v_p(x) \geq 0\} = \bigcap_{p \in A} V_p \tag{38}$$

where V_p is the valuation ring of K defined by the p-adic valuation v_p.

If p is a norm of a field K, then there exist absolute values f_i of K ($i \in I$) such that $f_i \leq p$. Indeed, Theorem 8 implies that p_{sp} has a form $p_{sp} = \sup\{f_i : i \in I\}$; moreover, $p_{sp} \leq p$, and consequently, $f_i \leq p$ holds for every $i \in I$.

DEFINITION 10 Let p be a nontrivial norm of a field K, where K is the quotient field of a Dedekind domain R. We say that p is of finite type if the following conditions are satisfied.

(i) There exist only finitely many pairwise nonequivalent, Archimedean absolute values of K subordinate to p. If there exists at least one such absolute value, let $f_1, \ldots, f_n$ be absolute values satisfying (i), which are taken from the classes of equivalent absolute values in such a way that $f_i|\mathbb{Q}$ is the ordinary absolute value. Put $F = \sup(f_1,\ldots,f_n)$. If there exists no absolute value satisfying (i), then we take $F = \varphi_0$ (the trivial absolute value). We call F the Archimedean part of p.

(ii) For every ideal $\mathfrak{a} \neq (0)$ of R there exists a real number $\delta(\mathfrak{a}) \geq 1$ such that for every $x \in K$ there is $a \in \mathfrak{a}$ satisfying

$$F(x - a) \leq \delta(\mathfrak{a})$$

(iii) There exists a constant $\gamma > 0$ such that for every $x \in R$,

$$p(x) \leq \gamma \max(F(x), 1)$$

THEOREM 9 Every nontrivial norm of $\mathbb{Q}$ and $\mathbb{F}_q(X)$ (X being transcendental over $\mathbb{F}_q$) is of finite type.

Suppose that $K = \mathbb{Q}$. Let $p_1 = p|\mathbb{Z}$.

Case 1. $p_1 \sim |\cdot|$ (the Archimedean absolute value) $\Leftrightarrow p_1(x) \geq 1$ for every $x \in \mathbb{Z}$, $x \neq 0 \Leftrightarrow p_1$ is trivial. Indeed, let $p_1(x_n) \to 0$; since $p_1(x_n) \geq 1$, then $p_1(x_n) = 0$ for $n \geq n_0$, i.e., $x_n = 0$ for $n \geq n_0$. This implies that $x_n \to 0$. Conversely, let $x_n \to 0$; then $x_n = 0$ for large values of n, i.e., $p_1(x_n) = 0$ if $n \to \infty$.

Case 2. $p_1 \not\sim \varphi_0$ (the trivial absolute value) $\Rightarrow \mathbb{Z}$ is bounded in the norm p_1. Indeed, since p_1 is nontrivial, there exists $n > 1$ such that $p_1(n) < 1$. Let $C = \max\{p_1(k) : |k| < n\}$. Let $N \in \mathbb{Z}$. Then

$$N = a_k n^k + \cdots + a_1 n + a_0$$

where $a_i \in \mathbb{Z}$ and $|a_i| < n$. Thus $p_1(N) \leq C \sum_{i=0}^{k} p_1(n)^i \leq C(1 - p_1(n))^{-1}$.

This implies that we can take $\gamma = p(1)$ in case 1, and $\gamma = C(1 - p(n))^{-1}$ in case 2, where $n > 1$ is such that $p(n) < 1$.

This implies that (iii) is fulfilled for $K = \mathbb{Q}$. Since condition (i) is obvious, we shall prove that (ii) is satisfied. Indeed, there is nothing to do in case 2; it is sufficient to take $\delta(\mathfrak{a}) = 1$ identically. Now suppose that case 1 holds. Let $a \in \mathbb{Z}$, $a \neq 0$, be arbitrary. If $x \in \mathbb{Q}$, then there exists $k \in \mathbb{Z}$, such that $|(x/a) - k| \leq 1/2$ (take, for example, $k = \min\{[x/a] - n : n \in \mathbb{Z}\}$). Since $F(x) = |x|$ in case 1, we obtain $F(x - ak) \leq |a|/2 \leq ((a))$. We take $\delta((a)) = |a|/2$ if $|a| \geq 2$, and $\delta((a)) = 1$ if $a = \pm 1$.

Now let p be a nontrivial norm of $\mathbb{F}_q(X)$. Since a field of nonzero characteristic admits no Archimedean absolute value, conditions (i) and (ii) obviously hold. Indeed, we take $\delta(\mathfrak{a}) = 1$, since $F = \varphi_0$.

We are going to prove (iii). Since $\mathbb{F}_q$ is finite, there is a constant b such that $p(a) \le b$ holds for every $a \in \mathbb{F}_q$. Moreover, since the norm p is nontrivial on $\mathbb{F}_q(X)$, there exists a topological nilpotent $x \in K^\times$, i.e., $p(x) < 1$. The element x is transcendental over $\mathbb{F}_q$, since otherwise x would be a root of unity, i.e., $X^N = 1$ with a suitable $N \in \mathbb{N}$. Thus $1 \le p(1) = p(x^N) \le p(x)^N$ would imply that $p(x) \ge 1$, a contradiction! Now, for every $z \in \mathbb{F}_q[x]$, $z = a_s x^s + \cdots + a_1 x + a_0$, $p(z) \le b \sum_{i=0}^{s} p(x)^i \le b(1 - p(x))^{-1} = C$, i.e., $\mathbb{F}_q[x]$ is bounded with respect to p.

Let R be the integral closure of $\mathbb{F}_q[x]$ in $\mathbb{F}_q(X)$. It is well known that R is a Dedekind domain (see, e.g., Narkiewicz [1, Th. 1.9, p. 22]; also Zariski and Samuel [1, Vol. I. Th. 19, Chap. V, Sec. 8, p. 281]). Indeed, $\mathbb{F}_q(X)$ is a finite extension of $\mathbb{F}_q(x)$ of degree max(deg(f),deg(g)), where $x = f(X)/g(X)$, $f,g \in \mathbb{F}_q[X]$, are relatively prime polynomials. Thus we can apply the theorem above (Narkiewicz [1]; Zariski and Samuel [1]). Moreover, there exist elements $w_1, \ldots, w_N \in \mathbb{F}_q(X)$ such that

$$R \subset w_1\mathbb{F}_q[x] + \cdots + w_N\mathbb{F}_q[x]$$

This implies that R is bounded with respect to p, being a subset of the bounded set $w_1\mathbb{F}_q[x] + \cdots + w_N\mathbb{F}_q[x]$. Consequently, every nontrivial norm of $\mathbb{F}_q(X)$ is of finite type.

We shall need the following.

LEMMA 10 Let K be the quotient field of a Dedekind domain R and p a nontrivial norm of finite type on K. Then for every $a \in K^\times$ satisfying $p(a) < 1$, there exists a number $\gamma(a) > 0$ satisfying

$$p(x) \le \gamma(a) \max(F(x),1) \qquad \text{for every } x \in R_{(a)}$$

Proof: Let $n(a) = p_1^{a_1} \cdots p_r^{a_r}$, $a_i \ge 1$.

Case 1. The Archimedean part of p is trivial. Let $x = y/z \in R_{(a)}$, where $y,z \in R$ and $v_{p_i}(z) = 0$ for $i = 1, \ldots, r$. Since $p(a) < 1$, a^n and xa^n tend to zero, i.e., $p(xa^t) \le 1$ with a suitable $t \in \mathbb{N}$.

There exists $u \in K$ satisfying

$$v_{p_i}(u - z^{-1}) \geq ta_i \qquad v_p(u) \geq 0 \text{ if } p \neq p_i \qquad (i = 1, \ldots, r)$$

by the strong approximation theorem (Chapter 2, Theorem 5). It is clear that $u \in R$ ($v_q(u) \geq 0$ for every prime ideal q of R) and moreover, $v_q((zu - 1)a^{-t}) \geq 0$ for every prime ideal q of R, implying that $v = (zu - 1)a^{-t} \in R$, i.e., $x(1 + va^t) = xzu = yu \in R$. Since $p(b) \leq \gamma$ for $b \in R$ (p is of finite type and F is trivial), we obtain, finally,

$$p(x) = p(yu - v(xa^t)) \leq \gamma + \gamma p(xa^t) \leq 2\gamma \text{ if } x \in R_{(a)}$$

Case 2. F is nontrivial. Let $x \in R_{(a)}$ and take $z \in R$, $z \neq 0$, with $xz \in R$. Again, as in case 1, applying the strong approximation theorem, we can find $u \in K$ satisfying

$$v_{p_i}(u - x) \geq a_i \qquad v_p(u) \geq 0 \text{ if } p \neq p_i \qquad (i = 1, \ldots, r)$$

It follows immediately that $u \in R$ and $(u - x)a^{-1} \in R_{(a)}$. Let $\gamma = \gamma(d(a))$. Definition 10 implies that there exists $b \in d(a)$ such that $F(b - (u - x)a^{-1}) \leq \gamma$. Define $u_1 = u - ab$, $x_1 = b - (u - x)a^{-1}$. Then $u_1 = x - ax_1$, with $u_1 \in R$, $x_1 z \in R$, $x_1 \in R_{(a)}$, $F(x_1) \leq \gamma$, and

$$\begin{aligned} p(x_1) &\leq p(x_1 z)p(z^{-1}) \leq \gamma p(z^{-1})\max(F(x_1 z), 1) \\ &\leq \gamma p(z^{-1})\max(\delta F(z), 1) \end{aligned}$$

It can be shown, by induction on t, repeating the construction above, that for every $t \in \mathbb{N}$ there exist $u_i \in R$ and $x_i \in R_{(a)}$ $(i = 1, 2, \ldots, t)$ satisfying $u_i = x_{i-1} - ax_i$ $(x_0 = x)$, $x_i z \in R$, where $F(x_i) \leq \delta$ and $p(x_i) \leq \gamma p(z^{-1})\max(F(z), 1)$. Moreover, we have

$$x = u_1 + u_2 a + \cdots + u_t a^{t-1} + x_t a^t \tag{39}$$

Since $p(a) < 1$, we can take t so large that $p(x_t a^t) \leq 1$.

Case 2a. Assume that $F(x) \leq 1$. Since $u_i = x_{i-1} - ax_i$, we obtain that $F(u_i) \leq F(x_{i-1}) + F(a)F(x_i) \leq \delta(1 + F(a))$ holds for every

$i = 1, 2, \ldots, t$. Then, in view of (39), one obtains

$$\begin{aligned} p(x) &\le \sum_{i=1}^{t} p(u_i a^{i-1}) + p(x_t a^t) \le \sum_{i=1}^{t} p(u_i) p(a)^{i-1} + 1 \\ &\le \gamma \sum_{i=1}^{t} p(a)^{i-1} \max(F(u_i),1) + 1 \\ &\le \gamma \sum_{i=1}^{t} p(a)^{i-1} \max(\delta(1 + F(a)),1) + 1 \end{aligned}$$

Since $\delta \ge 1$, we have

$$\begin{aligned} p(x) &\le 1 + \gamma\delta(1 + F(a)) \sum_{i=1}^{t} p(a)^{i-1} \\ &\le 1 + \gamma\delta(1 + F(a)) \frac{p(a)}{1 - p(a)} = \gamma_1(a) \end{aligned}$$

i.e., $F(x) \le 1$ implies that $p(x) \le \gamma_1(a)$ holds for every $x \in R_{(a)}$.

Case 2b. As F is nontrivial, so char(K) = 0. Let $p_1, \ldots, p_\ell$ be the primes for which there exists a prime ideal p_i $(i = 1, \ldots, r)$ such that p_i occurs in the decomposition of $p_j\mathbb{Z}_K$ into prime ideals $\mathbb{Z}_K$, the integral closure of $\mathbb{Z}$ in K. There exists a constant $\beta(a) > 1$ such that for every $\rho \ge 1$ there exists an integer n in the interval $[\rho, \beta(a)\rho]$ with $(n, p_1 \cdots p_\ell) = 1$.

Let now $F(x) \ge 1$. It follows from the above that we can find an integer n satisfying

$$F(x) \le n \le \beta(a)F(x) \qquad \text{and} \qquad (n, p_1, \cdots p_\ell) = 1$$

Since $F(w) = |w|$ (the ordinary absolute value), if $w \in \mathbb{Q}$, then $F(x/n) = F(x)n^{-1} \le 1$. Finally, $x/n \in R_{(a)}$, i.e., $p(x/n) \le \gamma_1(a)$, implying case 2a. Consequently,

$$p(x) \le p\left(\frac{x}{n}\right)p(n) \le \gamma_1(a)p(1)n \le \gamma_1(a)p(1)\beta(a)F(x)$$

This shows that the conclusion of Lemma 10 holds with $\gamma(a) = \gamma_1(a)\beta(a)p(1)$.

THEOREM 10 Let L be a finite field extension of a field K. Let K be the quotient field of a Dedekind domain R and S be the (classical) integral closure of R in L. If p' is a nontrivial norm of finite type on K, its arbitrary extension p to L is also of finite type.

Proof: We have to prove that conditions (i)-(iii) of Definition 10 are satisfied. It is evident that (i) holds true.

(ii) Let $f_1, \ldots, f_h$ be pairwise nonequivalent Archimedean absolute values of L, $f_i \leq p$, such that $f_i(a) = |a|$ for $a \in \mathbb{Q}$ ($i = 1, \ldots, h$) and let $f_i' = f_i|K$. Evidently, $f_i' \not\sim f_j'$ for every $i \neq j$, since $f_i' \neq f_j'$. If $(0) \neq a \lhd S$ is an ideal of S, we put $a' = a \cap R$. Assume that $[L:K] = n$ and let $x_1, \ldots, x_n \in S$ be a basis of L over K. If $x = b_1x_1 + \cdots + b_nx_n \in L$ ($b_j \in K$) is arbitrary, we can find elements $c_j \in a'$ satisfying

$$f_i'(b_j - c_j) \leq \delta(a') \qquad i = 1, \ldots, h;\ j = 1, \ldots, n$$

since p' is of finite type [Definition 10(ii)]. Putting $y = c_1x_1 + \cdots + c_nx_n$, we obtain $y \in a$, and moreover,

$$f_i(x - y) \leq \delta(a') \sum_{k=1}^{n} f_i'(x_k)$$

(iii) Let $w_i \leq p'$ ($i = 1, 2, \ldots, k$) be nonequivalent Archimedean absolute values. If $k = 0$, then $p'(x) \leq \gamma$ holds for every $x \in R$. Since the Archimedean part F of p is trivial, and S is a finitely generated module over R, say, by $s_1, \ldots, s_N$, then for every $x \in S$, $x = r_1s_1 + \cdots + r_Ns_N$ ($r_j \in R$), $p(x) \leq \gamma(p(s_1) + \cdots + p(s_N)) = \gamma'$, i.e., (iii) is fulfilled.

Now suppose that $k \geq 1$. We can assume that $w_i(x) = |x|$ for $x \in \mathbb{Q}$ ($i = 1, \ldots, k$). The absolute values $f_1, \ldots, f_h$ extend $w_1, \ldots, w_k$. Recall that by Corollary 2 of Theorem 2, for every Archimedean absolute value w of K there is an embedding g of K into $\mathbb{C}$ such that $w(x) = |g(x)|^t$ ($t > 0$), $x \in K$. We say that w is real if $g(K) \subset \mathbb{R}$, and complex otherwise. Let r_{i1} be the number of real and r_{i2} be the number of complex extensions of w_i to L. Put

$$s_i = \begin{cases} n & \text{if } w_i \text{ is complex} \\ r_{i1} + r_{i2} & \text{otherwise} \end{cases}$$

Let W_{ij} $(j = 1, 2, \ldots, s_i)$ be nonequivalent extensions of w_i to L. Define

$$I_j' = \{i : W_{ji} \le p\} \qquad I_j'' = \{1, \ldots, s_j\} \setminus I_j'$$

$$W = \sup\{W_{ij} : i = 1, \ldots, k;\ j \in I_i'\}$$

$$w = \sup\{w_i : i = 1, \ldots, k\}$$

$$V = \sup\{W_{ij} : i = 1, \ldots, k;\ j = 1, \ldots, s_i\}$$

It is clear that, by definition of f_i, $W = \sup\{f_i : 1 \le i \le h\}$. Since p is nontrivial, we can find $a \in K^\times$ with $p(a) < 1$. Consider the rings of fractions $R_{(a)}$ and $S_{(a)}$, respectively. If $n(a) = R$, then $R_{(a)} = K$ and $S_{(a)} = L$. If $n(a) \ne R$, then $R_{(a)}$ is a semilocal principal ideal domain (PID), as the intersection of finitely many local PIDs V_p (p being a prime ideal of R). Clearly, $S_{(a)}$ is the integral closure of $R_{(a)}$ in L. Since S is a finitely generated module over R (see Zariski and Samuel [1, Vol. I, Th. 7, p. 264]), $S_{(a)}$ is also free over $R_{(a)}$, of rank n. Assume that $S_{(a)} = R_{(a)}x_1 + \cdots + R_{(a)}x_n$. Lemma 10 implies that there exists a constant $\gamma(a) > 0$ such that

$$p'(x) \le \gamma(a) \max(w(x), 1) \qquad x \in R_{(a)} \tag{40}$$

and a constant $\alpha > 0$ with

$$x \in S_{(a)} \quad \text{and} \quad V(x) \le 1 \Rightarrow p(x) \le \alpha \tag{41}$$

First we shall prove that

> There exists an $\alpha' > 0$ such that $x \in L$ and $V(x) \le 1$, $x = c_1x_1 + \cdots + c_nx_n$ $(c_i \in K)$, imply that $w(c_i) \le \alpha'$ for $1 \le i \le n$. (42)

Let $g_1, \ldots, g_n$ be embeddings of L into $\mathbb{C}$, where $g_i|K = g$ is a fixed embedding of K into $\mathbb{C}$. Since $x = c_1x_1 + \cdots + c_nx_n$, we obtain Cramer's system of linear equations with respect to $g(c_1), \ldots, g(c_n)$:

$$g_j(x) = g(c_1)g_j(x_1) + \cdots + g(c_n)g_j(x_n) \qquad j = 1, \ldots, n \tag{43}$$

Since different embeddings are linearly independent over $g(K)$, this implies that $\det(g_i(x_j)) \neq 0$. Consequently, (43) is Cramer's system; i.e., there exist constants $d_{m1}, \ldots, d_{mn}, d_m \in \mathbb{C}$, such that

$$g(c_m) = d_{m1}g_1(x) + \cdots + d_{mn}g_n(x) + d_m \qquad m = 1, \ldots, n \tag{44}$$

Every Archimedean absolute value $\bar{w}$ of K has the form $\bar{w}(y) = |\bar{g}(y)|$, where $\bar{g}$ is an embedding of K into $\mathbb{C}$. If $\bar{W}$ is any of its extensions to L, then $\bar{W}(z) = |g_i(z)|$, $z \in L$, with a suitable index i. Taking absolute values of both sides in (44), one obtains

$$\begin{aligned}\bar{w}(c_m) &\leq |d_{m1}||g_1(x)| + \cdots + |d_{mn}||g_n(x)| + |d_m| \\ &\leq C_m(|g_1(x)| + \cdots + |g_n(x)| + 1) \\ &\leq C_m(nV(x) + 1) \leq C_m(n + 1)\end{aligned}$$

where $C_m = \max(|d_{m1}|, \ldots, |d_{mn}|, |d_m|)$. Finally,

$$w(c_m) \leq C = (n + 1) \max(C_1, \ldots, C_n)$$

as required in (42).

Now we can prove that (41) is fulfilled. Indeed, let $x = c_1x_1 + \cdots + c_nx_n \in S_{(a)}$, $V(x) \leq 1$. Since $c_j \in R_{(a)}$, then the assertion (41) follows from (40) and (42):

$$\begin{aligned}p(x) &\leq p(c_1)p(x_1) + \cdots + p(c_n)p(x_n) \\ &= p'(c_1)p(x_1) + \cdots + p'(c_n)p(x_n) \\ &\leq C(p'(c_1) + \cdots + p'(c_n)) \\ &\leq C\gamma(a)(\max(w(c_1),1) + \cdots + \max(w(c_n),1)) \\ &\leq nC\gamma(a) \max(\alpha',1)\end{aligned}$$

Let

$$n(a)S = P_1^{r_1} \cdots P_s^{r_s} \qquad P_i \neq P_j \text{ if } i \neq j$$

Let v_i be the P_i-adic exponential valuation of L. Let $i \in \{1,\ldots,k\}$

and $j \in I_i''$ be fixed. A definition of I_i'' implies that there exists $b_{ij} \in L$ satisfying

$$W_{ij}(b_{ij}) > 1 \qquad p(b_{ij}) < 1$$

Put $\ell = s_1 + \cdots + s_k$ and take an arbitrary $x \in S_{(a)}$. Then one can find $m(i,j) \in \mathbb{N}$ such that

$$W_{ij}(xb_{ij}^{-m(i,j)}) \leq \frac{1}{2\ell} \tag{46}$$

The approximation theorem for absolute values implies existence of an element $y_{ij} \in L$ satisfying

$$W_{ij}(y_{ij} - xb_{ij}^{-m(i,j)}) \leq \min\left\{\frac{1}{\ell W_{ij}(b_{ij}^{m(i,j)})}, \frac{1}{2\ell}\right\} \tag{47}$$

$$W_{pq}(y_{ij}) \leq \frac{1}{\ell W_{ij}(b_{ij}^{m(i,j)})} \qquad \begin{cases} q \in \{1,\dots,s_p\} \text{ if } p \neq i \\ q \in \{1,\dots,s_i\} \setminus \{j\} \text{ if } p = i \\ p = 1, \dots, k \end{cases} \tag{48}$$

$$v_m(y_{ij}) \geq \max(0, -v_m(b_{ij}^{m(i,j)})) \qquad m = 1, \dots, s \tag{49}$$

The last condition implies that $y_{ij} \in S_{(a)}$ and $y_{ij}b_{ij}^{m(i,j)} \in S_{(a)}$. Conditions (47)-(49) imply that

$$W_{pq}(y_{ij}) \leq \frac{1}{\ell} \qquad p = 1, \dots, k;\ q = 1, \dots, s_p \tag{50}$$

Since $y_{ij} \in S_{(a)}$ and $V(y_{ij}) \leq 1$ [by (50)], then applying (41), we have $p(y_{ij}) \leq \alpha$. As $p(b_{ij}) < 1$, then

$$p(y_{ij}b_{ij}^{m(i,j)}) \leq \alpha \tag{51}$$

Define

$$y = x - \sum_{i=1}^{k} \sum_{j \in I_i''} y_{ij}b_{ij}^{m(i,j)} \tag{52}$$

Obviously, $y \in S_{(a)}$, since every component of the sum (52) lies in $S_{(a)}$. If $p \in \{1,\dots,k\}$, $q \in I_p''$, then

$$W_{pq}(y) = W_{pq}\left(x - b_{pq}^{m(p,q)} y_{pq} - \sum_{i=1}^{k} \sum_{\substack{j \in I_i'' \text{ if } i \neq p \\ j \in I_p'' \setminus \{q\} \text{ if } i = p}} b_{ij}^{m(i,j)} y_{ij}\right)$$

$$\le \frac{1}{\ell} + \frac{\ell - 1}{1} = 1 \le 1 + W(x) \qquad \text{by (47) and (48)}$$

If $t \in \{1,\ldots,k\}$, $q \in I_t'$, then $W_{tq}(b_{ij}) < 1$, by definition of I_t', since $W_{tq} \le p$, by (45). Moreover, $W_{tq}(y_{ij}) \le 1/\ell$ [see (50)]. Thus we can evaluate y as follows:

$$W_{tq}(y) \le W_{tq}(x) + \sum_{i=1}^{k} \sum_{j \in I_i''} W_{tq}(y_{ij}) W_{tq}(b_{ij})^{m(i,j)}$$

$$\le W_{tq}(x) + \ell \frac{1}{\ell} = W_{tq}(x) + 1 \le W(x) + 1$$

Consequently, $W_{pq}(y) \le W(x) + 1$ holds for every pair p, q. Hence $V(y) \le W(x) + 1$.

There exists $m \in \mathbb{N}$ with $m^{-1} \in S_{(a)}$ satisfying

$$W(x) + 1 \le m \le \beta(a)(W(x) + 1)$$

(see Lemma 10, case 2b). This implies that

$$V\left(\frac{y}{m}\right) = \frac{1}{m} V(y) \le \frac{1}{m}(W(x) + 1) \le 1$$

Since $y/m \in S_{(a)}$, then $p(y/m) \le \alpha$ by (41). Finally, we have

$$p(y) = p\left(\frac{y}{m} m\right) \le \alpha m p(1) \le \alpha p(1)\beta(a)(W(x) + 1)$$

and applying (51) to (52), we obtain

$$p(x) \le p(y) + \alpha\ell \le \alpha(p(1)\beta(a)(W(x) + 1) + \ell)$$
$$\le \alpha(2p(1)\beta(a) + \ell)\max(W(x), 1)$$

This proves (iii) and the theorem.

LEMMA 11 Let K be the quotient field of a Dedekind domain R, p a nontrivial norm of finite type on K, and F the Archimedean part of p. Then the set of ideals

$$E = \{n(x) : x \in N_0(\mathcal{T}_p),\ x \neq 0\}$$

has greatest element e. If $e \neq R$, then $e = p_1 \cdots p_r$, $p_i \neq p_j$ for $i \neq j$ and $p_i \in \mathbb{P}(R)$. Every ideal $\mathfrak{a} \in E$ has the form

$$\mathfrak{a} = p_1^{n_1} \cdots p_r^{n_r}$$

If $E' = \{p_1, \ldots, p_r\}$, then

$$\{x \in K : p(x) < 1\} \subset N_0(T_p) \subset \mathcal{O}(E') \tag{53}$$

Proof: If $R \in E$, there is nothing to prove, since then $E' \neq \emptyset$ and $\mathcal{O}(E') = K$. If $R \notin E$, then take a maximal element $n(a) = \mathfrak{a} = p_1^{\alpha_1} \cdots p_r^{\alpha_r} \in E$. Define $\mathfrak{h} = p_1 \cdots p_r$. There exists $x \in K^\times$ satisfying $n(x) = \mathfrak{h}$. If $d = \max(d_1, \ldots, d_r)$, then $x^d a^{-1} \in R_{(a)}$. Since $a \in N_0(T_p)$, there exists $\ell \in \mathbb{N}$ with $p(a^\ell) < 1$.

First we shall prove that $\mathfrak{a} = \mathfrak{h}$.

Case 1. F is trivial. If $m \in N$ is arbitrary, write $m = dq + t$, $0 \leq t < d$. Then $p(x^m) = p((x^d/a)^q x^t a^q) \leq \gamma(a^\ell)p(x^t)p(a^q)$, by Lemma 10. If $m \to \infty$, then also $q \to \infty$, i.e., $x \in N_0(T_p)$. This implies that $\mathfrak{h} \in E$. Consequently, $\mathfrak{a} \subset \mathfrak{h}$ gives $\mathfrak{a} = \mathfrak{h}$, by the maximality of $\mathfrak{a}$.

Case 2. F is nontrivial. Take an $n \in \mathbb{N}$ such that ideals $\mathfrak{h}$ and nR are relatively prime and $F((x/n)a^{-1}) \leq 1$. Putting $y = x/n$, one has $ya^{-1} \in R_{(a)}$, and by the same arguments as in case 1,

$$p(y^m) \leq \gamma(a^\ell)p(y^t)p(a^q)$$

implying that $y \in N_0(T_p)$, i.e., $\mathfrak{h} \in E$, and $\mathfrak{a} = \mathfrak{h}$.

It remains to show that a greatest (i.e., a unique maximal) element exists in E. Indeed, let $\mathfrak{a}$ and $\mathfrak{h}$ be two maximal elements of E and $\mathfrak{a} = n(a)$, $\mathfrak{h} = n(b)$. Then $\mathfrak{a} = p_1 \cdots p_r$, $\mathfrak{h} = p_1 \cdots p_h q_1 \cdots q_s$ $(r > 0,\ h \geq 0,\ s > 0)$ with pairwise different prime ideals p_i and q_j. There exists an $x \in K^\times$ such that $x^{-1} \in R_{(a)}$ and $n(b/x) = p_1 \cdots p_h$.

Case 1. F is trivial. Then $p((b/x)^m) \leq p(x^{-1})^m p(b^m) \leq \gamma(a^\ell)p(b)^m$; here $\ell \in \mathbb{N}$ is taken in order to satisfy $p(a^\ell) < 1$. Thus $b/x \in N_0(T_p)$ and $\mathfrak{h} \subset n(b/x)$. This implies that $\mathfrak{h} = n(b/x)$, by the maximality of $\mathfrak{h}$, and consequently, $s = 0$. Since either $\mathfrak{a} \subset \mathfrak{h}$ (if

$r \geq h$) or $\mathfrak{b} \subset \mathfrak{a}$ (otherwise) and the both ideals are maximal elements of E, we obtain $\mathfrak{a} = \mathfrak{b}$.

Case 2. F is nontrivial. We can find $n \in \mathbb{N}$ such that the pairs of ideals $\{nR, \mathfrak{a}\}$ and $\{nR, \mathfrak{b}\}$ are relatively prime and $R(1/nx) \leq 1$. Consequently, $1/nx \in R_{(a)}$, $n(b/nx) = p_1 \cdots p_h$. Thus $n(b/nx) = \mathfrak{b}$, by the maximality of $\mathfrak{b}$ in E. We proceed in the same way as in case 1 in order to obtain $\mathfrak{a} = \mathfrak{b}$. The proof of Lemma 11 is finished.

THEOREM 11 Let K be the quotient field of a Dedekind domain R, p a nontrivial norm of finite type of K, E as in Lemma 11, and $e = p_1 \cdots p_r$ the greatest element of E. Then $T_p = T_q$, where $q = \sup(F, \varphi_{p_1}, \ldots, \varphi_{p_r})$. In other words, the norms p and q are equivalent.

Proof: We shall prove first that $T_p \leq T_q$. Let $a \in K^{\times}$, $p(a) < 1$ be such that $e = p_1 \cdots p_r$ divides $n(a)$ and $R_{(a)} = \mathcal{O}(\{p_1, \ldots, p_r\})$ (Lemma 11). Since a is a topological nilpotent of (K, T_p), then for every $\varepsilon > 0$ there exists $m \in \mathbb{N}$ with $p(a^m) < \varepsilon(\gamma(a))^{-1}$. Then there exists $\delta > 0$ such that for every $z \in K$ with $q(z) < \delta$, $za^{-m} \in R_{(a)}$ and $F(za^{-m}) \leq 1$. Indeed, it is sufficient to take

$$\delta = \min(F(a)^m, \varphi_{p_1}(a)^m, \ldots, \varphi_{p_r}(a)^m)$$

where φ_p denotes the p-adic absolute value of K. Lemma 10 implies that

$$\begin{aligned} p(z) &= p(za^{-m}a^m) \leq p(za^{-m})p(a^m) \leq \gamma(a) \max(F(za^{-m}), 1)p(a^m) \\ &\leq \gamma(a)\varepsilon(\gamma(a))^{-1} = \varepsilon \end{aligned}$$

Let $U_\varepsilon(N) = \{x \in K : N(x) < \varepsilon\}$, N being a norm of K. So we have proved that for every $\varepsilon > 0$ there is $\delta > 0$ such that $U_\delta(q) \subset U_\varepsilon(p)$. Hence $T_p \leq T_q$.

Conversely, it remains to show that $T_q \leq T_p$. First note that $q = \sup(f_1, \ldots, f_n, \varphi_{p_1}, \ldots, \varphi_{p_r})$, where $F = \sup(f_1, \ldots, f_n)$ is the Archimedean part of p. We have $f_i \leq p$ for every $i = 1, \ldots, n$ by Definition 10, and moreover, $\sup(T_{p_1}, \ldots, T_{p_r}) \leq T_p$, since $N_0(T_p)$ is open and bounded in (K, T_p) and $N_0(T_p) \subset V_{p_1} \cap \cdots \cap V_{p_r}$, by Lemma 11

(T_p denotes the p-adic topology of K). So

$$T_q = \sup(T_{f_1},\dots,T_{f_n},T_{p_1},\dots,T_{p_r}) \le T_p$$

This proves Theorem 11.

THEOREM 12 (Kiyek [6]; Mahler [3]) Let p be a nontrivial norm of a global field K. Then p is equivalent to a norm of the form $\sup(w_1,\dots,w_N)$, where w_i are nontrivial and nonequivalent absolute values of K.

LEMMA 12 If K is a finite field extension of $\mathbb{Q}$, and p is a nontrivial norm of K, then $p|\mathbb{Q}$ is also nontrivial.

Proof of Lemma 12: If $K = \mathbb{Q}$, there is nothing to do. Hence let $K \ne \mathbb{Q}$. Let $a \in K^\times$ satisfy $p(a) < 1$. Assume that $p|\mathbb{Q}$ is trivial, i.e., $p(x) \ge 1$ holds for every nonzero $x \in \mathbb{Q}$. This implies that $p_{sp}|\mathbb{Q}$ is also trivial, by definition of the spectral norm. Consequently, p_{sp} and p are equivalent on $\mathbb{Q}$ and both define the discrete topology of $\mathbb{Q}$. Then, by Theorem 8, $p_{sp}|\mathbb{Q} = \sup\{f_i : i \in I\}$, where I is a finite set and f_i are pairwise nonequivalent absolute values of $\mathbb{Q}$. Since $p_{sp}(x) \ge 1$, if $x \in \mathbb{Q}^\times$, we may assume that the set $\{f_i : i \in I\}$ contains the trivial absolute value φ_0. Consequently, applying Ostrowski's theorem (Chapter 2, Theorem 10), we can assume that

$$p_{sp}|\mathbb{Q} = \sup(\varphi_0,\varphi_1,\dots,\varphi_s,\varphi_\infty) \tag{54}$$

where φ_i is the p_1-adic absolute value, and φ_∞ the ordinary absolute value. Put $q = p_{sp}$ for brevity. Then $q(a) \le p(a) < 1$. Let $t \in \mathbb{N}$ be taken such that $p(a^t) < 1/(n+1)$. Thus $q(a^t) < 1/(n+1)$ (here $n = [K:Q]$). Since $a^t \in K$ and $1 < n = [K:Q] < \infty$, we can find integers $a_1, \dots, a_n$ such that $0 \ne b = a_n a^{nt} + \cdots + a_1 a^t \in \mathbb{Z}$. Let $m > \max(|a_1|,\dots,|a_n|,|b|)$ satisfy $(m, p_1\cdots p_s) = 1$. Then

$$1 \le q\left(\frac{b}{m}\right) = q\left(\sum_{j=1}^{n} \frac{a_j}{m} a^{tj}\right) \le \sum_{j=1}^{n} q\left(\frac{a_j}{m}\right) q(a^{tj}) \le \sum_{j=1}^{n} q(a^{tj})$$

since $q(a_j/m) \le 1$, by (54). Consequently,

$$1 \le q\left(\frac{b}{m}\right) \le \sum_{j=1}^{n} q(a^t)^j < \sum_{j=1}^{n}\left(\frac{1}{n+1}\right)^j < \sum_{i=1}^{\infty}\left(\frac{1}{n+1}\right)^j = \frac{1}{n} < 1$$

a contradiction! This implies that $p|\mathbb{Q}$ is nontrivial.

Proof of Theorem 12: Case 1. char(K) = 0. Then [K:Q] = n for some $n \in \mathbb{N}$. Let p be a nontrivial norm of K. Lemma 12 implies that $p' = p|\mathbb{Q}$ is nontrivial, and consequently, p' is of finite type (Theorem 9). Theorem 10 implies that p, an extension of p' to K, is also of finite type. Thus p has the desired form, by Theorem 11.

Case 2. char(K) = $q \neq 0$. Thus [K:F(t)] = n for some $n \in \mathbb{N}$, where F is a finite field and t is transcendental over F. Let p be a nontrivial norm of K. This implies that there exists $x \in K^\times$ with $p(x) < 1$, i.e., $x \in N_0(T_p)$. The element x is transcendental over F, since otherwise x would be a root of unity, i.e., $x^N = 1$ with a suitable $N \in \mathbb{N}$. This would imply that $1 \le p(1) \le p(x)^N$, i.e., $p(x) \ge 1$, a contradiction!

We shall prove that $[K:F(x)] < \infty$. If $x \in F(t)$, $x = f(t)/g(t)$, where f and g are relatively prime polynomials, then [F(t):F(x)] = max(deg(f),deg(g)), i.e., [K:F(x)] is finite, since [K:F(t)] is finite.

Otherwise, let $x \notin F(t)$. Note that dim transc[K:F] = 1, i.e., x and t are algebraically dependent over F. This implies that K is algebraic over F(x). Moreover, $[K:F(x)] < \infty$. Indeed, if [K:F(x)] = ∞, then also $[K:F(t,x)] = \infty$, since $[F(t,x):F(x)] < \infty$. This is impossible, since $F(t) \subset F(t,x) \subset K$, and K is finite over F(t), i.e., K is also finite over F(t,x).

Since $p' = p|F(x)$ is nontrivial, it must be of finite type (Theorem 9). Theorem 10 implies that p, an extension of p' to K, is also of finite type. Theorem 11 implies that p is equivalent to a norm of the form $\sup(w_1,\ldots,w_N)$, where w_i are nontrivial and non-equivalent absolute values of K.

Theorem 12 is proved.

THEOREM 13 Let L be an algebraic function field of one variable over a field F. Then every nontrivial norm p of L, bounded on F, is equivalent with a norm $\sup(w_1,\ldots,w_k)$, where w_i are nontrivial and pairwise nonequivalent absolute values of L, trivial on F.

Proof: A proof is similar to that of Theorem 12 (see Kiyek [6]).

In several cases the topologies T_{w_i} can be effectively determined. If $L = F(X)$, $T \in L_r(L)$, F is bounded in (L,T) and there exists a nonzero topological nilpotent of (L,T) lying in $F[X]$, then $J = N_0 \cap F[X]$ is a proper ideal of $F[X]$. The monic generator of J is the product of distinct prime elements $p_1, \ldots, p_s \in F[X]$ and $T = \sup T_i$, where T_i is p_i-adic topology of L (J.-A. Cohen [1, Th. 1]).

THEOREM 14 Let (K,T) be a nondiscrete locally bounded complete topological field with a nonzero topological nilpotent.

(i) If the characteristic of K is zero, then either $\mathbb{Q}$ lies discretely in K, or K contains as a closed topological subfield the reals $\mathbb{R}$ or a p-adic number field $\mathbb{Q}_p$.

(ii) If the characteristic of K is nonzero, then K contains as a closed topological subfield a field of formal Laurent series over a finite field.

Remark Theorem 14 remains true without the assumption that (K,T) has a nonzero topological nilpotent (see Mutylin [3, Th. 3] for case (i); case (ii) follows from Nichols and Cohen [1, Th. 3, p. 860] and from a result of J.-A. Cohen [2, Th. 2] stating that every locally bounded ring topology on $F(X)$, for which F is bounded, satisfies assumptions of Mutylin [3, Th. 3]). However, the original proofs are more complicated than the proof presented here, which is taken from Więsław [12].

Proof of Theorem 14: Theorem 4 of this chapter implies that T is induced on K by a nontrivial norm, say p.

(i) Let K have zero characteristic and suppose that $\mathbb{Q}$ is a nondiscrete subfield of K. This means that $p_1 = p|\mathbb{Q}$ is a nondiscrete norm inducing the original topology T on $\mathbb{Q}$. But then, applying Mahler's theorem (Theorem 12), we see that p_1 is equivalent to a norm of the form (1). Since K is complete at T, (K,T) must contain the closure of $\mathbb{Q}$ with respect to p_1, i.e., the completion of $\mathbb{Q}$ at p_1. Thus K contains as a topological subring the direct sum

$$\hat{\mathbb{Q}}_1 \oplus \hat{\mathbb{Q}}_2 \oplus \cdots \oplus \hat{\mathbb{Q}}_s$$

where $\hat{\mathbb{Q}}_i$ is the completion of $\mathbb{Q}$ at the absolute value φ_i, by Artin-Whaples' approximation theorem (Chapter 2, Theorem 3). However according to Ostrowski's theorem (Chapter 2, Theorem 10), the ordinary absolute value and the p-adic norms are the only nondiscrete absolute values on $\mathbb{Q}$. Since a field has no proper zero divisors, the direct sum reduces to one element: $\mathbb{Q}_p$ or $\mathbb{R}$.

(ii) Now let the characteristic of K be nonzero. As in case (i), the topology T is induced on K by a nontrivial norm p. Let $t \in K$ be a nonzero topological nilpotent. The element t is transcendental over the prime field $\mathbb{F}_p$ of K, since otherwise it would be algebraic over $\mathbb{F}_p$ and, consequently, a root of unity, which is impossible, t being a nilpotent. It proves that K contains a subfield $F = \mathbb{F}_p(t)$ of the rational functions of one variable t over $\mathbb{F}_p$. Since t is a nonzero nilpotent element, the norm $p_1 = p|F$ is nontrivial. Kiyek's theorem (Theorem 12) implies that p_1 is equivalent to a norm of a form (1). But every nondiscrete absolute value on F is equivalent either to a q-adic absolute value ($q \in \mathbb{F}_p[t]$, a prime element) or to the norm φ_∞, where $\varphi_\infty(f/g) = \exp(\deg(f) - \deg(g))$; $f,g \in \mathbb{F}_p[t]$ (see Chapter 2, Theorem 11). This implies that K contains the closure of F, i.e., the completion $\hat{F}$ of F at $\sup\{\varphi_i : i = 1, 2, \ldots, s\}$. The approximation theorem for absolute values gives $\hat{F} = \hat{F}_1 \oplus \hat{F}_2 \oplus \cdots \oplus \hat{F}_s$, where $\hat{F}_i$ is the completion of F at φ_i. Since every $\hat{F}_i$ is a formal Laurent series field over $\mathbb{F}_p$ and since K is a field, $n = 1$ and the proof is finished.

Now we shall prove that a nondiscrete locally bounded and complete topological field of zero characteristic, with nondiscrete rationals $\mathbb{Q}$, has a nonzero topological nilpotent.

Let (K,T) be a topological field having the above-mentioned properties. Suppose that $T|\mathbb{Z}$ is nondiscrete. Then $T|\mathbb{Z}$ being a locally bounded ring topology has an ideal base of the neighborhoods of zero (Arnautov [1, Th. 2]; Mutylin [3, Lemma 11]; see Dikranjan and Więsław [1] for more general results). Consequently, $\mathcal{T}|\mathbb{Z}$ is the supremum of a family of p-adic topologies on $\mathbb{Z}$. This family, however, consists of a single element, a p-adic topology, since otherwise, the closure of $\mathbb{Z}$ in K would have proper divisors of zero. This shows that $p \in \mathbb{Q}$ is a topological nilpotent of $(K,\mathcal{T})$.

The case when $\mathcal{T}|\mathbb{Z}$ is discrete is considered in Mutylin [3]. Theorem 14 can also easily be deduced from Shanks and Warner [2]. See Theorem 10, Chapter 9, Section 3, for the general case.

6. EXAMPLES OF FIELDS COMPLETE WITH RESPECT TO NORMS INEQUIVALENT TO ABSOLUTE VALUES

Let A be a Dedekind domain and K its quotient field. Suppose that A satisfies the following finite norm property:

For every prime ideal p of A, $N(p) = |A/p| < \infty$. (55)

Let $\mathbb{P}$ be the set of prime ideals of A, $m = |\mathbf{P}|$, v_p the p-adic exponential valuation and $\varphi_p(x) = N(p)^{-v_p(x)}$, $x \in K^\times$, the p-adic absolute value on K. The condition (55) implies that the residue class field $\bar{k}_p$ is finite, since $\bar{k}_p = A_p/pA_p \simeq A/p$. Let $\pi \in A_p \setminus pA_p$, $v_p(\pi) = 1$, be fixed. Put

$$K_0 = K \qquad K_n = K(\pi^{1/q^n}) \qquad F = \bigcup_{n=0}^{\infty} K_n$$

where q is a fixed prime number. In the sequel v_p (resp. φ_p) will denote an extension of v_p (resp. φ_p) to F, and M the completion of (F,φ_p).

Let $x \in M$. Then $x = \lim x_n$, where $x_n \in F$. Every element x_n has the form $\Sigma\, a_s\pi^{m_s}$, $a_s \in K$, $m_s = k_s/q^{n_s}$; $n_s, k_s \in \mathbb{N}$, $m_s < 1$, where the sum is finite. The finiteness of $\bar{k}_p$ guarantees that every a_s can be written as

$$a_s = \sum_{r=t_s}^{\infty} a_r^{(s)}\pi^r \qquad t_s \in \mathbb{Z} \tag{56}$$

where $a_r^{(s)}$ runs over a finite set S, satisfying $|S| = |\bar{k}_p|$. It implies that x_n has the form

$$\sum_{r,s} a_r^{(s)} \pi^{m(r,s)} \qquad a_r^{(s)} \in S,\ m(r,s) = r + \frac{k_s}{q^s} \tag{57}$$

We call the elements π^w ($w \in \mathbb{Q}$) monomials. Since $\varphi_p(x - x_n) \to 0$ for $n \to \infty$, we have $\varphi_p(x_{n+1} - x_n) \to 0$, which implies that in the representation of x_n as a sum of monomials, the coefficients are stable (when $n \to \infty$). It means that every $x \in M$ has the form

$$x = \sum_{j \in J} a_j \pi_j \qquad a_j \in S \tag{58}$$

where π_j are monomials and $|J| \le \aleph_0$. Our discussion implies that

$$\begin{aligned}&\text{For every } \varepsilon > 0 \text{ there exists } J_0 \subset J,\ |J_0| < \infty, \text{ such that}\\ &\text{for every } j \notin J_0,\ \varphi_p(\pi_j) < \varepsilon\end{aligned} \tag{59}$$

The uniqueness of representation (58) follows from (59). Now, put for a monomial $x \in M$,

$$\begin{aligned}p(x) &= \varphi_p(x)N(p)^n \quad \text{if } x \in K_n \setminus K_{n-1},\ p(0) = 0 \quad \text{and}\\ p(x) &= \max\{p(\pi_j) : j \in J\} \quad \text{for } x \text{ of the form (58)}\end{aligned} \tag{60}$$

Let

$$L = \{x = \Sigma\, a_i \pi_i \in M : x \text{ satisfies (59) with respect to } p\}$$

It can be shown in the same way as in Mutylin [2] that p is a non-Archimedean norm and L is complete topological field in the topology T_p induced by p. Since $p = \varphi_p$ on K, then (K_p, T_p), the completion of (K, T_p), is a closed subfield of (L, T_p). The topology T_p cannot be induced by an absolute value. Indeed, $p = \varphi_p$ on K_p. If T_p were induced by an absolute value, say φ, $T_p = T$, then we should have $\varphi = \varphi_p^\lambda$ with a suitable positive λ, since the absolute value φ_p can be uniquely extended from the complete field K_p to $K_p(\pi^{1/q^n})$, and finally, to M. But $T_p|L \ne T_p|L$, since for $x_n = \pi^n(\pi^{1/q^n})^{-1}$, $\varphi_p(x_n) = N(p)^{-n+q^{-n}} \to 0$, and $p(x_n) = \varphi_p(x_n)N(p)^n = N(p)^{q-n} \to 1$ with $n \to \infty$.

Clearly, T_p cannot be induced by a Krull valuation, since L contains nonzero topological nilpotents with respect to p (e.g., π).

The construction above proves

THEOREM 15 (Więsław [14, Th. 1]; Mutylin [2, the case K = $\mathbb{Q}$]) Let K be the quotient field of a Dedekind domain A, satisfying the finite norm property for ideals. If $\mathbb{P}$ is the set of prime ideals of A, $m = |\mathbb{P}|$, then K can be embedded in at least m nonisomorphic complete locally bounded and nonabsolute-valued (but normed) topological fields.

THEOREM 16 (Więsław [14, Th. 2]) Let K be a locally compact disconnected topological field. Then K can be topologically embedded in a field complete with respect to a locally bounded normed (but nonabsolute-valued) field topology.

Proof: Apply Theorem 15 and the classification theorem for locally compact disconnected topological fields (Chapter 6, Section 2).

COROLLARY 1 The pseudonorm p on the field L, defined in the proof of Theorem 15, is not regular.

Proof: Indeed, if p were regular, then Theorem 8 and the approximation theorem for absolute values would imply that $p \sim \psi$, ψ being an absolute value, which is impossible (compare the proof of Theorem 15).

COROLLARY 2 Every compact Dedekind domain can be openly embedded in a locally bounded complete normable (but not necessarily absolute valued) topological field.

Proof: Apply Theorem 16 and Theorem 3 from Warner [1].

COROLLARY 3 Let K be a locally bounded, complete, and disconnected topological field of characteristic 0 such that every closed subfield F of K has a finite group G(F) of its topological field automorphisms. Then K is a topological subfield of a complete locally bounded nonabsolute-valued field L, which has a closed subfield M such that the dimension of M over K (as a vector space) is infinite and G(M) is an infinite group.

Proof: The assumptions imply that K is a finite extension of a p-adic number field $\mathbb{Q}_p$ (Więsław [4]). Thus we can take as L the field constructed in the proof of Theorem 15. However, L is not locally compact; hence again applying Więsław [4], we obtain that there exists a closed subfield M of L having an infinite group G(M). It is clear that M is an infinite-dimensional extension of K, since otherwise M would be locally compact and G(M) would be finite.

If K is a nondiscrete normed field of zero characteristic and p is its norm such that $p(x) = |x|$ for every $x \in \mathbb{Q}$, then K is algebraically isomorphic to a subfield of the complex number field (Aurora [3,4]). The last statement remains valid if $p(x) = |x|$ holds for infinitely many $x \in \mathbb{N}$ (Aurora [5]). If there exists $n_0 \in \mathbb{N}$, $n_0 > 1$ with $p(n^2) = p(n)^2$ for all $n \geq n_0$, $p(n_0) > 1$, then there is a continuous embedding of $(K, \mathcal{T}_p)$ into $(\mathbb{C}, \mathcal{T}_\infty)$ (Aurora [5]). This embedding is, generally, not topological. Indeed, take a discontinuous automorphism g of $\mathbb{C}$ and put $p(x) = \max(|x|, |g(x)|)$ for $x \in \mathbb{C}$. Then the identity embedding of $(\mathbb{C}, \mathcal{T}_p)$ into $(\mathbb{C}, \mathcal{T}_\infty)$ is continuous but not topological.

A norm p on a field K is said to be multiplication monotone if $p(x) < p(x')$ and $p(y) < p(y')$ imply $p(xy) \leq p(x'y')$. It can be shown that a multiplication monotone norm p of a field K is equivalent to an absolute value, say v, that is $\mathcal{T}_p = \mathcal{T}_v$ (see Aurora [6]).

Relations between absolute values and spectral norms are studied in Güntzer and Remmert [1]. For other results about norma in the case of rings, see Rhodes [1,2].

NOTES

Now we shall present an example of a norm on F(X), nonequivalent to a power-multiplicative norm, such that the completion of F(X) with respect to this norm is a field, and such that F is a discrete but unbounded subset of this completion. This example is due to J.-A. Cohen [5].

Let F be an infinite field. It is easy to see that F can be written as a countable union of an increasing sequence of its subrings F_n, $n \geq 0$. For every $a \in F$ let $f(a) = \min\{n \geq 0 : a \in F_n\}$. Obviously, $f(a + b) \leq \max(f(a),f(b))$ and $f(ab) \leq \max(f(a),f(b))$ hold for every $a,b \in F$. Let $F((X))$ be the field of formal Laurent series of a variable X over F. Let

$$D = \left\{a = \sum_{k=m}^{\infty} a_k X^k \in F((X)) : \lim_{n\to\infty} 2^{f(a_n)-n} = 0\right\}$$

It can be shown (J.-A. Cohen [5]) that D is a subfield of $F((X))$ containing $F(X)$. Now define a function $N: D \to \mathbb{R}_+$ by putting

$$N(a) = \begin{cases} 0 & \text{if } a = 0 \\ \max\{2^{f(a_k)-k} : k \geq m\} & \text{if } a = \sum_{i=m}^{\infty} a_i X^i \neq 0 \end{cases}$$

Cohen [5] proved that N is a norm of D and D is complete with respect to the topology T_N induced by N. Moreover, F is a discrete but unbounded subset of D.

We formulate a classical theorem of Mazur [1]. An absolute-valued algebra A over the reals satisfying $\|ax\| = |a|\|x\|$ for every $a \in \mathbb{R}$ and $x \in A$ is isomorphic with $\mathbb{R}$, $\mathbb{C}$, or $\mathbb{H}$ (the real quaternions). It is not assumed that A has unit element. The original Mazur's proof appeared for the first time in Żelazko [5].

Let A be a normed ring. An element $a \in A$ is said to be the generalized divisor of zero if there exists a sequence $b_n \in A$ such that $\|b_n\| = 1$, and $\lim_{n\to\infty} \max(\|ab_n\|,\|b_n a\|) = 0$. Kaplansky proved (see, e.g., Belfi and Doran [1]) the following result. If a normed associative algebra A over the reals has no nonzero generalized divisors of zero, then A is isomorphic with $\mathbb{R}$, $\mathbb{C}$, or $\mathbb{H}$. F. B. Wright [1] proved that every absolute-valued (not necessarily commutative or associative) division algebra over the reals, with unity, is algebraic. Albert's theorem [4] implies that A is one of the following algebras: $\mathbb{R}$, $\mathbb{C}$, $\mathbb{H}$, or $\mathbb{D}$ (the Dickson-Cayley algebra). Urbanik and Wright [1,2] extended this result, showing that the only absolute-valued algebras over $\mathbb{R}$ with unity are those mentioned above. Finally, Urbanik [2] proved that if A is an absolute-valued algebra over the reals, not necessarily associative or commutative, such that all except at most countably many elements are quasi-invertible, then A is isomorphic with $\mathbb{R}$, $\mathbb{C}$, $\mathbb{H}$, or $\mathbb{D}$. For classical results in this case, see Albert [3,4].

Let A be a normed algebra (not necessarily associative) over $\mathbb{R}$. Let (x,y,z) be the associator of x, y, z, i.e., $(x,y,z) = x(yz) - (xy)z$. An algebra A is called *elastic* if for every $x,y \in A$ we have $(x,y,x) = 0$. For example, an alternative (or associative) algebra is elastic.

The following theorem generalizes results of Albert [3,4] and Urbanik and Wright [1,2].

THEOREM (El-Mallah and Micali [1]) Every elastic normed algebra over the reals is finite-dimensional. If A is a normed algebra over the reals, then each of the following conditions implies that A is finite-dimensional.

(i) For every $x \in A$, $(x,x,x) = 0$ and there exists nonzero $z \in A$ such that for every $x \in A$, $xz = zx$.

(ii) There exists a nonzero $a \in A$ such that a and a^2 commute with every $x \in A$, and for every $x \in A$, $(x,a,x) = 0$.

Recently El-Mallah [1] obtained another result.

THEOREM Let A be an absolute-valued algebra (not necessarily associative) over the reals. Assume that A is pre-Hilbert (the norm of A is induced by a scalar product in A, i.e., $\|x\|^2 = (x,x)$ holds for every $x \in A$). If $(x,x,x) = 0$ holds for every $x \in A$ and $\dim_{\mathbb{R}} A < \infty$, then A is isomorphic with one of the classical normed division algebras (i.e., $\mathbb{R}$, $\mathbb{C}$, $\mathbb{H}$, or $\mathbb{D}$).

Let A be a ring regular in the sense of von Neumann. This means that for every $a \in A$ there exists $b \in A$ satisfying $aba = a$. An element b is called an inner inverse of a. By a generalized division algebra we shall mean a von Neumann regular algebra.

The following theorem is due to Hirschfeld [1].

THEOREM Let A be a von Neumann regular algebra over $\mathbb{C}$ (with or without unity). Then A is finite-dimensional if and only if the following conditions are satisfied.

(i) A is complete metric algebra.

(ii) There is no proper field in A that contains $\mathbb{C}$.

(iii) There exists a neighborhood of zero in A without idempotents.

An old problem in mathematics is to equip n-dimensional euclidean space $\mathbb{R}^n$ with a multiplication in order to obtain a real algebra which has some prescribed properties, such as (1) associativity, (2) commutativity, and (3) continuity of multiplication with respect to the euclidean norm. We shall present a construction of a multiplication in $\mathbb{R}^n$, due to Lucas [1]. Let $u, w \in \mathbb{R}^{n+1}$, $I = (1,1,\ldots,1) \in \mathbb{R}^{n+1}$. It is easy to see that the relation $u \sim w$ if and only if there exists $k \in \mathbb{R}$ such that $u - w = kI$ is an equivalence relation. Let us consider the quotient structure: $\mathbb{R}_n^* = \mathbb{R}^{n+1}/\sim$, with operations

$$(a_0,a_1,\ldots,a_n) + (b_0,b_1,\ldots,b_n) = (a_0 + b_0, \ldots, a_n + b_n)$$

$$(a_0,a_1,\ldots,a_n) * (b_0,b_1,\ldots,b_n) = (c_0,c_1,\ldots,c_n)$$

where

$$c_j = \sum_{\substack{s,t \\ s+t\equiv j(\mathrm{mod}(n+1))}} a_s b_t$$

$$c(a_0,a_1,\ldots,a_n) = (ca_0,ca_1,\ldots,ca_n) \qquad c \in \mathbb{R}$$

It can be shown that $\mathbb{R}_n^*$ and $\mathbb{R}[X]/(X^n + X^{n-1} + \cdots + X + 1)$ are isomorphic as $\mathbb{R}$-algebras. Let $a = (a_0,a_1,\ldots,a_n) \in \mathbb{R}_n^*$. Define $N(a) = ((n + 1)^2(1/n)\mathrm{Var}(a))^{1/2}$, where

$$\mathrm{Var}(a) = \frac{1}{n+1}\sum_{i=0}^{n+1}\left(a_i - \sum_{k=0}^{n+1}\frac{a_k}{n+1}\right)^2$$

Lucas [1] proved that the addition and multiplication in $\mathbb{R}_n^*$ are continuous with respect to N. The function N fails to satisfy the condition $N(a * b) \le N(a)N(b)$. Since $\mathbb{R}_n^*$ is an n-dimensional normed vector space over the reals, there exists an isometry f from $\mathbb{R}_n^*$ onto the euclidean normed space $\mathbb{R}^n$. Now we can carry the multiplication from $\mathbb{R}_n^*$ onto $\mathbb{R}^n$ by putting

$$a \cdot b = f(f^{-1}(a) * f^{-1}(b)) \qquad a,b \in \mathbb{R}^n \tag{61}$$

THEOREM $\mathbb{R}^n$ is a commutative and associative algebra over $\mathbb{R}$ with respect to componentwise addition and the multiplication defined by (61). The multiplication in $\mathbb{R}^n$ is continuous with respect to the Cartesian product topology in $\mathbb{R}^n$.

Connections of arithmetic and topology in division rings are described by means of norms (see Benz [1,2]). Let M be a topological division ring $(K,\mathcal{T})$, where the topology $\mathcal{T}$ is induced by a norm, say p. A subset A of M is called *absolutely convex* if for every $a_1, \ldots, a_k \in A$ and every $x_1, \ldots, x_k \in K$, $p(x_1) + \cdots + p(x_k) \le 1$ implies that $x_1a_1 + \cdots + x_ka_k \in A$. Vodinčar [1] proved the following

THEOREM Let M be a topological algebra over a topological division ring $(K,\mathcal{T})$, $\mathcal{T}$ being induced by a norm. Then the following conditions are equivalent.

(i) The topology of M is induced by an algebra norm.

(ii) There exists a bounded and absolutely convex neighborhood of zero in M.

It is not known whether Cohn's characterization of topologies induced by norms has an analog for noncommutative fields.

The original Aurora's theorem was proved for associative, not necessarily commutative rings. As a corollary from Aurora [2], we obtain that a power-multiplicative norm on a division ring is equivalent to a norm of the form $\sup\{w_i : i \in I\}$, where w_i are absolute values of the division ring. Related results in the case of normed algebras over $\mathbb{R}$ were obtained by Strzelecki [1,2].

Arnautov [1] gave the following criterion of normability of associative rings. A topology on a ring K can be induced by a norm if and only if K is a countable union of bounded sets and there exists a countable base $\{U_i\}$ of symmetric neighborhoods of zero in K such that $U_iU_j \subset U_{i+j}$ and $U_{i+1} + U_{i+1} \subset U_i$ hold for every $i,j \in \mathbb{N}$.

Arens [1] proved that the set of functions holomorphic on an open subset of $\mathbb{C}$ is a complete, convex, and metrizable algebra over

$\mathbb{C}$, which is not normable. Recall that a B_0 algebra is a completely metrizable locally convex topological algebra over $\mathbb{R}$ or $\mathbb{C}$. Arens [1] proved that if in a B_0 division algebra A the inversion $x \mapsto x^{-1}$ is continuous ($x \neq 0$), then A is topologically isomorphic either with $\mathbb{R}$, or with $\mathbb{C}$, or with $\mathbb{H}$. Żelazko [1] proved that the assumption of the continuity of inversion is superfluous. For another proof of Arens' theorems, see Isidro [1]. Some fundamental properties of normed division rings can be found in the book by Arnautov, Vodinčar, and Mihalev [1].

Cohn and Mahler [1] described norms on the rings of integers of algebraic number fields. For more general results in this direction, see J.-A. Cohen [3,4]. Properties of absolute-valued fields via spectral norm are described in Güntzer and Remmert [1].

The presentation of results in Section 5 is based mostly on Kiyek [6]. Theorem 6 has an analog in real Banach algebras (see Arens and Hoffman [1]).

4

Krull Valuations

In this chapter we summarize only the fundamental properties of valuations necessary for further investigation. A more detailed exposition of valuation theory can be found in Endler [7], Ribenboim [10], Schilling [6], and Zariski and Samuel [1, Vol. II].

1. FUNDAMENTAL PROPERTIES OF VALUATIONS

The fundamental notion of a valuation was introduced by Krull in his fundamental paper (Krull [1]).

DEFINITION 1 Let Γ be any multiplicative linearly ordered group with added zero such that $0 \cdot \gamma = \gamma \cdot 0 = 0$ and $0 \leq \gamma$ for every $\gamma \in \Gamma$. A function $v\colon K \xrightarrow{\text{onto}} \Gamma \cup \{0\}$ is called a *valuation of the field* K if the following conditions are satisfied.

(V_1) $v(x) \geq 0$, $v(x) = 0 \Leftrightarrow x = 0$.

(V_2) $v(xy) = v(x)v(y)$.

(V_3) There exists a $\lambda \in \Gamma$ such that

$$v(x + y) \leq \lambda \max(v(x), v(y)) \qquad \text{for all } x, y \in K$$

A valuation v is called *non-Archimedean* if $\lambda = 1$, where 1 is the unit element of Γ. The group Γ is called the *value group of a valuation* v.

A valuation v of a field K defines a field topology T_v on K: We take sets of the form

$$U(\gamma) = \{x \in K : v(x) < \gamma\} \qquad \gamma \in \Gamma \tag{1}$$

as a base of the neighborhoods of zero in (K, T_v).

As in the case of a non-Archimedean absolute value, we define a valuation ring V_v and a residue class field $\bar{k}_v$ of a non-Archimedean valuation v.

DEFINITION 2 Let Γ be an additive linearly ordered group. A function $v: K \xrightarrow{\text{onto}} \Gamma \cup \{\infty\}$ ($\infty + \infty = \infty$, $\infty + \gamma = \infty$ for $\gamma \neq \infty$; $\gamma < \infty$ for every $\gamma \in \Gamma$) is called an *(additive) valuation of the field* K if the following conditions are satisfied.

(V_1') $v(x) \neq \infty$ for $x \in K^\times$; $v(0) = \infty$.

(V_2') $v(xy) = v(x) + v(y)$.

(V_3') $v(x + y) \geq \min(v(x), v(y))$.

If we take a partially ordered ℓ-group $(\Gamma, \leq)$ in Definition 2, we obtain the definition of a semivaluation.

LEMMA 1 Definitions 1 and 2 are equivalent for non-Archimedean valuations. In other words, there is a one-to-one correspondence between the set of all non-Archimedean valuations on a field K (in the sense of Definition 1) and the additive valuations on K.

Proof: Let $v: K \xrightarrow{\text{onto}} \Gamma \cup \{0\}$ be a valuation of K. Write the group operation of Γ additively and denote this (isomorphic with Γ) group by Δ. Define the mapping $f: \Gamma \cup \{0\} \to \Delta \cup \{\infty\}$ by $f(g) = -\bar{g}$ ($\bar{g} \in \Delta$, the image of g in Δ) and $f(0) = \infty$. Put

$$v_{ad}(x) = f(v(x)) \qquad \text{for every } x \in K \tag{2}$$

It is immediate that v_{ad} satisfies conditions of Definition 2, i.e., v_{ad} is an additive valuation of K.

The same construction assigns to each additive valuation of K a non-Archimedean valuation, by changing the roles of Γ and Δ.

DEFINITION 3 Let v_i be a valuation on a field K with the value group $\Gamma_i = v_i(K^\times)$. We say that v_1 and v_2 are equivalent, $v_1 \sim v_2$, if there exists an order isomorphism $f: \Gamma_1 \xrightarrow{\text{onto}} \Gamma_2$ such that the diagram

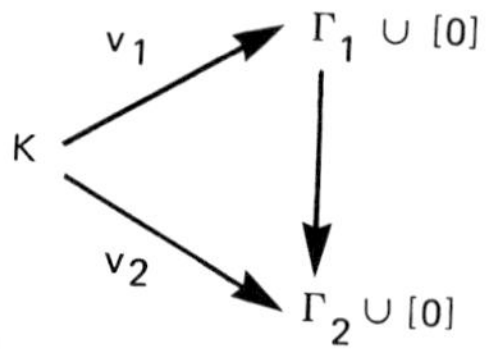

is commutative, i.e., $v_2 = f \circ v_1$. It follows immediately that $v_1 \sim v_2$ implies that $T_{v_1} = T_{v_2}$.

We shall see later that $T_{v_1} = T_{v_2}$ does not imply, generally, that $v_1 \sim v_2$. In the sequel, by valuation we shall often mean a class of equivalent valuations.

THEOREM 1 A valuation v on a field K is equivalent to an absolute value if and only if v has rank ≤ 1.

Proof: This follows immediately from Theorem 1 and Lemma 2 of the Appendix. Note that only the trivial valuation (inducing the discrete topology) has rank 0.

THEOREM 2 Let v be a valuation on a field K of rank ≥ 1. Then $T_v \in L_r(K)$ is a minimal and locally bounded topology. Moreover, for every neighborhood U of zero, $(K \setminus U)^{-1}$ is bounded and $T_v \in L_f(K)$.

Proof: To prove that T_v is a minimal element of $L_r(K)$, it is sufficient to repeat the proof of Theorem 13 of Chapter 2. The local boundedness of T_v is obvious, since every $U(\gamma) \in B(T_v)$ defined by the formula (1) is bounded. Finally, if $U = U(\gamma)$, then $(K \setminus U)^{-1} = U(\gamma^{-1}) \setminus \{0\}$. Its closure is also bounded. It follows immediately that T_v is a field topology.

Example 1 Let $K = k(X_1, X_2, \ldots, X_n)$ be a rational function field of indeterminates $X_1, X_2, \ldots, X_n$ with coefficients from a field k. First, we shall consider a ring of all polynomials $R = K[X_1, X_2, \ldots, X_n]$. We agree to assign to the monomial $X_1^{m_1} X_2^{m_2} \cdots X_n^{m_n}$ the value

$$v(X_1^{m_1} X_2^{m_2} \cdots X_n^{m_n}) = (e^{-m_1}, e^{-m_2}, \ldots e^{-m_n}) \in (\mathbb{R}_+^\times)^n \tag{3}$$

Next, we define a multiplication in $\Gamma = (\mathbb{R}_+^\times)^n$:

$$(r_1,\ldots,r_n)(s_1,\ldots,s_n) - (r_1s_1,\ldots,r_ns_n) \tag{4}$$

We order n-tuples of positive reals lexicographically so that

$$(r_1,\ldots,r_n) > (s_1,\ldots,s_n) \Leftrightarrow \exists\ k \quad (0 \le k \le n) \tag{5}$$

$$r_i = s_i \text{ for } i = 1, \ldots, k, \text{ and } r_{k+1} > s_{k+1}$$

Each polynomial $f(X_1,\ldots,X_n)$ can be expressed as

$$f(X_1,\ldots,X_n) = X_1^{m_1}F_1(X_1)$$

where $F_1(X_1) \in k[X_2,\ldots,X_n][X_1]$. Let $f_0(X_2,\ldots,X_n) = F_1(0)$ be a non-zero polynomial in X_2, ..., X_n. In the same way we can write

$$f_0(X_2,\ldots,X_n) = X_2^{m_2}F_2(X_2)$$

where $F_2(X_2) \in k[X_3,\ldots,X_n][X_2]$ and $F_2(0) \neq 0$. Similarly, we can obtain the numbers m_3, ..., m_n. Let us put

$$v(f) = (e^{-m_1},e^{-m_2},\ldots,e^{-m_n})$$

Since every element of K is of the form f/g, where f,g $\in$ R and g $\neq$ 0, we can extend this definition of v to K by putting $v(f/g) = v(f)f(g)^{-1}$ and $v(0) = 0$. It is easy to verify that v is a valuation on K, of rank n.

Example 2 Let K be an arbitrary field and Γ an arbitrary linearly ordered abelian group. We consider the set $L = K\{t^\Gamma\}$ of all formal sums

$$x = a_{\alpha_1}t^{\alpha_1} + a_{\alpha_2}t^{\alpha_2} + \cdots \tag{6}$$

where $a_{\alpha_i} \in K$, $\alpha_i \in \Gamma$ and $\alpha_1 < \alpha_2 < \cdots < \alpha_\rho < \cdots$, $\rho < \sigma$, and σ is a fixed ordinal number. We define multiplication by the formula

$$t^\alpha t^\beta = t^{\alpha+\beta} \qquad \text{for all } \alpha,\beta \in \Gamma \tag{7}$$

Next we define addition and multiplication for expressions of the form (6) in the usual formal fashion. We again obtain a field. Finally, we set $v(x) = \alpha_1$ for $x \neq 0$ defined by formula (6). Thus we obtain a valuation on L. The valuation ring V_v consists of all series (6) having $\alpha_1 \geq 0$ (recall that Γ is additively written) and the maximal prime ideal P_v of V_v contains all series with $\alpha_1 > 0$. The residue class field $\bar{k}_v \simeq V_v/P_v$ is isomorphic to K, and $\Gamma = v(L^\times)$ is the value group of v.

Thus we have proved

THEOREM 3 Let K be an arbitrary field and Γ any linearly ordered abelian group. Then there exist a field extension L of K and a valuation v on L such that $v(L^\times) = \Gamma$ and $\bar{k}_v \simeq K$.

Note that the field $K\{t^\Gamma\}$, where K is an algebraically closed field and Γ is a linearly ordered divisible abelian group, is universal in the sense that for every field F of the same characteristic and cardinality, there exists an isomorphism of F into $K\{t^\Gamma\}$ (MacLane [4, Th. 3]).

DEFINITION 4 Two valuations v and w of a field K are said to be independent if $K = V_v + V_w$.

DEFINITION 5 A valuation ring A of a field K is a subring of K such that

$$K^\times = A^\times \cup (A^\times)^{-1} \tag{8}$$

The importance of the foregoing notion follows from:

THEOREM 4 (Endler [7, Th. (7.1), p. 46]) Let v be a non-Archimedean valuation on a field K and V_v its valuation ring. The mapping $v \mapsto V_v$ is a one-to-one correspondence between the set of all equivalence classes of non-Archimedean valuations of K and the set of all valuation rings of K.

Proof: For every valuation v on a field K, the ring V_v is a valuation ring of the field K. Moreover, if v and w are non-Archimedean valuations of K, and $v \sim w$, then $V_v = V_w$, as for every $x \in K$,

$v(x) \leq 1 \Leftrightarrow w(x) \leq 1$. It is easily seen that $v \not\sim w$ implies that $V_v \neq V_w$.

Conversely, let A be a valuation ring of K. We have to construct a non-Archimedean valuation v on K such that $A = V_v$. Let U(A) be the group of units of A. Put $\Gamma_A = K^\times/U(A)$. The quotient group Γ_A becomes a linearly ordered group with respect to the order $xU(A) \leq yU(A) \Leftrightarrow xy^{-1} \in A$. Adjoin the zero 0 to Γ_A and put $0 \cdot g = 0$ and $0 < g$ for every $g \in \Gamma_A$. Define a function $v_A: K \to \Gamma_A \cup \{0\}$ by putting $v_A(x) = xU(A)$ if $x \in K^\times$ and $v_A(0) = 0$. It is easy to verify the conditions of Definition 1. Indeed, (V_1) holds by definition of v_A. Condition (V_2) is satisfied, since v_A is a group homomorphism. Note that $x \in A \Leftrightarrow v_A(x) \leq 1$, i.e., $V_{v_A} = A$. Now let $v_A(x) \leq v_A(y)$. Assume that $y \neq 0$. Then $xy^{-1} \in A$, and as A is a ring, $1 + xy^{-1} \in A$. Consequently, $v_A(x + y) = v_A(y)v_A(1 + xy^{-1}) \leq v_A(y) = \max(v_A(x),v_A(y))$, proving (V_3).

Finally, we prove that if $A = V_v$, then $v \sim v_A$. Indeed, $v: K^\times \to \Gamma$ is an epimorphism, and $\mathrm{Ker}(v) = U(V_v) = U(A)$. Thus we obtain a commutative diagram:

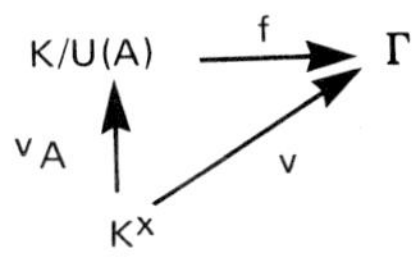

where f is an order isomorphism of groups. This means, however, that $v = f \circ v_A$, i.e., $v \sim v_A$.

THEOREM 5 (Krull [1]) Let K be a field with a non-Archimedean valuation v, Γ its value group, and R the residue class field of v. Then

$$|K| \leq |R|^{|\Gamma|}$$

Proof: For $\gamma \in \Gamma$ define

$$Q_\gamma = \{x \in K : v(x) \leq \gamma\} \quad \text{and} \quad P_\gamma = \{x \in K : v(x) < \gamma\}$$

Regard these sets as abelian groups under addition. Each quotient group $V_\gamma = Q_\gamma/P_\gamma$ has a natural structure as a vector space over

$R = Q_1/P_1$ of dimension 1. Indeed, we define multiplication by scalars $\bar{a} = a + P_1 \in R$ as $(a + P_1)(x + P_\gamma) = ax + P_\gamma$, $v(a) \le 1$, $v(x) \le \gamma$. Take an $a_\gamma \in Q_\gamma \setminus P_\gamma$. Then $v(a_\gamma) = \gamma$ and $V_\gamma = R\bar{a}_\gamma$, i.e., $\dim_R V_\gamma = 1$. If $\bar{b} = b + P_\gamma$ $(b \in Q_\gamma)$ is arbitrary, then write $b + P_\gamma = (c + P_1)(a_\gamma + P_\gamma)$. Taking $c \in Q_1$ such that $v(c) = v(b)v(a_\gamma)^{-1}$, we obtain $v(c) \le 1$, since $v(b)v(a_\gamma)^{-1} \le 1$ (such a choice of c is possible since v maps $K^\times$ onto Γ). Consequently, $\dim_R V_\gamma = 1$ and $|V_\gamma| = |R|$ holds for every $\gamma \in \Gamma$. Thus $Q_\gamma = \bigcup_{r \in R} (a_r + P_\gamma)$ and the cosets are pairwise disjoint, $a_r \in Q_\gamma$.

For each $\gamma \in \Gamma$ the additive group of K is the union of distinct cosets of Q_γ, i.e., $K = \bigcup_a (a + Q_\gamma)$. Each of these cosets is itself a union of distinct cosets of the smaller group P_γ. Consequently, $K = \bigcup_a \bigcup_{r \in R} (a + a_r + P_\gamma)$. We are now in a position to set up an injection from K to R^Γ. For $x \in K$ define $\bar{x} \in R^\Gamma$ by setting $\bar{x}(\gamma) = r$, where $x \in a_r + a + P_\gamma$. The function $\bar{x}$ is well defined, since the cosets of P_γ whose union is K are distinct. Note that for $x,y \in K$, $v(x - y) = \gamma \Leftrightarrow \bar{x}(\gamma) \ne \bar{y}(\gamma)$. Indeed, $v(x - y) = \gamma \Leftrightarrow v(x - y) \le \gamma$ and $v(x - y) \nless \gamma$, but this means that $x + Q_\gamma = y + Q_\gamma$ and $x + P_\gamma \ne y + P_\gamma$. Thus there must be different indices attached to the cosets of P_γ in which x and y, respectively, lie.

Let $\bar{x}(\gamma) = \bar{y}(\gamma)$ hold for every $\gamma \in \Gamma$. Consequently, for every $\gamma \in \Gamma$, $v(x - y) \ne \gamma$. Thus $v(x - y) = 0$, and $x = y$. The theorem now follows immediately.

The proof presented is due to Gravett [1] (see Rayner [1]).

2. INDEPENDENCE THEOREMS

THEOREM 6 (Endler [7, Th. (11.16), p. 80]) Let $A_1, \ldots, A_n$ be pairwise independent valuation rings of a field K, and let v_i be a non-Archimedean valuation corresponding to A_i with value group Γ_i $(i = 1, 2, \ldots, n)$. Then for any $(x_1,\ldots,x_n) \in K^n$ and any $(\gamma_1,\ldots,\gamma_n) \in \Gamma_1 \times \cdots \times \Gamma_n$, there exist infinitely many $x \in K$ such that

$$v_i(x - x_i) = \gamma_i \qquad i = 1, 2, \ldots, n \tag{9}$$

For this result and related theorems, see Jaffard [2].

As a corollary we obtain

THEOREM 7 (independence theorem) Under the assumptions of Theorem 6 for every $(\gamma_1,\ldots,\gamma_n) \in \Gamma_1 \times \cdots \times \Gamma_n$, there exist infinitely many $x \in K$ such that $v_1(x) = \gamma_1, \ldots, v_n(x) = \gamma_n$.

Let $v_1, \ldots, v_n$ be valuations of a field K and $\Gamma_1, \ldots, \Gamma_n$ the corresponding value groups. Let $(\gamma_1,\ldots,\gamma_n) \in \Gamma_1 \times \cdots \times \Gamma_n$ and $(b_1,\ldots,b_n) \in K^n$ be fixed elements. Suppose that Δ_{ij} (resp. Δ_{ji}) is the isolated subgroup of Γ_i (resp. Γ_j) which corresponds to the largest prime ideal p common to the valuation rings A_i and A_j of v_i and v_j, respectively. Krull proved that the groups Γ_i/Δ_{ij} and Γ_j/Δ_{ji} are order isomorphic and represent (after identification) the value group of the valuation ring $(A_i)_p = (A_j)_p$. Let $g_{ij}: \Gamma_i \to \Gamma_i/\Delta_{ij}$ be the canonical epimorphism.

DEFINITION 6 A set $(\gamma_1,\ldots,\gamma_n) \in \Gamma_1 \times \cdots \times \Gamma_n$ is called *compatible* if $g_{ij}(\gamma_i) = g_{ji}(\gamma_j)$ for every pair $i \neq j$.

Now we are able to formulate the next theorem.

THEOREM 8 (Ribenboim [1]) Suppose that the values $(\gamma_1,\ldots,\gamma_n) \in \Gamma_1 \times \cdots \times \Gamma_n$ are compatible. Let $(b_1,\ldots,b_n) \in K^n$ be arbitrary and satisfy $v_i(b_i - b_j) < \alpha_i$ for every $i \neq j$. Then the system of equations

$$v_i(x - b_i) = \gamma_i \qquad i = 1, 2, \ldots, n \tag{10}$$

has a solution $x \in K$ if and only if $\gamma_i - v_i(b_i - b_j) \in \Delta_{ij}$ holds for each pair $i \neq j$.

In Chapter 7 we look again at the independence of valuations from a more general point of view.

3. TOPOLOGICAL CHARACTERIZATION OF FIELDS WITH VALUATIONS

If K is a field and R is an order in K (i.e., a subring of K with identity whose quotient is K) we may topologize K by taking the sets aR, $a \in K^\times$, as neighborhoods of zero. The topology T obtained in this way is a ring topology. On the other hand, R can be used to define a kind of a valuation on K, called a half-valuation (see Chapter 9, Definition 1) in the following way. Let $U = U(R)$ be the group of units of R. We may order $K^\times/U$ by

$$aU \leq bU \Leftrightarrow ab^{-1} \in R \tag{11}$$

This makes $K^\times/U$ a partially ordered group. It is easy to see that $K^\times/U$ is a directed group. The canonical epimorphism $v: K^\times \to K^\times/U$ has the following properties:

$$v(xy) = v(x)v(y) \tag{12}$$

$$v(x), v(y) \leq \gamma \Rightarrow v(x - y) \leq \gamma \tag{13}$$

Let $\Gamma = K^\times/U$. We adjoin the symbol 0 to Γ with the claim that $0 \cdot \gamma = \gamma \cdot 0 = 0$, $0 < \gamma$ for every $\gamma \in \Gamma$. We extend v to K, putting $v(0) = 0$. Thus

$$v(x) = 0 \Leftrightarrow x = 0 \tag{14}$$

Now we formulate

LEMMA 2 Let K be a field. Then there is one-to-one correspondence between orders in K and classes of equivalent half-valuations of K.

(We define an equivalence of half-valuations in the same way as for valuations.)

Proof: Follows from the construction above. If v is a semi-valuation of K, $\Gamma = v(K^\times)$ and 1 is the unit element of Γ, then $R = \{x \in K : v(x) \leq 1\}$ is the corresponding order in K.

Note that the topology defined by an order is a field topology.

LEMMA 3 Let R be an order in a topological field (K,T). Then the following conditions are equivalent.

(i) The topology induced by R is equal to T.

(ii) If v is the half-valuation corresponding to R, then $T_v = T$.

(iii) R is a bounded neighborhood of zero for (K,T).

Proof: (i) $\Leftrightarrow$ (ii) If R defines T, $v(x) = xU \in K^\times/U$, then $\{y \in K : v(y) \leq 1\} = R$, and conversely.

(i) $\Rightarrow$ (iii) R is bounded and $R \in \mathcal{B}(T)$.

(iii) $\Rightarrow$ (i) Since $R \in \mathcal{B}(T)$, $aR \in \mathcal{B}(T)$ for every $a \in K^{\times}$, R being bounded, then for every $V \in \mathcal{B}(T)$ there exists $W \in \mathcal{B}(T)$ with $RW \subset V$. This implies that $aR \subset V$ for $a \in R \cap W$, $a \neq 0$. Thus $T \leq S$, where $S \in L_f(K)$ is the topology defined by R. Conversely, since $R \in \mathcal{B}(T)$ and T is locally bounded, $\mathcal{B}(T) = \{aR : a \in K^{\times}\}$ (Chapter 1, Theorem 7). Finally, $T = S$.

COROLLARY Let (K,T) be a topological field. The following conditions are equivalent.

(i) There exists $V \in \mathcal{B}(T)$ such that the additive group $gp(V) \subset K^+$ is bounded.

(ii) There exists a bounded order $R \in \mathcal{B}(T)$.

(iii) There is an order R in K defining T.

(iv) The topology T is induced by a half-valuation.

DEFINITION 7 A *subset* A of a topological field (K,T) is *bounded away from 0* if there exists $W \in \mathcal{B}(T)$ satisfying $W \cap A = \emptyset$.

Now we may formulate an important theorem due to Zelinsky [2].

THEOREM 9 Let (K,T) be a proper topological field. Then there exists a non-Archimedean valuation v satisfying $T = T_v$ if and only if (K,T) satisfies the following conditions.

(i) There exists $V \in \mathcal{B}(T)$ such that the additive group gp(V) generated by V in K^+ is bounded.

(ii) If $A \subset K$ is bounded away from zero, then A^{-1} is bounded.

Proof: Necessity follows immediately from Theorem 2.

To prove sufficiency, let R be any bounded order in (K,T), $R \in \mathcal{B}(T)$. Put

$$X = \{x \in K : x \notin R \text{ and } x^{-1} \notin R\} \tag{15}$$

The set X is bounded since $X \cap R = \emptyset$, $R \in \mathcal{B}(T)$, and $X = X^{-1}$, by (ii).

Note that a set A is bounded in (K,T) exactly when $A \subset Rb$ for some $b \in K^{\times}$. Indeed, $\mathcal{B}(T) = \{bR : b \in K^{\times}\}$, by Lemma 3. Moreover, Lemma 12 of Chapter 1 implies that A is bounded if and only if for every $b \in K^{\times}$ there exists $c \in K^{\times}$ satisfying $Ac \subset Rb$. Thus $A \subset Rbc^{-1}$.

We shall call such an element b for which $A \subset Rb$, a bound for A. If b is a bound for A, so is bc^{-1} for every $c \in R^{\times}$. Indeed, $Ac \subset Rbc = (Rc)b \subset Rb$, or $A \subset Rbc^{-1}$. Hence we may choose a bound p for X with the property that $p^{-1} \in R$. (Note that $p \notin R$, since otherwise $X \subset pR \subset R$, contradicting $X \cap R = \emptyset$.) Let

$$Y = \{p, p^2, \ldots, p^n, \ldots\} \tag{16}$$

Consider first the case when Y is not bounded. Then Y^{-1} is not bounded away from zero. If Y^{-1} were bounded away from zero, say $Y^{-1} \cap aR = \emptyset$, $(Y^{-1})^{-1} = Y$ would be bounded by (ii), a contradiction. Thus for every $a \in K^{\times}$ there exists $n \in \mathbb{N}$ with $p^{-n} \in Ra$. But if $N \geq n$, then $p^{-N} \in Rp^{-n} \subset Ra$. This proves that p^{-n} tends to zero.

Observe that the set N_0 of topological nilpotents of $(K,\mathcal{T})$ is open. Indeed, N_0 is nonzero since $p^{-1} \in N_0$. The openness of N_0 follows from Lemma 3 of Chapter 3. Finally, note that since N_0 is open and condition (ii) holds, we can apply Theorem 14(v) of Chapter 2 to obtain $\mathcal{T} = \mathcal{T}_v$, where v is an absolute value of K. This absolute value is in fact non-Archimedean since the additive group generated by 1, being a subset of R, is bounded. This concludes the proof in case that Y is not bounded.

Suppose now that Y is bounded. Let $S = R[X]$. Since $R \subset S$, $S \in \mathcal{B}(\mathcal{T})$ and is an order. The order S is bounded. Let $a \in R$, $x_1, \ldots, x_n \in X$ and let q be a bound for Y (i.e., $Y \subset Rq$). Then

$$ax_1 \cdots x_n \in R(Rp)^n \subset Rp^n \subset Rq$$

Since Rq is closed under addition, $S \subset Rq$. Thus S is bounded.

Now we are able to define a valuation v satisfying $\mathcal{T} = \mathcal{T}_v$. Let v be the half-valuation corresponding to the order S. Lemma 3 implies that $\mathcal{T} = \mathcal{T}_v$. It only remains to show that v is a valuation, i.e., that $\Gamma = v(K^{\times})$ is linearly ordered. This follows from an obvious (by definition of X) property of S: S is a valuation ring, i.e., $K^{\times} = S^{\times} \cup (S^{\times})^{-1}$. Thus if $x,y \in K^{\times}$, then either $x^{-1}y \in S$ or $y^{-1}x \in S$. Hence either $v(y) \leq v(x)$ or $v(x) \leq v(y)$. This completes the proof.

4. EXTENSIONS OF VALUATIONS

Note first that Theorems 18, 19, 21, and 22 of Chapter 2 remain true for arbitrary non-Archimedean valuations (i.e., for valuations of arbitrary rank). Hasse showed in 1925 that for an algebraic number field and integers $e_1, \ldots, e_g$, $f_1, \ldots, f_g$, there exists an extension L of degree $n = \Sigma_{i=1}^{g} e_i f_i$ over K such that a given prime ideal p of the ring R_K of algebraic integers of K may be factored in R_L as $p = P_1^{e_1} P_2^{e_2} \cdots P_g^{e_g}$, where the prime ideal P_i has the relative degree f_i. This result holds for a wide class of non-Archimedean valuations (see Ribenboim [3,4,6,10]).

We state the following theorem without proof.

THEOREM 10 (Fukawa [1]) Let K_0 be a field, v_0 a valuation on K_0 with the value group Γ_0 and the residue class field Δ_0. Let Γ_1 be a linearly ordered abelian group containing Γ_0, and Δ_1 a field containing Δ_0. Then v_0 can be extended to a valuation v_1 on a field K_1 containing K_0 with the value group Γ_1 and the residue class field Δ_1.

Now we prove

THEOREM 11 (Więsław [5]) Let K be a field with a valuation v. Then K has only a finite number of nonisomorphic valuated extensions if and only if $K \simeq \mathbb{C}$ or K is a real closed field such that $K(i) \simeq \mathbb{C}$, both isomorphisms being topological.

Proof: ($\Leftarrow$) By Theorem 5 of Mutylin [3], $\mathbb{R}$ and $\mathbb{C}$ are the only locally bounded extensions of the reals $\mathbb{R}$. Since every topology induced by a valuation is locally bounded (Theorem 2 of this chapter), the result follows (see also Chapter 9, Theorem 13).

($\Rightarrow$) (S. Warner) Suppose that the topology is induced on K by a valuation, say v, having the value group Γ. We claim that the order of Γ must be Archimedean.

Suppose to the contrary that Γ is a non-Archimedean ordered group. Let $K = K_0$, $K_{i+1} = K_i(X_{i+1})$, X_{i+1} being transcendental over K_i ($i = 0, 1, \ldots$). Extend v to v_1 on K_1, then to v_2 on K_2, by, for $f \in K_i[X_{i+1}]$, $f = \Sigma_{j=0}^{N} a_j X_{i+1}^j$, $v_{i+1}(f) = \min\{v_i(a_j) : j = 0, 1, \ldots, N\}$, extending to $K_{i+1} = K_i(X_{i+1})$. Since dim transc$[K_i:K] = i$, no two K_i are K-isomorphic. Note that all valuations have their values in the divisible group generated by Γ.

Hence it follows that Γ is an Archimedean ordered group, so is order-isomorphic to a subgroup of the multiplicative group of the positive reals $\mathbf{R}_+^\times$ (see Theorem 1 in the Appendix), i.e., v has rank 1, or equivalently, v is an absolute value. Now the result follows from Knopfmacher and Sinclair [1].

For other results on this subject, see Fukawa [2], Nagata, Nakayama, and Tuzuku [1], Nakano [1], Neukirch [1], and M. J. Wright [1].

NOTES

We present a kind of valuation, defined by Mathiak [1-3], and some of its fundamental properties. Let K be a division ring and W a linearly ordered set with the least element zero. A function $m: K \to W$ from K onto W is called a *Mathiak valuation* (or an *M-valuation*) if the following conditions are satisfied.

(M_1) $m(x) = 0$ if and only if $x = 0$.
(M_2) $m(x + y) \leq \max(m(x), m(y))$ for every $x, y \in K$.
(M_3) For every $x \in K$ there exists an order-preserving mapping $v(x): W \to W$ such that

$$m(xy) = v(x)(m(y))$$

holds for every $y \in K$.

A subring B of a division ring K is called the *valuation ring* of K if $K = B \cup (B^\times)^{-1}$ and $B \neq K$. It is easy to see that the set $B_m = \{x \in K : m(x) \leq m(1)\}$ is a valuation ring of K. The converse is also true: Every valuation ring of K defines an M-valuation of K (Mathiak [1]). Every valuation ring B of a division ring K defines a topology of K. If B is a valuation ring of K, the sets xB $(x \in K^\times)$ can be taken as a base of the neighborhoods of zero in K. Every topology defined by a valuation ring is a group topology of K^+

(the additive group of K) but not necessarily a ring (a division ring) topology. The following theorem due to Nathiak [2] characterizes ring topologies among the topologies defined by M-valuations.

THEOREM Let B be a valuation ring of a division ring K. Then the following conditions are equivalent.

(i) B defines a proper division ring topology.

(ii) The right multiplication $x \mapsto xa$ is continuous with respect to the topology defined by B.

(iii) For every $a \in K^\times$ the valuation ring $B_a = a^{-1}Ba$ defines the same topology as B.

(iv) For every $c \in K^\times$ there exists $d \in K^\times$ such that $dB \subset Bc$.

As a corollary it follows from the theorem that every valuation ring B of a division ring K which contains an invariant valuation ring B_1 (i.e., $B_1 = a^{-1}B_1a$ holds for every $a \in K^\times$) defines a proper division ring topology (see Schilling [6]; there, invariant valuation rings play an essential role in the study of the classical valuations on division rings). Related results dealing with the classical valuations can be found in Kowalsky and Dürbaum [1].

Assume that m is an M-valuation of a division ring K. Mathiak observed that a topology defined by m is a proper division ring topology if and only if for every $\varepsilon > 0$ there exists $\delta > 0$ such that for every $x \in K^\times$, $m(x) \leq \delta$ implies that $m(x^{-1}) > \varepsilon$. It can be shown (Mathiak [3]) that if a topology defined by an M-valuation is a division ring topology, then this topology is minimal.

For related results on valuations of division rings, see Dürbaum [1,2] and Štipelman [1]. For new results, see Pank and Roquette [1] and Saltman [1]. A systematic exposition of functional analysis over topological division rings can be found in Prolla [1]. Extensions of valuations on division rings are studied in Cohn and Mahdavi-Hezavehi [1]. For generalizations on near-fields, see Wefelscheid [3,4].

5

Topologies of Type V

1. FUNDAMENTAL PROPERTIES

DEFINITION 1 Let K be a field and $\mathcal{T} \in \mathcal{L}_r(K)$. We say that $\mathcal{T}$ is of *type V* (or that $\mathcal{T}$ is *locally retrobounded*) if for every $U \in \mathcal{B}(\mathcal{T})$ the set $(K \setminus U)^{-1}$ is bounded.

Example 1 Let v be a valuation or an absolute value on a field K and $\mathcal{T} = \mathcal{T}_v$. Then $\mathcal{T}$ is of type V.

Indeed, for $U = \{x \in K : v(x) < a\}$, $(K \setminus U)^{-1} = \{y \in K : v(y) \leq a^{-1}\}$ is a bounded subset of $(K,\mathcal{T})$.

Example 2 (see Chapter 1, Example 7) Let field K be linearly ordered by the relation $\leq$ and let $\mathcal{T}$ be the order topology defined by this relation. We have already seen that $\mathcal{T}$ is a field topology. Moreover, the topology $\mathcal{T}$ is of type V: If $U = \{x \in K : |x| < \varepsilon\}$, then $(K \setminus U)^{-1} = \{y \in K : |y| \leq \varepsilon^{-1}\}$; (here $|\cdot|$ denotes the absolute value defined by the order $\leq$).

LEMMA 1 Let K be a field with a proper ring topology $\mathcal{T}$. Then there exists $U \in \mathcal{B}(\mathcal{T})$ satisfying

$$(K \setminus U) \cup (K \setminus U)^{-1} = K^{\times} \tag{1}$$

Proof: Suppose to the contrary that for every $V \in \mathcal{B}(\mathcal{T})$, $(K \setminus V) \cup (K \setminus V)^{-1}$ is a proper subset of $K^{\times}$. Thus for every V there exists $x_V \notin (K \setminus V) \cup (K \setminus V)^{-1}$. Thus $x_V, x_V^{-1} \in V$. Let $U \in \mathcal{B}(\mathcal{T})$ be arbitrary. Take $V \in \mathcal{B}(\mathcal{T})$ satisfying $VV \subset U$ (Chapter 1,

Theorem 1). Then we should have $1 = x_V x_V^{-1} \in VV \subset U$, contradicting the assumption that T is a proper topology.

DEFINITION 2 Let k be a field and $T \in L_r(K)$. A *set* $A \subset K$ is said to be *restricted* if $0 \notin \overline{(A \setminus \{0\})^{-1}}$ (the bar denotes the closure of a set).

LEMMA 2 Let K be a field and $T \in L_r(K)$. Then the following conditions are equivalent.

(i) $A \subset K$ is restricted.

(ii) There exists $U \in \mathcal{B}(T)$ such that $1 \notin AU$.

Proof: (i) ⇒ (ii) Suppose that for every $U \in \mathcal{B}(T)$, $1 \in AU$. It means that there exist $u \in U$ and $a_u \in A$ such that $1 = a_u u$, i.e., $u = a_u^{-1} \in (A \setminus \{0\})^{-1}$. If U runs over $\mathcal{B}(T)$, then the net (a_u^{-1}) tends to zero, i.e., $0 \in \overline{(A \setminus \{0\})^{-1}}$, contradicting (i).

(ii) ⇒ (i) Suppose that A is not restricted: $0 \in \overline{(A \setminus \{0\})^{-1}}$. Thus for every $U \in \mathcal{B}(T)$, $U \cap (A \setminus \{0\})^{-1} \neq \emptyset$. Let $0 \neq x_u \in U \cap (A \setminus \{0\})^{-1}$. Then $x_U = a_U^{-1}$, where $a_U \in A$, i.e., $1 = x_U a_U \in AU$, a contradiction.

LEMMA 3 Let K be a field and $T \in L_r(K)$ a proper topology. The following conditions are equivalent.

(i) T is of type V.

(ii) For every $A \subset K$ bounded away from zero, A^{-1} is bounded.

(iii) For every $A, B \subset K$ bounded away from zero, AB is bounded away from zero.

(iv) For every $U \in \mathcal{B}(T)$ there exists $V \in \mathcal{B}(T)$ such that if $ab \in V$, then $a \in U$ or $b \in U$.

(v) Every restricted subset of K is bounded.

Proof: (i) ⇒ (ii) Let $A \cap V = \emptyset$, where $V \in \mathcal{B}(T)$. Then $A \subset K \setminus V$ and $A^{-1} \subset (K \setminus V)^{-1}$ is bounded as a subset of a bounded set.

(ii) ⇒ (i) Let $U \in \mathcal{B}(T)$ and take $A = K \setminus U$ in (ii).

(ii) ⇒ (iii) Suppose that A and B are bounded away from zero. Then A^{-1} and B^{-1} are bounded, and $(AB)^{-1} = A^{-1}B^{-1}$ is also bounded. We must prove that there exists $V \in \mathcal{B}(T)$ for which $AB \cap V = \emptyset$ holds. Suppose to the contrary that for every $W \in \mathcal{B}(T)$, $W \cap AB \neq \emptyset$. Let

$z_W = a_W \cdot b_W \in W \cap AB$, where $a_W \in A$ and $b_W \in B$. Since $(AB)^{-1}$ is bounded, then for every $U \in \mathcal{B}(T)$ there exists $V \in \mathcal{B}(T)$ such that $(AB)^{-1}V \subset U$. Then

$$(a_V b_V)^{-1} z_V = 1 \in (AB)^{-1} V \subset U$$

a contradiction, since U was arbitrary and T was supposed to be a Hausdorff topology. Thus AB is bounded away from zero.

(iii) ⇒ (iv) Take $U \in \mathcal{B}(T)$ arbitrary and $A = B = K \setminus U$. Then there exists $V \in \mathcal{B}(T)$ satisfying $AB \cap V = \emptyset$. Let $ab \in V$. Then either $a \in U$ or $b \in U$. Indeed, if $a \notin U$ and $b \notin U$, then $a \in A$, $b \in B$, and $ab \in AB$, so $ab \notin V$, contrary to our assumption.

(iv) ⇒ (i) Let $U,V \in \mathcal{B}(T)$. By assumption, there exists $W \in \mathcal{B}(T)$ such that $(K \setminus V \cap U)(K \setminus V \cap U) \subset K \setminus W$. This implies that for $Z = V \cap U$, $W(K \setminus Z)^{-1} \subset Z$. Otherwise, there would be $w \in W$ and $z \in K \setminus Z$ with $wz^{-1} \in K \setminus Z$. Thus $(wz^{-1})z = w \in K \setminus W$, a contradiction. Now $W(K \setminus U)^{-1} \subset W(K \setminus Z)^{-1} \subset Z \subset V$.

(v) ⇒ (ii) Let A be bounded away from zero, i.e., $A \cap V = \emptyset$ for a suitable $V \in \mathcal{B}(T)$. We can assume that $\bar{A} \cap V = \emptyset$, taking, if necessary, a smaller V. Thus $0 \notin \bar{A} = \overline{(A^{-1})^{-1}}$. A^{-1} is bounded, by assumption (v).

(ii) ⇒ (v) Let A be restricted: $0 \notin \overline{(A \setminus \{0\})^{-1}}$. This means that there exists $V \in \mathcal{B}(T)$ such that $V \cap B = \emptyset$, where $B = \overline{(A \setminus \{0\})^{-1}}$. Thus B^{-1} is bounded, by (ii). Since $A \subset B^{-1}$, A is also bounded.

LEMMA 4 Let K be a field and T a type V topology on K. Then T is locally bounded.

Proof: There exists $U \in \mathcal{B}(T)$ such that (1) holds, by Lemma 1. This implies that $U^\times \subset (K \setminus U)^{-1}$ and $U \subset (K \setminus U)^{-1} \cup \{0\}$ is a bounded neighborhood of zero.

THEOREM 1 Let K be a field. Then every topology of type V on K is a field topology.

Proof: Let T be a proper topology of type V on K. To prove Theorem 1, it is sufficient to show that for every $U \in \mathcal{B}(T)$ there exists $V \in \mathcal{B}(T)$ satisfying $(1 + V)^{-1} \subset 1 + U$ (by Theorem 1 of Chapter 1).

First take $V_1 \in \mathcal{B}(T)$ with $-1 \notin V_1 + V_1$. Since $(K \setminus U)^{-1}$ is bounded, there exists $V_2 \in \mathcal{B}(T)$ such that $V_2(K \setminus U)^{-1} \subset V_1$. Let $V \in \mathcal{B}(T)$, $V \subset V_1 \cap V_2$. Then $-1 \notin V + V(K \setminus U)^{-1}$; thus for every $v \in V^{\times}$,

$$-1 \notin v + v(K \setminus U)^{-1} \quad \text{or} \quad 1 + v \in -v(U^{\times})^{-1}$$

Finally, $(1 + v)^{-1}(-v) = (1 + v)^{-1} - 1 \in U$, or $(1 + V)^{-1} \subset 1 + U$.

THEOREM 2 An almost-order R defines the topology of type V on a field K if and only if the following condition is satisfied:

$$\exists\, c \in R^{\times}\ \forall\, x \in K^{\times} \qquad x \in R \text{ or } x^{1} \in c^{-1}R \tag{2}$$

Proof: (⇒) By our assumption $(K \setminus R)^{-1}$ is bounded. Thus there exists $c \in K^{\times}$ with the property that $c(K \setminus R)^{-1} \subset R$. Since $c = ab^{-1}$, where $a, b \in R^{\times}$, $a(K \setminus R)^{-1} \subset bR \subset R$, we can suppose, without restricting generality, that $c \in R^{\times}$. It follows that $(K \setminus R)^{-1} \subset c^{-1}R$. But this last inclusion is a statement equivalent to (2).

(⇐) Conversely, suppose that (2) holds. Then $c(K \setminus R)^{-1} \subset R$, i.e., $(K \setminus R)^{-1}$ is bounded, since R is bounded. We have $\mathcal{B}(T) = \{bR : b \in K^{\times}\}$ (Chapter 1, Theorem 8). Now

$$(K \setminus bR)^{-1} = b^{-1}(K \setminus R)^{-1} \subset (bc)^{-1}R$$

implies that for every $U \in \mathcal{B}(T)$ the set $(K \setminus U)^{-1}$ is bounded.

LEMMA 5 Let (K,T) be a topological field of type V, R an almost-order inducing T, and $N_0 = N_0(T)$ the set of topological nilpotents of (K,T). If $x \in R \setminus N_0$, then the set

$$\{x^{-n} : n \in \mathbb{N}\} \text{ is bounded in } (K,T)$$

Proof: Note that one can find $U \in \mathcal{B}(T)$ such that $\{x^n : n \in \mathbb{N}\} \subset K \setminus U$. Indeed, otherwise for every $u \in R^{\times}$ there exists n_0 such that $x^{n_0} \in uR$. This implies that $x^{n_0+1} = x^{n_0}x \in uRx \subset uR$, since $x \in R^{\times}$. Thus $x^n \in uR$ for every $n \geq n_0$. Since u is arbitrary, x^n tends to zero and $x \in N_0$, a contradiction. Since $\{x^n : n \in \mathbb{N}\} \subset K \setminus U$, $\{x^{-n} : n \in \mathbb{N}\} \subset (K \setminus U)^{-1}$ is bounded.

2. CHARACTERIZATION OF TOPOLOGIES OF TYPE V

THEOREM 3 Let $(K,\mathcal{T})$ be a topological field of type V having a nonzero topological nilpotent. Then $\mathcal{T}$ is induced by a nontrivial absolute value.

Proof: Since $N_0(\mathcal{T}) \neq \{0\}$, $N_0(\mathcal{T})$ is open and $\mathcal{T}$ is a proper field topology (Chapter 3, Lemma 3). Lemma 3 shows that for every $A \subset K$ bounded away from 0, A^{-1} is bounded. Now applying Theorem 14(v) of Chapter 2, the result follows.

THEOREM 4 Let $(K,\mathcal{T})$ be a proper topological field of type V. If $N_0(\mathcal{T}) = \{0\}$, then there is an almost-order S defining $\mathcal{T}$, which is, moreover, a valuation ring in K.

Proof: Let R be an almost-order in K which induces $\mathcal{T}$ and satisfies (2) (Theorem 2). Let

$$\tilde{R} = \{x \in K : xR \subset \{c^{-n} : n = 0, 1, 2, \ldots\}R\} \tag{3}$$

Since $RR \subset R$, $1 \in R$, we find that $R \subset \tilde{R}$ and $\tilde{R} \subset RM$, where $M = \{c^{-n} : n = 0, 1, 2, \ldots\}$. Lemma 5 proves that M is bounded, since $c \in R^{\times}$ and $c \notin N_0$ ($N_0 = \{0\}$, by assumption). This proves that RM, and hence $\tilde{R}$, is bounded. Since $\tilde{R} \in \mathcal{B}(\mathcal{T})$, it induces, by Theorem 8 of Chapter 1, the topology $\mathcal{T}$. We shall prove that $\tilde{R}$ is an almost-order.

Let $x, y \in \tilde{R}$. Then $x, y \in RM$, i.e., $x \in Rc^{-r}$, $y \in Rc^{-s}$ with suitably chosen $r, s \in \mathbb{N}$. It follows that

$$xR \subset Rc^{-r}, \quad yR \subset Rx^{-s}, \quad \text{and} \quad (xy)R \subset Rc^{-(r+s)}$$

Thus $xy \in \tilde{R}$. This proves that $\tilde{R}\tilde{R} \subset \tilde{R}$, i.e., condition ($O_2$) of Definition 9 of Chapter 1 is fulfilled. The remaining conditions of this definition clearly hold.

Now we shall prove that $\tilde{R}$ is a full almost-order in K: $K^{\times} = \tilde{R}^{\times} \cup (\tilde{R}^{\times})^{-1}$, or equivalently, $(K \setminus \tilde{R})^{-1} \subset \tilde{R}$. In fact, $R \subset \tilde{R}$ implies that $K \setminus \tilde{R} \subset K \setminus R$ and $(K \setminus \tilde{R})^{-1} \subset (K \setminus R)^{-1}c^{-1}R \subset R$. Let $z \in \tilde{R}^{\times}$ satisfy $z(\tilde{R} \pm \tilde{R}) \subset \tilde{R}$. Then

$$\tilde{R} \pm \cdots \pm \tilde{R} \text{ } (2^n \text{ times}) \subset z^{-n}\tilde{R} \tag{4}$$

Inclusion (4) shows that if S is the ring generated by R in K, then $S \subset \bigcup_{n=0}^{\infty} z^{-n}\tilde{R}$. Thus $S \subset \tilde{R}[z^{-1}]$ and $S \in \mathcal{B}(\mathcal{T})$. The set $\{z^{-n} : n = 0, 1, 2, \ldots\}$ is bounded, by Lemma 5, since $z \in \tilde{R} \setminus N_0$; but $\tilde{R}$ is bounded and S is bounded as a subset of the product of two bounded sets. Since the ring S contains $\tilde{R}$ and $\tilde{R}$ is a full almost-order, S also is a valuation ring. Thus we have found a valuation ring S in K, which induces topology $\mathcal{T}$.

We are now able to generalize Theorem 9 of Chapter 4.

THEOREM 5 (Fleischer [2]; Kowalsky and Dürbaum [1]) Let $(K,\mathcal{T})$ be a proper topological field. Then the following conditions are equivalent.

(i) $\mathcal{T}$ is the topology of type V.

(ii) There exists either an absolute value or a non-Archimedean valuation v satisfying $\mathcal{T} = \mathcal{T}_v$.

Proof: Since the implication (ii) ⇒ (i) is obvious (see Example 1), it suffices to prove the converse implication. If $N_0(\mathcal{T}) \neq \{0\}$, the result follows from Theorem 3. If $N_0(\mathcal{T}) = \{0\}$, Theorem 4 works: There is an order S in K which is a valuation ring and induces $\mathcal{T}$. We can apply Theorem 9 of Chapter 4 since $\mathcal{T}$ is of type V, i.e., satisfies condition (ii) of this theorem, and $S \in \mathcal{B}(\mathcal{T})$ is a bounded additive subgroup of K^+. Thus $\mathcal{T} = \mathcal{T}_v$, where v is a non-Archimedean valuation on K.

Remark Since S is a valuation ring inducing the topology on K, we can produce a valuation satisfying $\mathcal{T} = \mathcal{T}_v$, in the same manner as at the beginning of Section 3 of Chapter 4.

Now we are going to present another characterization of topologies of type V, following Kiltinen [2].

Let $(K,\mathcal{T})$ be a topological field, $f \in K[X]$, $\deg(f) \geq 2$, and let $A = f(K)$. Put $\tilde{K} = \{F \subset K : |F| < \infty\}$. For $f \in K[X]$ we define f^{-1}: $A \to \tilde{K}$ in the obvious way. In order to talk about continuity for f^{-1}, we will consider a natural uniform structure on $\tilde{K}$. For every $U \in \mathcal{B}(\mathcal{T})$, let

$$\tilde{U} = \{(F_1,F_2) \in \tilde{K}^2 : \forall\, a \in F_i\ \exists\, b \in F_j,\ a - b \in U;\ i \neq j\} \tag{5}$$

Let F be the filter generated by sets (5). It can be checked that $(\tilde{K},F)$ is a uniform structure. The set A will have the relative topology.

LEMMA 6 Let (K,T) be a topological field of type V over which the polynomial $f(X) = X^n - 1$ factors into linear polynomials. Then the nth root function f^{-1} is uniformly continuous.

Proof: Put $m = 1$ if $\mathrm{char}(K) = 0$ and $m = p^k$, where $n = p^k r$, $(p,r) = 1$ if $\mathrm{char}(K) = p$. Then $X^n - 1 = (X^r - 1)^m$, where $n = mr$, and one can see that $X^r - 1$ has no multiple roots. Since $X^n - 1$ factors completely in $K[X]$, there are r distinct nth roots of unity, $e_1 = 1, e_2, \ldots, e_r$. Let $b \in K$. Then $X^n - b^n = (X - b)^m(X - e_2b)^m \cdots (X - e_rb)^m$.

To prove the uniform continuity of f^{-1}, we must show that for every $U \in B(T)$ there exists $V \in B(T)$ such that for every $a,b \in K$, $a^n - b^n \in V$ implies that $(f^{-1}(a^n),f^{-1}(b^n)) \in \tilde{U}$. Let $U \in B(T)$. Since every finite set is bounded, there exists $U_1 \in B(T)$ such that $\{e_i : i = 1, \ldots, r\}U_1 \subset U$. Applying an induction argument to Lemma 3(iv), we assert that there exists $V \in B(T)$ such that for every $x_1, \ldots, x_n \in K$, if $x_1x_2\cdots x_n \in V$, then some x_i lies in U_1, $i = 1, 2, \ldots, n$. Let

$$a^n - b^n = (a - b)^m(a - e_2b)^m\cdots(a - e_rb)^m \in V$$

Since the topology T is of type V, there exists $i \in \{1,\ldots,r\}$ satisfying $a - e_ib \in U_1$.

Suppose that $c \in f^{-1}(a^n)$. Then $c = ae_j$ for suitable j. Also, $e_ibe_j \in f^{-1}(b^n)$ and $c - e_ibe_j = (a - e_ib)e_j \in U_1e_j \subset U$. Similarly, if $c \in f^{-1}(b^n)$, there exists $d \in f^{-1}(a^n)$ such that $c - d \in U$. Thus we have proved that if $a^n - b^n \in V$, then $(f^{-1}(a^n),f^{-1}(b^n)) \in \tilde{U}$.

COROLLARY For every topological field (K,T) of type V, the square root function is uniformly continuous.

Proof: It is sufficient to apply Lemma 6 to $f(X) = X^2 - 1$.

THEOREM 6 (Kiltinen [2, Th. 2]) Let $(K,\mathcal{T})$ be a proper topological field and let $\mathrm{char}(K) \neq 2$. Suppose that $f \in K[X]$ and $\deg(f) = 2$. Then the following conditions are equivalent.

(i) f^{-1} is uniformly continuous.

(ii) The topology $\mathcal{T}$ is of type V.

Proof: Let $f(X) = aX^2 + bX + c \in K[X]$, $a \neq 0$. Note that since $\mathrm{char}(K) \neq 2$, f can be transformed into a polynomial of the form $a(Y^2 + k)$, $k \in K$, using the substitution $Y = X - b/2a$

Moreover, we can omit the factor a, since division is continuous. Thus, without losing generality, let $f(X) = X^2 + k$. Note that for $y = f(a) \in A$, $f^{-1}(y) = \{a,-a\}$. The function f^{-1} is uniformly continuous exactly when

$$\begin{gathered} \forall\ U \in \mathcal{B}(\mathcal{T})\ \exists\ V \in \mathcal{B}(\mathcal{T}) \\ f(a) - f(b) \in V \Rightarrow a - b \in U \text{ or } a + b \in U \end{gathered} \tag{6}$$

(ii) $\Rightarrow$ (i) $f(a) - f(b) = (a - b)(a + b)$. Since $\mathcal{T}$ is of type V, for every $U \in \mathcal{B}(\mathcal{T})$ there exists $V \in \mathcal{B}(\mathcal{T})$ such that $a^2 - b^2 \in V$ implies that $a - b \in U$ or $a + b \in U$, by Lemma 3(iv). Thus (6) is satisfied.

(i) $\Rightarrow$ (ii) Assume that (6) is true. Let $U \in \mathcal{B}(\mathcal{T})$ be arbitrary and let $V \in \mathcal{B}(\mathcal{T})$ satisfy (6). Suppose that $cd \in V$. Since $\mathrm{char}(K) \neq 2$, there are elements $a,b \in K$ such that $a - b = c$, $a + b = d$, namely, $a = (1/2)(d + c)$, $f = (1/2)(d - c)$. But then $f(a) - f(b) = (a + b) \times (a - b) = cd \in V$, so by (6), $c = a - b \in U$ or $d = a + b \in U$.

Remark The assumption $\mathrm{char}(K) \neq 2$ cannot be omitted, since there are nondiscrete topological fields of characteristic 2 which are not type V, but in which the (single-valued) square-root function is uniformly continuous (see Kiltinen [2]).

It is not known under what conditions, if any, the uniform continuity of f^{-1} for f a polynomial of degree 3 or more implies that the topology is of type V.

3. MINIMAL LOCALLY BOUNDED RING TOPOLOGIES ON FIELDS

Let K be a field. Recall that a topology T on K is said to be a *minimal ring (field) topology* if $T \in L_r(K)$ ($L_f(K)$) and the antidiscrete topology $\mathbb{O}_K$ is the only topology of $L_r(K)$ properly weaker than T.

We start this section with the following.

THEOREM 7 Let K be a field and $T \in L_r(K)$ be a proper topology. Then the following conditions are equivalent.

(i) T is a minimal element of $L_r(K)$.

(ii) $T \in L_f(K)$ and is a minimal element of $L_f(K)$.

Proof: (i) ⇒ (ii) Theorem 4 of Chapter 1 implies that there is a topology $T_1 \in L_f(K)$ satisfying $\mathbf{0}_K \neq T_1 \leq T$. Since $T_1 \in L_r(K)$ and T is minimal, $T_1 = T$ and thus $T \in L_f(K)$. The topology T is obviously a minimal element of $L_f(K)$.

(ii) ⇒ (i) Suppose that T is not minimal in $L_r(K)$. Then there is an $S \in L_r(K)$ with $\mathbf{0}_K \neq S < T$. By Theorem 4 of Chapter 1, there is $S_1 \in L_f(K)$, $\mathbf{0}_K \neq S_1 \leq S$, a contradiction.

THEOREM 8 Let K be a field and $T \in L_r(K)$ a proper ring topology. Then the following conditions are pairwise equivalent.

(i) T is a locally bounded topology and is a minimal element in $L_r(K)$.

(ii) T is a locally bounded topology and is a minimal element in $L_f(K)$.

(iii) T is a locally bounded topology and is a minimal element in the subset of $L_f(K)$ consisting of all locally bounded field topologies.

(iv) There exists an absolute value or a non-Archimedean valuation v on K for which $T = T_v$.

(v) T is a topology of type V.

(vi) For every $A \subset K$ bounded away from zero in (K,T), A^{-1} is bounded in (K,T).

(vii) For every $A,B \subset K$ bounded away from zero in (K,T), AB is bounded away from zero in (K,T).

Proof: (i) ⇔ (ii) follows by Theorem 7. (ii) ⇒ (iii) is obvious. (iv) ⇔ (v) is implied by Theorem 5, and (v) ⇔ (vi) ⇔ (vii) by Lemma 3. Moreover, (iv) ⇒ (i) results from Theorem 13 of Chapter 2 and Theorem 2 of Chapter 4.

It remains to prove (iii) ⇒ (iv) (the equivalence (iii) ⇔ (iv) is due to Fleischer [1]). In order to do it, the following lemma is necessary.

LEMMA 7 Let p be a nontrivial norm on a field K and let F be the set of all nontrivial norms q on K satisfying $q \le p$ [i.e., $q(x) \le p(x)$ for every $x \in K$]. Then the partially ordered set $(F,\le)$ contains a minimal element. Moreover, every minimal element q of F is an absolute value: $q(xy) = q(x)q(y)$ holds for all $x,y \in K$.

Proof of Lemma 7: We apply the Kuratowski-Zorn lemma. Let $F_0 \subset F$ be a linearly ordered subset. Then q_0 defined by $q_0(x) = \inf\{q(x) : q \in F_0\}$, $q_0 \in F$. Let q be any minimal element of F. Then for a fixed $a \in K^\times$, q_a defined by the formula

$$q_a(x) = \inf\{q(a^n x)q(a)^{-n} : n = 0, 1, 2, \ldots\} \tag{7}$$

satisfies $q_a \in F$ and, moreover, $q_a \le q$. Since q is minimal, by assumption, we obtain $q_a = q$, i.e., $q(x) = q(a^n x)q(a)^{-n}$ holds for every $n \ge 0$. In particular, for $n = 1$, $q(ax) = q(a)q(x)$ for arbitrary $x \in K$. But $a \in K^\times$ was arbitrarily chosen. Thus $q(ab) = q(a)q(b)$ holds for every $a,b \in K$.

Remark One can also proceed as follows. Since $p_{sp} \le p$, and by Theorem 8 of Chapter 3, $p_{sp} = \sup \varphi_i$, φ_i being absolute values; then F contains an absolute value.

Now we can prove the implication (iii) ⇒ (iv). Suppose that T is a minimal element in the subset of $L_f(K)$ consisting of all locally bounded field topologies on a field K.

Case 1. $N_0(T) \ne \{0\}$, i.e., (K,T) has a nonzero topological nilpotent. Since $T \in L_f(K)$ is locally bounded and $N_0(T) \ne \{0\}$, it is open (Chapter 3, Lemma 3), hence T is induced by a norm, say p (by Theorem 4 of Chapter 3). Lemma 7 implies that there exists a nonempty set A of absolute values w on K, satisfying $w \le p$. It follows that $\mathcal{T}_w \le \mathcal{T}_p = \mathcal{T}$. Since $\mathcal{T} \in \mathcal{L}_f(K)$ is minimal, $\mathcal{T} = \mathcal{T}_w$.

Case 2. $N_0(\mathcal{T}) = \{0\}$, i.e., $(K,\mathcal{T})$ has no nonzero topological nilpotents. Let R be an almost-order defining the topology $\mathcal{T}$. Put $R_a = \{a^n : n = 0, 1, 2, \dots\}$. Suppose first that there exists $c \in K^\times$ such that R_c is not bounded. Let $R^* = \{a \in K : R_a \text{ is bounded}\}$. Let $z \in R^\times$ satisfy $z(R \pm R) \subset R$. Take arbitrary $a,b \in R^*$. Since R_a and R_b are bounded, R_aR_b is also bounded; thus there is $V \in \mathcal{B}(\mathcal{T})$ such that $R_aR_bV \subset R$. Then, for arbitrary $n \in \mathbb{N}$,

$$(ab)^nV = a^nb^nV \subset R_aR_bV \subset R$$

and

$$(z(a + b))^nV \subset z^n(a^n + \cdots + b^n)V \subset z^n(R + \cdots + R)(2^n \text{ times}) \subset R$$

This implies that $ab \in R^*$ and $z(a + b) \in R^*$, that is, $R^*R^* \subset R^*$, $z(R^* \pm R^*) \subset R^*$. Moreover, $R \subset R^*$ and $R^* \ne K$ $(c \notin R^*)$ is an almost-order defining a topology $\mathcal{S} \in \mathcal{L}_r(K)$, $0_K \ne \mathcal{S} \le \mathcal{T}$. Since $\mathcal{T}$ is minimal, R^* is bounded and defines $\mathcal{T}$. Let $Z = \{z \in K : (R^* + R^*)z \subset R^*\}$. There exists no Z-almost-order R_1 $(\ne K)$ in K, containing R^*, since otherwise R_1 would define a proper ring topology on K, weaker than $\mathcal{T}$, which contradicts minimality of $\mathcal{T}$. Thus R^* is a maximal almost-order in K. Applying Dürbaum's theorem (Chapter 2, Theorem 15), it results that R^* is an almost-order of a nontrivial absolute value, contradicting an assumption that $N_0(\mathcal{T}) = \{0\}$.

Now suppose that R_a is bounded for all $a \in K$. Let A be a subring of K generated by R. A is an order defining $\mathcal{T}$, since $\mathcal{T}$ is minimal. Lemma 13 of Chapter 1 proves that A is bounded. Let $S = A[X]$, $X = \{x \in K : x \notin A \text{ and } x^{-1} \notin A\}$. The ring S contains A, is bounded (Chapter 1, Lemma 13), and thus S defines $\mathcal{T}$. It is not difficult to see that S is a valuation ring in K: $K^\times = S^\times \cup (S^\times)^{-1}$. It follows that a half-valuation v defining $\mathcal{T}$ on K is a valuation (see the last lines of the proof of Theorem 9 of Chapter 4).

Contrary to the situation described in Theorem 8, it is not known whether there exist locally unbounded and minimal field or ring topologies on fields. It would be interesting to answer this important question.

4. STRICTLY MINIMAL TOPOLOGICAL FIELDS

All results in this section were obtained by Nachbin [1].

DEFINITION 3 Let (K,T) be a topological field and let S be an arbitrary topology on K. The *topology* S is said to be *admissible* with respect to T if K equipped with S is a topological vector space over (K,T).

This implies that $S \leq T$; indeed, the multiplication $(x,y) \mapsto xy$ is a continuous mapping $(K,T) \times (K,S) \to (K,S)$. In particular, putting $y = 1$ in the last condition, we see that the identity mapping is a continuous mapping $(K,T) \to (K,S)$, that is, $S \leq T$.

DEFINITION 4 A topological field (K,T) is said to be *strictly minimal* if there exists no topology S on K admissible with respect to T such that $S < T$. In other words, the only topology on K admissible with respect to T is T itself.

LEMMA 8 Let (K,T) be a strictly minimal topological field. Then (K,T) is minimal. If (E,T_E) is a topological vector space over (K,T), then every one-dimensional vector subspace of (E,T_E) is isomorphic (as a topological vector space over (K,T)) with (K,T).

Proof: Let S be a field topology on K, $S \leq T$. Then S is admissible with respect to T; but T is strictly minimal. Therefore, $S = T$, that is, T is minimal.

To prove the last statement of Lemma 8, take arbitrary $x_0 \in E$, $x_0 \neq 0$. Then the mapping $a \mapsto ax_0$ of (K,T) onto (Kx_0,T_E) is continuous, by definition of topological vector space. Define a topology S on K in the following way: A subset U of K is S-open $\Leftrightarrow$ there exists an open subset V of Kx_0 satisfying $U = \{a \in K : ax_0 \in V\}$. In this way (K,S) becomes a topological vector space over (K,T), isomorphic with (Kx_0,T_E). Thus $S \leq T$, i.e., S is admissible with respect to T. Since T is strictly minimal, $S = T$, and the mapping $a \mapsto ax_0$ $(a \in K)$ is a homeomorphism of (K,T) and (Kx_0,T_E). It is clear that this map is a linear bijection of the foregoing vector spaces over K.

THEOREM 9 Every topological field of type V is strictly minimal.

Proof: Let $(K,\mathcal{T})$ be a topological field of type V and let $\mathcal{S}$ be a topology on K, admissible with respect to $\mathcal{T}$. In particular, $\mathcal{S} \le \mathcal{T}$. Since scalar multiplication $(x,y) \mapsto xy$ is a continuous function from $(K,\mathcal{T}) \times (K,\mathcal{S})$ into $(K,\mathcal{S})$, then for every $U \in \mathcal{B}(\mathcal{S})$ there exist $V_1 \in \mathcal{B}(\mathcal{T})$ and $V_2 \in \mathcal{B}(\mathcal{S})$ satisfying $V_1V_2 \subset U$. Suppose that $1 \notin U$. Then $1 \notin V_1V_2$. Write $x = y^{-1}$ for every $x \in V_2$, $x \neq 0$. Obviously, $y \notin V_1$ since otherwise $1 = xy$ would lie in V_1V_2, contrary to our assumption. Thus $1 \notin V_1V_2$ implies that $V_2 \setminus \{0\} \subset (K \setminus V_1)^{-1}$. Since $(K,\mathcal{T})$ is of type V, $V_2 = (V_2 \setminus \{0\}) \cup \{0\}$ is bounded in $(K,\mathcal{T})$, as it is a union of two bounded sets. Let $V \in \mathcal{B}(\mathcal{T})$ be arbitrary. Thus there exists $b \in K^\times$ with $bV_2 \subset V$. Note that $bV_2 \in \mathcal{B}(\mathcal{S})$. This shows that $\mathcal{T} \le \mathcal{S}$, that is, $\mathcal{T} = \mathcal{S}$. In this way we have proved that $(K,\mathcal{T})$ is a strictly minimal topological field.

We are going to describe all topological fields $(K,\mathcal{T})$ having the property that every finite-dimensional vector space over K admits only one admissible topology, i.e., the Cartesian product topology. Before doing it, we shall need some lemmas.

LEMMA 9 Let $(K,\mathcal{T}_K)$ be a topological field. Let $(E,\mathcal{T}_E)$ and $(F,\mathcal{T}_F)$ be topological vector spaces over $(K,\mathcal{T}_K)$, $\dim_K F = 1$, and $f\colon E \to F$ be a nonzero linear transformation. Let $(K,\mathcal{T}_K)$ be strictly minimal. Then the following conditions are equivalent.

(i) f is continuous.

(ii) $\operatorname{Ker}(f) = f^{-1}(0)$ is closed.

If $(K,\mathcal{T}_K)$ is not strictly minimal, the result is not true.

Proof: (i) $\Rightarrow$ (ii) is obviously true, without any assumption on $(K,\mathcal{T}_K)$.

(ii) $\Rightarrow$ (i) Since $\dim_K F = 1$ and f is a nonzero transformation, then f is onto. Define the direct image $\mathcal{S}_F$ of the topology $\mathcal{T}_E$ in the following way: $Y \subset F$ is open with respect to $\mathcal{S}_F \Leftrightarrow f^{-1}(Y)$ is open in $(E,\mathcal{T}_E)$. It is easy to see that $(F,\mathcal{S}_F)$ is a topological vector space over $(K,\mathcal{T}_K)$. Lemma 8 implies that $\mathcal{T}_F = \mathcal{S}_F$. It means, however, that f is continuous.

Suppose that (K,T_K) is not strictly minimal and let T be a topology on K admissible with respect to T_K such that $\mathbf{0}_K \neq T < T_K$. Put $E = F = K$ and consider (E,T) as a topological vector space over (K,T_K) and (F,T_K) as the space over (K,T_K). Then under the identity $f\colon E \to F$, $f^{-1}(0) = \{0\}$ is closed in E, but f is not continuous.

LEMMA 10 Let $E = E_1 \oplus E_2$ be a topological vector space over a strictly minimal topological field (K,T_K), where E_1 and E_2 are vector subspaces of E and $\dim_K E_1 = 1$. Let T be an admissible topology on E and $T_i = T|E_i$ $(i = 1, 2)$. Then the following conditions are equivalent.

(i) $T = T_1 \times T_2$ (the product topology).

(ii) E_2 is closed in E.

Proof: (i) ⇒ (ii) since $T = T_1 \times T_2$ implies that E_1 and E_2 are closed in E.

(ii) ⇒ (i) Assume that E_2 is closed in E. Put $T^* = T_1 \times T_2$. Then T^* is an admissible topology on E. For every neighborhood W of zero in (E,T) there exists another one V satisfying $V + V \subset W$. Then $V_i = V \cap E_i$ is a neighborhood of zero in (E_i,T_i) $(i = 1, 2)$. Moreover, $V_1 + V_2 \subset W$. Since $V_1 + V_2$ is a neighborhood of zero in (E,T^*), then $T \leq T^*$. We shall prove the opposite inequality, i.e., $T^* \leq T$. Let $x = x_1 + x_2 \in E_1 \oplus E_2$ and let $p_i\colon E \to E_i$ be the projection of E onto E_i, $p_i(x) = x_i$. Since $E_2 = p_1^{-1}(0)$ is closed in E, Lemma 9 implies that $p_1\colon (E,T) \to (E_1,T_1)$ is continuous. Since $p_2(x) = x - p_1(x)$, then $p_2\colon (E,T) \to (E_2,T_2)$ is also continuous.

Now let U be a neighborhood of zero in (E,T^*). Then there exists a neighborhood U_i of zero in (E_i,T_i) $(i = 1, 2)$ satisfying $U_1 + U_2 \subset U$. Since $p_i\colon E \to E_i$ is continuous, there exists a neighborhood V_i of zero in (E,T) such that $p_i(V_i) \subset U_i$ $(i = 1, 2)$. If $V = V_1 \cap V_2$, then $V \subset U_1 + U_2 \subset U$. Consequently, $T^* \leq T$. This proves that $T = T^*$.

LEMMA 11 Let (K,T_K) be a topological field. If (K,T_K) is strictly minimal, then every n-dimensional vector space over K admits exactly one admissible topology with respect to which every $(n - 1)$-dimensional vector subspace of E is closed.

Proof: Let $\mathcal{T}_E$ be an admissible topology on E. Since every finite-dimensional subspace of E is an intersection of (n - 1)-dimensional vector subspaces, all vector subspaces of E are closed. Let $E = E_1 \oplus \cdots \oplus E_n$, $\dim_K E_i = 1$ for i = 1, 2, ..., n and $\tilde{E}_i = E_i \oplus \cdots \oplus E_n$. Let $\mathcal{T}_i = \mathcal{T}|E_i$ and $\tilde{\mathcal{T}}_i = \mathcal{T}|\tilde{E}_i$. Since $\tilde{E}_2$ is closed in E, $\mathcal{T} = \mathcal{T}_1 \times \tilde{\mathcal{T}}_2$, by Lemma 10. Since $\tilde{E}_3$ is closed in E and $\tilde{E}_3 \subset \tilde{E}_2$, thus $\tilde{E}_3$ is closed in $\tilde{E}_2$, and consequently $\tilde{\mathcal{T}}_2 = \mathcal{T}_2 \times \tilde{\mathcal{T}}_3$, by Lemma 10. By induction we obtain $\mathcal{T} = \mathcal{T}_1 \times \cdots \times \mathcal{T}_n$. Since $(K,\mathcal{T}_K)$ is strictly minimal, every $\mathcal{T}_i$ is the unique admissible topology on E_i, and finally, $\mathcal{T}$ is the unique admissible topology on E.

To finish the proof it suffices to repeat arguments given at the end of the proof of Lemma 9.

THEOREM 10 Let $(K,\mathcal{T})$ be a topological field. Then the following conditions are equivalent.

(i) Every finite-dimensional vector space over K has only one admissible topology (as a topological vector space over $(K,\mathcal{T})$).

(ii) $\mathcal{T}$ is strictly minimal and $(K,\mathcal{T})$ is complete.

Proof: (i) ⇒ (ii) It was shown (Lemma 8) that every one-dimensional vector space over K admits exactly one admissible topology if and only if $(K,\mathcal{T})$ is strictly minimal. Assume that $(K,\mathcal{T})$ is not complete and let $\hat{K}$ be the completion of $(K,\mathcal{T})$. Since, by assumption, $\mathcal{T}$ is a proper field topology, K lies densely in the topological ring $(\hat{K},\hat{\mathcal{T}})$, $\hat{\mathcal{T}}$ being the unique prolongation of $\mathcal{T}$ to $\hat{K}$. Let $z \in \hat{K} \setminus K$ and $E = Kz + K$ be the vector subspace of $\hat{K}$ generated over K by z and 1. Let $\mathcal{T}_1 = \hat{\mathcal{T}}|E$. Since K lies densely in E, the closure of K in E, $\bar{K} = E$. Since E and K^2 are isomorphic vector spaces and K^2 is a topological vector space over $(K,\mathcal{T})$, we may transfer its topology to an admissible topology $\mathcal{T}_2$ of E. It is clear that the closure of K with respect to $\mathcal{T}_2$ equals K since $(K,\mathcal{T})$ is strictly minimal and K is a one-dimensional subspace of E, consequently closed (Lemma 8). Thus we have obtained two topologies $\mathcal{T}_1$ and $\mathcal{T}_2$ on E, admissible with respect to $(K,\mathcal{T})$, $\mathcal{T}_1 \neq \mathcal{T}_2$, a contradiction! Thus it follows that $(K,\mathcal{T})$ is complete.

(ii) ⇒ (i) By assumption, every one-dimensional vector space over K possesses only one admissible topology T and (K,T) is complete. We proceed by induction on n. Assume that every (n - 1)-dimensional vector space over K has exactly one admissible topology with respect to (K,T), (K,T) being complete. Let E be a vector space over K, $\dim_K E = n$, and T_E be an admissible topology with respect to (K,T) on E. If $F \subset E$ is an $(n - 1)$-dimensional subspace of E, then $T_E|F$ is the product of $n - 1$ copies of T. Thus F is complete, and consequently closed in E. Lemma 11 implies that T_E is unique, being the product of n copies of T.

THEOREM 11 Let (K,T) be a proper topological field. Then the following conditions are equivalent.

(i) Every automorphism of every finite-dimensional topological vector space over (K,T) is continuous (in fact, is a topological isomorphism).

(ii) (K,T) is strictly minimal and complete.

Proof: It suffices to show, in view of Theorem 10, that every finite-dimensional vector space over K has only one admissible topology with respect to (K,T). Let $E = K^n$ and let T_1 and T_2 be two admissible topologies on E. Put $F = E^2$. Then the topology $T_1 \times T_2$ on F is admissible. Since $(x,y) \overset{f}{\longmapsto} (y,x)$ is an automorphism of F, then f is continuous, and consequently, $T_1 = T_2$.

(ii) ⇒ (i) In view of Theorem 10 and by the assumption, the Cartesian product topology $T \times \cdots \times T$ (n times) is the only vector space topology on every n-dimensional vector space E over (K,T). It is obvious that every automorphism of $E^n \simeq E$ is continuous. Indeed, if f is an automorphism of K^n, then $f(x) = (y_1,\ldots,y_n) = Ax^T$, $x = (x_1,\ldots,x_n)$, and $y_i = a_{i1}x_1 + \cdots + a_{in}x_n$, $\det(A) \neq 0$. Since field operations are continuous, y_i is continuous as a function of x, and consequently, f is continuous. The proof is finished.

NOTES

Fraedrich [1, Satz 1, p. 83] has proved that a proper locally bounded field topology T of $\mathbb{Q}$ is the supremum of V-topologies if and only if the set $N_0(T)$ is nonzero and bounded. Thus every norm of $\mathbb{Q}$ is power-multiplicative (see Chapter 3, Theorem 8).

A notion of V-topology is defined for division rings in the same way as for fields. The topologies defined either by valuations or absolute values are easily seen to be of type V. A complete description of topologics of type V on division rings is still unknown.

Recall that a proper subset A of a division ring K is called an almost-order if the following conditions are satisfied: $0, \pm 1 \in A$; $A \subset A$; there exist nonzero elements $z, z' \in A$ such that $z(A \pm A) \subset A$ and $(A \pm A)z' \subset A$; $K = A(A^{\times})^{-1} \cup (A^{\times})^{-1}A$, where $A^{\times}$ denotes nonzero elements of A. An almost-order A is invariant if $d^{-1}Ad = A$ holds for every $d \in K^{\times}$. An almost-order A is called full if it satisfies $K^{\times} = A^{\times} \cup (A^{\times})^{-1}$. Two almost-orders A_1 and A_2 are left (resp. right) equivalent if there exist $c, d \in K^{\times}$ such that $cA_1 \subset A_2$ and $dA_2 \subset A_1$ (resp. $A_1c \subset A_2$ and $A_2d \subset A_1$).

THEOREM (Kowalsky and Dürbaum [1]) Every valuation v on a division ring defines an invariant full almost-order A_v by putting

$$A_v = \{a \in K : v(a) \leq 1\} \tag{8}$$

Moreover, there is one-to-one correspondence between the classes of equivalent invariant full almost-orders and the classes of equivalent valuations of K.

Maximality of an almost-order in a division ring is defined in the same manner as for fields.

THEOREM (Dürbaum [1]) An almost-order A of an absolute value v of a division ring K [i.e., of the form (8)] is an invariant and maximal almost-order of K.

Moreover, there exist maximal and invariant almost-orders in division rings which cannot be presented in the form (8).

Now we shall give a partial description of topologies of type V on division rings.

THEOREM (Kowalsky and Dürbaum [1]) (i) Let T be a V-topology on a division ring K such that $N_0(T) = \{0\}$. Then T is induced by a Krull-

Schilling valuation if and only if there exists a bounded and invariant neighborhood of zero in $(K,\mathcal{T})$.

(ii) Let $\mathcal{T}$ be a V-topology on a division ring K such that $N_0(\mathcal{T}) \neq \{0\}$. Then $\mathcal{T}$ is induced by an absolute value if and only if $N_0(N \cup N_0) \subset N_0$, where $N_0 = N_0(\mathcal{T})$ and $N = N(\mathcal{T})$.

General results on ordered division rings can be found in Neumann [1]. Many new results on ordered fields are contained in Priess-Crampe [1] and Pestov [1]. Nachbin's results presented in Section 4 are true for topological division rings.

Let $(K,\mathcal{T})$ be a topological field. We say that the zeros of polynomials over K are continuous functions of their coefficients if for every $f(X) = a_nX^n + \cdots + a_0 \in K[X]$ $(a_n \neq 0,\ n \geq 1)$ completely reducible over K for every zero a of f, and for every $U \in \mathcal{B}(\mathcal{T})$ there exists $V \in \mathcal{B}(\mathcal{T})$ such that for every $b_i \in a_i + V$ $(i = 0, 1, \ldots, n;\ b_n \neq 0)$, if $g(X) = b_nX^n + \cdots + b_0$ is completely reducible over K, then there exists $b \in a + U$ satisfying $g(b) = 0$. It can be shown (Marchisa [1]) that in every V-topological field the zeros of polynomials are continuous functions of the coefficients. Moreover, every topological field $(K,\mathcal{T})$, $\mathcal{T}$ being a supremum of V-topologies, has this property. Related results can be found in Endler [3] and in Schiffels [1].

6
Locally Compact Topological Fields

1. EXAMPLES OF LOCALLY COMPACT FIELDS

Let $\mathbb{F}_q$ be a Galois field with $q = p^n$ elements and $K = \mathbb{F}_q(X)$ the rational function field in the variable X over $\mathbb{F}_q$. It is easy to see (Chapter 2, Theorem 10) that every nontrivial absolute value w on K has either the form $w(a) = |a|_\infty$, where for $0 \neq a = f(X)/g(X)$ $(f,g \in \mathbb{F}[X])$ and $N = \deg(g) - \deg(f)$, $|a|_\infty = e^{-N}$, $|0|_\infty = 0$, or the form $w(a) = |a|_P$, where $|\cdot|_P$ is a P-adic absolute value on K induced by a prime element $P \in \mathbb{F}_q[X]$. By completion of K at w we obtain the field K_w, the formal Laurent series in X^{-1} over $\mathbb{F}_q$ in the case $w = |\cdot|_\infty$ and the so-called "P-adic fields of prime characteristic" in the second case (compare Example 3 of Chapter 1).

Now let K be a finite algebraic extension of $\mathbb{Q}$. Since K is the quotient field of the ring $\mathbb{Z}_K$ of elements of K integral over $\mathbb{Z}$, and $\mathbb{Z}_K$ is a Dedekind domain, for a fixed prime ideal p of $\mathbb{Z}_K$ we have

$$(a) = aR_K = p^{v_p(a)}\,\mathfrak{a} \quad \text{with} \quad p \nmid \mathfrak{a} \qquad a \in R_K$$

This gives the possibility of topologizing the field K (see Chapter 1, Example 9): For $d \in K^\times$, $d = ab^{-1}$, where $a,b \in \mathbb{Z}_K$, put

$$|d|_p = \frac{|a|_p}{|b|_p} \qquad |a|_p = e^{-v_p(a)} \qquad |0|_p = 0$$

(We can take any constant c, $0 < c < 1$, instead of e^{-1}; often one takes $c + N(p)^{-1}$, $N(p) = |R_K/p|$, to preserve the product formula for absolute values.) By completion of K with respect to the p-adic norm $|\cdot|_p$, we obtain the so-called *p-adic fields*. The first easy theorem states:

THEOREM 1 Let K be a finite field extension of $\mathbb{Q}$ or $\mathbf{F}_q(X)$. All fields obtained from K by the completion with respect to nontrivial absolute values of K are locally compact.

For $\mathbb{R}$ and $\mathbb{C}$ this is obvious and for disconnected fields (arising from K by completion with respect to nontrivial non-Archimedean absolute values) it follows from the fact that the compact sets aV_w, $a \in K_w^\times$, $V_w = \{b \in K_w : w(b) \leq 1\}$ form a base of zero neighborhoods for (K_w, T_w). For detailed proofs, see Borewicz and Šafarevič [1], Narkiewicz [1], and Weil [1].

Van Dantzig [1] was first to characterize the locally compact topological fields. Assuming the second axiom of countability, he proved that every locally compact topological field can be obtained as described in Theorem 1. Pontrjagin [1] showed that the only locally compact and connected topological fields are $\mathbb{R}$ and $\mathbb{C}$ in their usual topologies. Several results in this direction in the disconnected case were proved by Jacobson [1], and Hasse and Schmidt [1]. A key to the classification theorem lies in the following:

THEOREM 2 (Loonstra [3]) Let w be a nontrivial and non-Archimedean absolute value on a field K. Then (K, T_w) is locally compact if and only if the following conditions hold.

(i) (K, T_w) is a complete field.

(ii) (K, T_w) is a discretely valued field [i.e., the value group $w(K^\times)$ of w is a discrete subgroup of $\mathbf{R}_+^\times$].

(iii) (K, T_w) has a finite residue class field: $\bar{k}_w = V_w/P_w = \{x \in K : w(x) \leq 1\}/\{x \in K : w(x) < 1\}$ is finite.

For a detailed proof, see Narici, Beckenstein, and Bachman [1].

In a locally compact (and separable) totally disconnected field, the topology is induced by a non-Archimedean absolute value (Jacobson [1]). The theorem of Šafarevič-Kaplansky (Chapter 2, Theorem 14) implies that a topology on a locally compact field is induced by a nontrivial absolute value, and this implies [applying results of

Ostrowski (Chapter 2, Theorems 10 and 11) and Loonstra (Theorem 2)] that every locally compact field is a finite extension of either $\mathbb{R}$, $\mathbb{Q}_p$, of $\mathbb{F}_q((X))$.

A first complete proof of the classification theorem for locally compact fields was given by Kowalsky [1]. His proof is based on the properties of topologies of type V. Other proofs of the van Dantzig-Pontrjagin-Jacobson-Kowalsky theorem were given by several authors (see Endo [3]; Fraedrich [1]; Jacobson [1]; Moriya [6]; Otobe [1]; Volk [1]; Weiss and Zierler [1]; Warner [2]; and Więsław [12]).

Braconnier [1] defined the module of an automorphism of a locally compact group (for a full exposition, see Weil [1]). Let G be a locally compact abelian group, μ its Haar measure, and $f \in \mathrm{Aut}(G)$. Then for every measurable set $A \subset G$, $\mu(A)$ and $\mu(f(A))$ differ by a multiplicative constant independent of A; hence $\mu(f(A)) = v(f)\mu(A)$, $v(f)$ a constant. In particular, taking for G the additive group K^+ of the locally compact topological field K and $f(x) = ax$, $a \in K^\times$, and putting $v(0) = 0$, we obtain a function $v\colon K \to \mathbb{R}_+$, satisfying:

(1°) $v(x) > 0$ for $x \in K^\times$, $v(0) = 0$.

(2°) $v(xy) = v(x)v(y)$.

(3°) There exists a positive number M such that for every $x \in K$, $v(x) \leq 1$ implies that $v(1 + x) \leq M$.

In order to have the triangle inequality $v(x + y) \leq v(x) + v(y)$, instead of ($3^o$), one should eventually take v^s instead of v, with s satisfying $M^s \leq 2$ (compare Lemma 1 of Chapter 2).

The method of description of the locally compact fields by exploring the Haar measure was sketched in Weiss and Zierler [1] and details were presented in Endo [3]. Moriya [6] proved directly that the topology on a locally compact field is induced by an absolute value. Namely, he proved that conditions of Theorem 14 of Chapter 2 are satisfied.

In the next section we present some characterizations of locally compact fields, based on Więsław [1,3,4,12].

2. CHARACTERIZATIONS OF LOCALLY COMPACT FIELDS

We shall start with the following

LEMMA 1 Every locally compact topological vector space over a non-trivial absolute-valued complete field is finite-dimensional.

For an elementary proof in the case of $K = \mathbb{R}$, see Schaefer [1, Th. 3.6, p. 23]. We reproduce his proof here with some modifications.

Proof: Let $(E,\mathcal{S})$ be a locally compact topological vector space over a complete topological field $(K,\mathcal{T}_v)$, where v is an absolute value on K. Then every one-dimensional vector subspace of $(E,\mathcal{S})$ is isomorphic with $(K,\mathcal{T}_v)$, as a topological vector space over $(K,\mathcal{T}_v)$, by Lemma 8 of Chapter 5, since the field topology $\mathcal{T}_v$ is strictly minimal (Chapter 5, Theorem 9). This implies that every one-dimensional vector subspace of $(E,\mathcal{S})$ is closed, hence locally compact. In particular, $(K,\mathcal{T}_v)$ is a locally compact field.

Let V be any compact and balanced (i.e., $a \in K$, $v(a) \le 1$, $x \in V$ imply $ax \in V$) neighborhood of zero in $(E,\mathcal{T})$. Let $t \in K^\times$ be a topological nilpotent. We will prove that $\{t^nV : n \in \mathbb{N}\}$ is a base of the neighborhoods of zero in $(E,\mathcal{S})$. Indeed, let $\mathcal{B}$ be a base of the neighborhoods of zero for $(E,\mathcal{S})$. Take $U \in \mathcal{B}$ arbitrarily. Then we can find balanced $W \in \mathcal{B}$ with $W + W \subset U$. Since V is compact, there exist $x_1, \ldots, x_k \in V$ satisfying

$$V \subset (x_1 + W) \cup \cdots \cup (x_k + W)$$

Moreover, there exists $b \in K^\times$ with $bx_i \in W$ for every $i = 1, 2, \ldots, k$ and $v(b) \le 1$, since scalar multiplication is continuous. Finally, if $v(t^n) < v(b)$, then

$$t^nV \subset bV \subset (bx_1 + bW) \cup \cdots \cup (bx_k + bW) \subset W + W \subset U$$

This proves that $\{t^nV : n \in \mathbb{N}\}$ is a base of the neighborhoods of zero for $(E,\mathcal{S})$.

Let $\varepsilon \in \mathbb{R}_+$ satisfy $\varepsilon < 1/2$ and let $\alpha \in K^\times$ be such that $v(\alpha) \le \varepsilon$. By assumption, there exist $y_i \in V$ $(i = 1, 2, \ldots, n)$ such that

$$V \subset (y_1 + \alpha V) \cup \cdots \cup (y_n + \alpha V)$$

Let $M = Ky_1 + \cdots + Ky_n$. Thus $V \subset M + \alpha V$. We will prove that $M = E$. Suppose to the contrary that $M \neq E$. Let $c \in E \setminus M$. We may assume that $y_1, \ldots, y_n$ form a basis of M over K. Note that M is closed since every Ky_i is closed. Indeed, let $x \in \bar{M}$. Then there exists a sequence $x_m \in M$, satisfying $x = \lim x_m$. Write $x_m = \Sigma_{s=1}^n \alpha_{sm} y_s$, $\alpha_{sm} \in K$. Then, by assumption on x_m, (x_m) is a Cauchy sequence in M. Since (K, T_v) is complete, the unique vector space topology on M over (K, T_v) is the Cartesian product topology (Chapter 5, Theorem 10); one can also consult Narici, Beckenstein, and Bachman [1, Chap. 3], and Bachman [1]. Consequently, $(\alpha_{sm})_{m \in \mathbb{N}}$ is a Cauchy sequence for $s = 1, 2, \ldots, n$. Since (K, T_v) is complete, $\alpha_s = \lim_m \alpha_{sm} \in K$. Finally, $x = \lim_m x_m = \alpha_1 y_1 + \cdots + \alpha_n y_n \in M$, i.e., $\bar{M} = M$. Set

$$L = \{\lambda \in K^\times : (c + \lambda V) \cap M \neq \emptyset\}$$

Then $d = \inf\{v(\lambda) : \lambda \in L\} > 0$, since $\{c + \lambda V : \lambda \in K^\times\}$ is a base of the neighborhoods of c. This implies that there exist $y \in M$ and $\beta \in K^\times$ such that $c - y \in \beta V$ and $d \leq v(\beta) \leq d(1 + \varepsilon)$; put $x_0 = \beta^{-1}(c - y)$. Since $x_0 \in V$ and $V \subset M + \alpha V$, it implies that there are $z \in M$ and $u \in V$ with $x_0 = z + \alpha u$, or $\beta^{-1}(c - y) = z + \alpha u$. Hence $c = y + \beta z + \alpha\beta u$, and $c - \alpha\beta u = y + \beta z \in M$. We have $v(\alpha\beta) = v(\alpha)v(\beta) \leq \varepsilon(1 + \varepsilon)d < (1/2)(1 + 1/2)d = (3/4)d < d$, contradicting the definition of d. Finally, $E = M$, and $\dim_K E < \infty$.

We shall also need the following:

LEMMA 2 Let G be a locally compact Hausdorff space with a group structure. If the group operation $(x,y) \mapsto xy$ is continuous, then G is the topological group; i.e., the inverse operation $x \mapsto x^{-1}$ is also continuous.

Proof: The case of compact spaces is straightforward. Recall a standard fact that a function $f: X \to Y$ from a topological space X into a compact space Y is continuous if and only if the graph $\{(x, f(x)) \in X \times Y : x \in X\}$ is closed in $X \times Y$. Now let G be a group, that is a compact space with continuous multiplication $g(x,y) = xy$.

Then, by a previous remark, the inversion $x \mapsto x^{-1}$ is continuous, since its graph $\{(x,x^{-1}) : x \in G\} = g^{-1}(e)$ is closed. For a general case, see Ellis [1].

THEOREM 3 Let K be a field and $(K,\mathcal{T})$ be a nondiscrete topological ring. Then the following conditions are equivalent.

(i) $(K,\mathcal{T})$ is a locally compact topological ring.

(ii) $(K,\mathcal{T})$ is a locally compact topological field.

(iii) K is a finite extension either of the reals, or of the p-adic numbers, or else of a formal Laurent series field in one variable over a finite field $\mathbb{F}_q$.

LEMMA 3 Let $(K,\mathcal{T})$ be a nondiscrete locally compact topological field. Then:

(i) $N_0(\mathcal{T}) \neq \{0\}$.

(ii) $\mathcal{T}$ is a topology of type V.

(iii) If $\mathrm{char}(K) = 0$, then $\mathcal{T}|\mathbb{Q}$ is nondiscrete.

Proof: Let $\mathcal{B}$ be a base of the neighborhoods of zero in $(K,\mathcal{T})$.

(i) Let us take relatively compact $U \in \mathcal{B}$ and let $V \in \mathcal{B}$ be a proper subset of U. Since $\bar{U}$ is compact, $\bar{U}$ is bounded (Chapter 1, Lemma 10), so there exists $W \in \mathcal{B}$ satisfying $UW \subset \bar{U}W \subset V$. Because $(K,\mathcal{T})$ is nondiscrete, we can find a nonzero $x \in W$ with

$$\bar{U} \supset U \supsetneq \bar{U}x \supsetneq \cdots \supsetneq \bar{U}x^n \supset Ux^n \supsetneq \bar{U}x^{n+1} \supset \cdots \tag{1}$$

Let y be a cluster point of the set $\{x^n : n \in \mathbb{N}\}$. Then $\bigcap_{n\in\mathbb{N}} \bar{U}x^n = \bar{U}y$. We are going to show that $y = 0$. Otherwise, since $\bar{U}y$ is compact, we should have

$$\bar{U}y = \bigcap_{n=1}^{\infty} \bar{U}x^n = \bigcap_{n=0}^{\infty} (\bar{U}x^n)x = (\bar{U}y)x \tag{2}$$

Thus $\bar{U}xy = \bar{U}y$ and $y \neq 0$ would imply $\bar{U}x = \bar{U}$, which contradicts (1). Thus zero is the unique cluster point of the set $\{x^n : n \in \mathbb{N}\}$, that is, $x \in N_0(\mathcal{T})$, $x \neq 0$.

(ii) Since $(K,\mathcal{T})$ is a locally bounded topological field, then for every $U \in \mathcal{B}$, $\{aU : a \in K^{\times}\}$ is a base of zero neighborhoods. We can suppose that U is an almost-order, by Theorem 8 of Chapter 1. Hence it is sufficient to show that $(K \setminus U)^{-1}$ is bounded. Moreover, $\{x^nU : n \in \mathbb{N}\}$ is a base of neighborhoods of zero in (K,T). Let $x \in N_0(T)$, $x \neq 0$. To prove the boundedness of $(K \setminus U)^{-1}$, it is sufficient to show that there exists an $n \in \mathbb{N}$ with $(K \setminus U)^{-1}x^n \subset U$. Suppose the contrary. Then there exists a sequence (a_n) such that both $a_n^{-1}x^n$ and a_n lie in $K \setminus U$. Putting $b_n = a_n^{-1}x^n$, we get $a_nb_n = x^n$. Let us note that for every $a \in K^\times$ there exists the smallest number $t(a)$ with $ax^{t(a)} \in U$. Then $ax^{t(a)-1} \in K \setminus U$ and $ax^{t(a)} \in K \setminus xU$. Finally, $ax^{t(a)} \in U \cap (K \setminus xU)$. Now let us take a_n (resp. b_n) for a and put

$$c_n = a_nx^{t(a_n)} \qquad d_n = b_nx^{t(b_n)} \qquad A = U \cap (K \setminus xU) \tag{3}$$

Since U is bounded, $\bar{A}$ is compact and the sequences (c_n) and (d_n) have the cluster points in $\bar{A}$, say c and d. Because $c_n, d_n \in K \setminus xU$, the cluster points c and d are both nonzero. But the sequence

$$c_nd_n = (a_nx^{t(a_n)})(b_nx^{t(b_n)}) = x^{n+t(a_n)+t(b_n)}$$

tends to zero, which gives a contradiction with the continuity of the multiplication in (K,T) (see Kowalsky [1, Satz 8, p. 265]).

(iii) Suppose that $T|\mathbb{Q}$ is discrete. Let $x \in N_0(T)$, $x \neq 0$, U a relatively compact (hence bounded) neighborhood of zero in (K,T). We can suppose that U is an almost-order defining the topology T in K. Then the sets x^nU, $n \in \mathbb{N}$, form a base of the neighborhoods of zero in (K,T). If $T|\mathbb{Q}$ were discrete, then $\mathbb{Q}^\times$ would be disjoint from some x^nU, i.e., $\mathbb{Q}^\times \subset K \setminus x^nU$. It means that $\mathbb{Q}^\times = (\mathbb{Q}^\times)^{-1} \subset (K \setminus x^nU)^{-1}$. Since the topology T is of type V, hence $(K \setminus x^nU)^{-1}$ is bounded, so $\mathbb{Q}^\times$ is bounded. Thus there exists $m \in \mathbb{N}$ with $\mathbb{Q}^\times x^m \subset Ux$, $\mathbb{Q}^\times \subset Ux^{1-m}$, or $\mathbb{Q}^\times \subset \bar{U}x^{1-m}$. Since $\bar{U}$ is compact, the inclusion $\mathbb{Q}^\times \subset \bar{U}x^{1-m}$ would imply that the set $\mathbb{N} \subset \mathbb{Q}^\times$ has a convergent subsequence. This contradicts the assumption that $T|\mathbb{Q}$ is discrete (see Fraedrich [1, Lemma 5, p. 66]).

As a corollary we obtain

LEMMA 4 Let (K,T) be a nondiscrete locally compact topological field. Then there exists a nontrivial absolute value w on K, inducing T, that is, $T = T_w$.

Proof: Since there is a nonzero topological nilpotent in (K,T) and T is of type V, by Lemma 3, then Theorem 3 of Chapter 5 implies the result.

Proof of Theorem 3: (i) $\Leftrightarrow$ (ii) Indeed, the implication (ii) $\Rightarrow$ (i) is trivial and (i) $\Rightarrow$ (ii) follows from Lemma 2, since the multiplicative group $K^\times$ of K is a locally compact space.

(iii) $\Rightarrow$ (ii) Since $\mathbb{R}$, $\mathbb{Q}_p$, and $\mathbb{F}_q((t))$ (the Laurent series field over a finite field $\mathbb{F}_q$) are all locally compact in the topologies induced by suitable absolute values, and they are complete, the only topologies on their finite extensions, induced by the absolute values extending the absolute values on the ground fields, are the Cartesian product topologies (compare Theorem 10 of Chapter 5). This implies that the finite extensions of the above-mentioned fields are locally compact fields.

(ii) $\Rightarrow$ (iii) Lemma 4 implies that the topology T is induced on K by a nontrivial absolute value. This absolute value is nontrivial on $\mathbb{Q}$ in the case of zero characteristic by Lemma 3, case (iii). Since (K,T) is locally compact, it must be complete. Applying Theorem 14 of Chapter 3 we obtain that K contains $\mathbb{R}$, $\mathbb{Q}_p$, or $\mathbb{F}_q((t))$ as a closed topological subfield. The field K, being a topological vector space over this closed subfield, is finite dimensional, since it is locally compact, as follows from Lemma 1.

The proof is finished.

We formulate below another characterization of locally compact fields of zero characteristic.

DEFINITION 1 Let (K,T) be a topological field. We write G(K) for the *group of* all its *topological automorphisms*. Thus $g \in G(K)$ exactly when $g \in \mathrm{Aut}(K)$ and g and g^{-1} are continuous functions from (K,T) into itself.

THEOREM 4 (Więsław [4]) Let $(K,\mathcal{T})$ be a proper topological field. Then the following conditions are equivalent.

(i) $(K,\mathcal{T})$ is a locally bounded, complete field and for every closed subfield F of $(K,\mathcal{T})$, $G(F)$ is finite.

(ii) $(K,\mathcal{T})$ is a locally compact field of characteristic zero.

(iii) K is a finite extension of the reals $\mathbb{R}$ or of some p-adic number field $\mathbb{Q}_p$ with the usual locally compact topologies.

We finish this chapter with some remarks on the Cauchy functional equation

$$f(x + y) = f(x) + f(y) \tag{4}$$

in a locally compact field (see Figiel [1]).

Let μ be the Haar measure of the additive group K^+ of a locally compact topological field K.

THEOREM 5 Let K be a locally compact field and $f\colon K \to K$ a function measurable with respect to the Haar measure μ of K. If f satisfies (4), then f is continuous.

Proof: It is easy to see that if a field K has the property formulated in Theorem 5, then every finite extension of K also has this property. Indeed, if L is of degree n over K, $w_1, \ldots, w_n$ is a basis of L over K, and $f\colon L \to L$ is a measurable function satisfying (4), then for every $x_i \in K$:

$$\begin{aligned} f(x_1w_1 + \cdots + x_nw_n) &= f(x_1w_1) + \cdots + f(x_nw_n) \\ &= x_1c_1 + \cdots + x_nc_n \end{aligned}$$

where $c_i \in L$ is a constant depending only on f and w_i. In view of Theorem 3 it is sufficient to consider only the fields $\mathbb{R}$, $\mathbb{Q}_p$, and $\mathbb{F}_q((X))$. Since the case of the reals is well known, we consider only the case $K = \mathbb{Q}_p$. A proof in the case $K = \mathbb{F}_q((X))$ is similar to that in the case $K = \mathbb{Q}_p$.

Let $K = \mathbb{Q}_p$. Denote by $S(a,\varepsilon)$ the ball in $\mathbb{Q}_p$ with center a and radius ε. It is easy to see that

$$S(0,p|x|_p) = \bigcup_{n=0}^{p-1} S\left(\frac{n}{p}\,x, |x|_p\right) \tag{5}$$

where the sets on the right side of (5) are pairwise disjoint. Since μ is regular, (5) implies that for every μ-measurable set A in $\mathbb{Q}_p$ the following equality holds:

$$\mu(pA) = \frac{1}{p}\,\mu(A) \tag{6}$$

Now let f: $\mathbb{Q}_p \to \mathbb{Q}_p$ be a μ-measurable solution of (4). Put $g_t(x) = tf(x) - xf(t)$; $x,t \in \mathbb{Q}_p$. It is immediate that

$$g_t(rx) = rg_t(x) \tag{7}$$

and

$$g_t(x + rt) = g_t(x) \qquad \text{for every } r \in \mathbb{Q} \text{ and } t,x \in \mathbb{Q}_p \tag{8}$$

Let $A_t(u) = \{x \in \mathbb{Q}_p : |g_t(x)|_p < u\}$. Then (8) implies that

$$\mu(A_t(pu) \cap S(0,|t|_p) = \mu\left(A_t(pu) \cap S\left(\frac{n}{p}\,t,|t|_p\right)\right) \qquad n \in \mathbb{N} \cup \{0\}$$

Applying (5), we obtain

$$\mu(A_t(pu) \cap S(0,|t|_p) = \frac{1}{p}\,\mu(A_t(pu) \cap S(0,p|t|_p))$$

Consequently, conditions (6) and (7) and the definition of the p-adic absolute value imply that

$$\begin{aligned}\mu(A_t(u) \cap S(0,|t|_p) &= \frac{1}{p}\,\mu\left(\frac{1}{p}\,A_t(u) \cap \frac{1}{p}\,S(0,|t|_p)\right)\\ &= \frac{1}{p}\,\mu(A_t(pu) \cap S(0,p|t|_p))\end{aligned}$$

and finally,

$$\mu(A_t(pu) \cap S(0,|t|_p)) = \mu(A_t(u) \cap S(0,|t|_p)) \tag{9}$$

Thus

$$\begin{aligned}&\mu(\{x \in S(0,|t|_p) : g_t(x) \neq 0\})\\ &= \mu\left(\bigcup_{n=-\infty}^{\infty} (A_t(p^n) \setminus A_t(p^{n-1})) \cap S(0,|t|_p)\right)\\ &= \sum_{n=-\infty}^{\infty} \mu((A_t(p^n) \setminus A_t(p^{n-1})) \cap S(0,|t|_p))\end{aligned}$$

$$= \sum_{n=-\infty}^{\infty} (\mu(A_t(p^n) \cap S(0,|t|_p)) - \mu(A_t(p^{n-1}) \cap S(0,|t|_p))) = 0$$

by (9) and by inductive arguments. This implies that $g_t = 0$ almost everywhere on $S(0,|t|_p)$. Thus $g_t = 0$ almost everywhere, by (8). Since $g_0 = 0$, then for every $t \in \mathbb{Q}_p$, the equality

$$0 = tg_1(x) - g_t(x) = tf(x) - txf(1) - tf(x) + xf(t)$$
$$= (f(t) - tf(1))x$$

holds for almost all $x \in \mathbb{Q}_p$. Thus $f(t) = tf(1)$, i.e., f is continuous.

NOTES

Generally, a study of alternative division rings is reduced to a study of associative division rings. Indeed, it is well known that every alternative division ring of characteristic $\neq 2$ is either an associative division ring or is isomorphic to an eight-dimensional Cayley-Dickson division algebra over its center. For this reason we are interested mainly in associative locally compact division rings. The results on nonassociative locally compact division rings can be found in Buchanan and Hahl [1], Peterson [1-4], and Bourne [2].

Theorem 3 remains true if we take an associative division ring instead of a field (see, e.g., Endo [3]). The connected case of this theorem was first proved by Pontrjagin [1].

7

Independent Topologies on Fields

1. EXAMPLES OF INDEPENDENT TOPOLOGIES

DEFINITION 1 Let $\mathcal{T}_1, \mathcal{T}_2 \in \mathcal{L}_i(K)$, $\mathcal{T}_1 \neq \mathcal{T}_2$ (i = r, f), where K is a field. We say that $\mathcal{T}_1$ and $\mathcal{T}_2$ are independent if one of the following equivalent conditions holds:

(I_1) $\forall\, a_1, a_2 \in K\ \forall\, U_i \in \mathcal{B}_{a_i}(\mathcal{T}_i)$ $(i = 1, 2)$, $U_1 \cap U_2 \neq \emptyset$.

(I_2) $\forall\, a_1, a_2 \in K\ \forall\, U_i \in \mathcal{B}(\mathcal{T}_i)$ $(i = 1, 2)$, $(a_1 + U_1) \cap (a_2 + U_2) \neq \emptyset$.

(I_3) $\forall\, a \in K\ \forall\, U_i \in \mathcal{B}(\mathcal{T}_i)$ $(i = 1, 2)$, $(a + U_1) \cap U_2 \neq \emptyset$.

Otherwise, we say that $\mathcal{T}_1$ and $\mathcal{T}_2$ are dependent. For arbitrary finite sets of topologies $\mathcal{T}_1, \ldots, \mathcal{T}_n \in \mathcal{L}_i(K)$ (i = r, f), we say that $\mathcal{T}_1, \ldots, \mathcal{T}_n$ are independent if for every $a_1, \ldots, a_n \in K$ and every $U_i \in \mathcal{B}_{a_i}(\mathcal{T}_i)$ $(i = 1, 2, \ldots, n)$, $U_1 \cap U_2 \cap \cdots \cap U_n \neq \emptyset$.

This notion goes back to Mahler [1, Def. 3, p. 103]. He called (pseudo)norms $W_1, \ldots, W_n$ of a ring R independent if for every $a^{(1)}, \ldots, a^{(n)} \in R$ there exists a sequence $a_m \in R$ satisfying $\lim_{m\to\infty} W_k(a^{(k)} - a_m) = 0$ for every $k = 1, 2, \ldots, n$. He remarked that if $W_1, \ldots, W_n$ are independent norms and W_i' are norms on R such that $W_i \sim W_i'$ holds for every $i = 1, 2, \ldots, n$, then $W_1', \ldots, W_n'$ are independent. This means, however, that this notion is defined rather for topologies than for norms.

Remark Definition 1 works also for rings and groups (Weber [2]).

Example 1 Let $\mathcal{T}_w \in \mathcal{L}_f(K)$ be a topology induced by a nontrivial absolute value w on a field K. Then for two absolute values w, w' with

$\mathcal{T}_w \neq \mathcal{T}_{w'}$, $\mathcal{T}_w$ and $\mathcal{T}_{w'}$ are independent. Indeed, since $\mathcal{T}_w, \mathcal{T}_{w'} \in L_f(K)$ are both minimal elements (Chapter 2, Theorem 13), $0_K \neq \mathcal{T}_w \nleq \mathcal{T}_{w'}$ and $0_K \neq \mathcal{T}_{w'} \nleq \mathcal{T}_w$, they must be nonequivalent. Thus the approximation theorem for absolute values (Chapter 2, Theorem 3) implies that $\mathcal{T}_w$ and $\mathcal{T}_{w'}$ are independent.

Example 2 Let $\mathcal{T}_p$ be a p-adic topology on a field K of fractions of a unique factorization domain A (compare Example 3 of Chapter 1). Then for any four nonassociate prime elements p, q, r, s $\in$ A, $\sup(\mathcal{T}_p,\mathcal{T}_q)$, $\sup(\mathcal{T}_r,\mathcal{T}_s)$ are independent, but $\sup(\mathcal{T}_p,\mathcal{T}_q)$ and $\sup(\mathcal{T}_p,\mathcal{T}_r)$ are dependent field topologies. Indeed, let $\mathcal{T}_1 = \sup(\mathcal{T}_p,\mathcal{T}_q)$ and $\mathcal{T}_2 = \sup(\mathcal{T}_r,\mathcal{T}_s)$. Then $\mathcal{B}(\mathcal{T}_1) = \mathcal{B}(\mathcal{T}_p) \cap \mathcal{B}(\mathcal{T}_q)$ and $\mathcal{B}(\mathcal{T}_2) = \mathcal{B}(\mathcal{T}_r) \cap \mathcal{B}(\mathcal{T}_s)$. Let $U_i \in \mathcal{B}(\mathcal{T}_i)$ for i = p, q, r, s and a_1, a_2 be arbitrary. We have

$$(a_1 + U_p \cap U_q) \cap (a_2 + U_r \cap U_s)$$
$$= (a_1 + U_p) \cap (a_1 + U_q) \cap (a_2 + U_r) \cap (a_2 + U_s) \neq \emptyset$$

by the approximation theorem for absolute values.

It remains to show that $\sup(\mathcal{T}_p,\mathcal{T}_q)$ and $\sup(\mathcal{T}_p,\mathcal{T}_r)$ are dependent field topologies. Let us take $a_1 = 0$, $a_2 = 1$, $U_p \in \mathcal{B}(\mathcal{T}_p)$ with $1 \notin U_p$. There exists $V_p \in \mathcal{B}(\mathcal{T}_p)$ such that $V_p + V_p \subset U_p$. Since $\mathcal{T}_p < \sup(\mathcal{T}_p,\mathcal{T}_q)$ and $\mathcal{T}_p < \sup(\mathcal{T}_p,\mathcal{T}_r)$, we have $V_p \in \mathcal{B}(\sup(\mathcal{T}_p,\mathcal{T}_q))$ and $V_p \in \mathcal{B}(\sup(\mathcal{T}_p,\mathcal{T}_r))$ and $(a_1 + V_p) \cap (a_2 + V_p) = V_p \cap (1 + V_p) = 0$. Indeed, if $V_p \cap (1 + V_p) \neq \emptyset$, then for $x \in V_p \cap (1 + V_p)$, we should have $x = 1 + y$, $y \in V_p$, or $1 = x - y \in V_p - V_p = V_p + V_p \subset U_p$, a contradiction.

Example 3 As in Example 2, let $v_1, \ldots, v_n, v_{n+1}, \ldots, v_m$ be nontrivial and pairwise nonequivalent absolute values on a field K. Then $\sup\{\mathcal{T}_{v_i} : i \in I\}$ and $\sup\{\mathcal{T}_{v_j} : j \in J\}$ are independent if and only if $I \cup J \subset \{1, 2, \ldots, m\}$ and $I \cap J = \emptyset$.

LEMMA 1 Let A be a ring. Let $\mathcal{T}_i'$ and $\mathcal{T}_i$ be ring topologies on A satisfying $\mathcal{T}_i' \leq \mathcal{T}_i$ for $i = 1, \ldots, n$. If $\mathcal{T}_1, \ldots, \mathcal{T}_n$ are independent, then for every subset $I \subset \{1,\ldots,n\}$, $\mathcal{T}_i'$ $(i \in I)$ are also independent. Moreover,

(i) If $T_1 \neq \mathbb{1}_A$ is Hausdorff, then $\sup(T_1,\ldots,T_n) \neq \mathbb{1}_A$.

(ii) If $n \geq 2$, then $\inf(T_1,\ldots,T_n) = \mathbf{0}_A$.

Proof: (i) The proof of the first statement is straightforward. Since $T_1 \neq \mathbb{1}_A$, there exists an $x_0 \in A$ such that $\{x_0\}$ is not open. Suppose that $\sup(T_1,\ldots,T_n) = \mathbb{1}_A$. Thus there exist sets O_i, open in (A,T_i) $(i = 1, \ldots, n)$ such that $O_1 \cap \cdots \cap O_n = \{x_0\}$. Since $\{x_0\}$ is not open and (A,T_1) is Hausdorff, there exists an open set U_1, $\emptyset \neq U_1 \subset O_1 \setminus \{x_0\}$. This implies that $U_1 \cap O_2 \cap \cdots \cap O_n = \emptyset$, a contradiction.

(ii) It is sufficient to consider only the case $n = 2$. For arbitrary $n > 2$ this follows by a simple induction. Put $T = \inf(T_1,T_2)$, $B = B(T)$, $B_i = B(T_i)$, $i = 1, 2$. Let $U \in B$. Take an arbitrary $a \in A$. Since T is a ring topology, there exists $V \in B(T)$ with $V + V \subset U$. The definition of T implies that there exists $U_i \subset V$, $U_i \in B_i$ $(i = 1, 2)$. Since T_1 and T_2 are independent, there exists $z \in U_1 \cap (a + U_2)$. Thus $z = a - y$, with $y \in U_2$, i.e., $a = z + y \in U_1 + U_2 \subset V + V \subset U$. This implies that $U = A$, i.e., $T = \mathbf{0}_A$.

THEOREM 1 Let v and w be two nondiscrete valuations on a field K. The following conditions are equivalent:

(i) v and w are independent valuations.

(ii) T_v and T_w are independent field topologies.

Proof: (i) ⇒ (ii) follows immediately from Theorem 6 of Chapter 4, Section 2.

(ii) ⇒ (i) Suppose that the topologies induced on K by v and w, T_v and T_w, are independent. Then for every $V \in B(T_v)$ and every $W \in B(T_w)$, $(a + V) \cap (b + W) \neq \emptyset$ for all $a,b \in K$. Let us take $V = V_v$ and $W = V_w$, $b = 0$. Then there exists $x \in (a + W) \cap V$ satisfying $x = a - y$, $y \in W$, or $a = x + y \in V_v + V_w$. It means (since element $a \in K$ was arbitrarily chosen) that $K = V_v + V_w$, i.e., the valuations v and w are independent.

THEOREM 2 If $T_1, T_2 \in L_f(K)$ are independent, then T_1 and T_2 are incomparable (i.e., $T_1 \nleq T_2$ and $T_2 \nleq T_1$). The converse is not true.

Proof: It is a general fact, which obviously requires no underlying algebraic structure. Since there exist, in the weaker of two comparable T_2 topologies on a set, disjoint nonempty open sets with more than one point, such topologies are not independent. For the last statement of Theorem 2, see Example 2.

THEOREM 3 Suppose that the proper topologies $\mathcal{T}_1$, $\mathcal{T}_2 \in \mathcal{L}_f(K)$ are independent. Let $\mathcal{T} = \sup(\mathcal{T}_1,\mathcal{T}_2)$. Then $(K,\mathcal{T})$ is densely embeddable in the product $(K,\mathcal{T}_1) \times (K,\mathcal{T}_2)$, and conversely.

Proof: It is sufficient to show that $f: (K,\mathcal{T}) \to (K,\mathcal{T}_1) \times (K,\mathcal{T}_2)$ defined by $f(x) = (x,x)$ is continuous and its inverse is also continuous. Let $\tilde{U} = U_1 \times U_2$ be a neighborhood of zero for $(K,\mathcal{T}_1) \times (K,\mathcal{T}_2)$. Then $f(x) \in \tilde{U} \Leftrightarrow x \in U_1 \cap U_2$. We shall show that $(K,\mathcal{T})$ is dense in $(K,\mathcal{T}_1) \times (K,\mathcal{T}_2)$. Let a_1, $a_2 \in K$ and $U_i \in \mathcal{B}(\mathcal{T}_i)$ for $i = 1, 2$. Since $\mathcal{T}_1$ and $\mathcal{T}_2$ are independent, we can find $a \in K$ satisfying $a \in (a_1 + U_1) \cap (a_2 + U_2)$. This means, however, that $f(a) - (a_1,a_2) \in \tilde{U} = U_1 \times U_2$.

COROLLARY If $\mathcal{T}_1$, $\mathcal{T}_2 \in \mathcal{L}_r(K)$ are independent, the completion of K at $\sup(\mathcal{T}_1,\mathcal{T}_2)$ equals the direct sum of completions $\hat{K}_i$ of $(K,\mathcal{T}_i)$ ($i = 1, 2$).

2. SOME RESULTS ON INDEPENDENT FIELD TOPOLOGIES

THEOREM 4 Let $\mathcal{T}_i$, $\mathcal{S}_j \in \mathcal{L}_f(K)$ ($i \in I$, $j \in J$) be families of field topologies on a field K. If every finite subset of $\{\mathcal{T}_i : i \in I\} \cup \{\mathcal{S}_j : j \in J\}$ is independent, the suprema of these families (in $\mathcal{L}_f(K)$), $\mathcal{T} = \sup\{\mathcal{T}_i : i \in I\}$ and $\mathcal{S} = \sup\{\mathcal{S}_j : j \in J\}$, are independent field topologies. Conversely, if $\mathcal{T}$ and $\mathcal{S}$ are independent, then $\mathcal{T}_i$ and $\mathcal{S}_j$ are independent for every pair $(i,j) \in I \times J$. The same holds if $\mathcal{T}_i$, $\mathcal{S}_j \in \mathcal{L}_r(K)$.

Proof: Let $U \in \mathcal{B}(\mathcal{T})$ and $V \in \mathcal{B}(\mathcal{S})$ be arbitrary. Then (by Lemma 4 of Chapter 1) U and V have the form $U = U_{i_1} \cap \cdots \cap U_{i_m}$, $V = V_{j_1} \cap \cdots \cap V_{j_k}$, where $U_1 \in \mathcal{B}(\mathcal{T}_i)$ and $V_j \in \mathcal{B}(\mathcal{S}_j)$ for $i \in I$ and $j \in J$. Let $a,b \in K$. Then

$$(a + U) \cap (b + V) = (a + U_{i_1} \cap \cdots \cap U_{i_m}) \cap (b + V_{j_i} \cap \cdots \cap V_{j_k})$$

$$= \bigcap_{s=1}^{n} (a + U_{i_s}) \cap \bigcap_{r=1}^{k} (b + V_{j_r}) \neq \emptyset$$

by our assumption. The last part of Theorem 4 is immediate from the following remark: If T, $S \in L_f(K)$ are independent and $T' \leq T$, $S' \leq S$ with T', $S' \in L_f(K)$, then T' and S' are independent (see Lemma 1).

THEOREM 5 (Weber [2, Folgerung (2.3)]; this result was obtained by M. E. Shanks in 1974 (unpublished)) If $T_i \in L_f(K)$ $(i = 1, 2)$ are proper field topologies, T_2 is minimal and locally bounded and $T_2 \nleq T_1$, then T_1 and T_2 are independent. In particular, $\sup(T_1, T_2) \neq \mathbb{1}_K$.

Proof: Note first that T_2 is of type V, by Theorem 8 of Chapter 5. Suppose to the contrary that T_1 and T_2 are dependent. Hence there exist U_2 and U_1 satisfying $U_2 \cap U_1 = \emptyset$, where U_2 is T_2-open and U_1 is T_1-open. We can assume, by Definition 1, that $U_2 \in \mathcal{B}(T_2)$. Moreover, U_1 can be supposed to lie in $\mathcal{B}_1(T_1)$. Indeed, if $1 \notin U_2$, for arbitrary $x \in U_1$, $x^{-1}U_2 \cap x^{-1}U_1 = \emptyset$ and $1 \in x^{-1}U_1$. Since $T_1 \in L_f(K)$, suppose that $U_1 = U_1^{-1}$. Thus U_1^{-1} is bounded away from zero in (K, T_2), i.e., U_1 is. The assumption that T_2 is of type V gives the boundedness of U_1. Let $W_2 \in \mathcal{B}(T_2)$ be arbitrary. Then there exists $V_2 \in \mathcal{B}(T_2)$ with $V_2 + V_2 \subset W_2$. Let $y \in V_2$ satisfy $yU_1 \subset V_2$. [It is possible to find a y "near zero" in (K, T_2) as U_1 is bounded.] It follows that

$$yU_1 - y \in \mathcal{B}(T_1) \qquad \text{and} \qquad yU_1 - y \subset V_2 - V_2 = V_2 + V_2 \subset W_2$$

This shows that $T_2 \leq T_1$, contrary to our assumption.

Thus T_1 and T_2 are independent field topologies.

THEOREM 6 Let T, $T_i \in L_f(K)$ be proper topologies such that:

(i) T_i is a V-topology for $i = 1, \ldots, n$.

(ii) For every $i \neq j$, $T_i \neq T_j$.

(iii) $T_i \nleq T$ for every $i = 1, \ldots, n$.

Then T, T_1, ..., T_n are independent. In particular,

$$\sup(T,T_1,\ldots,T_n) \neq \mathbb{1}_K$$

Proof: We proceed by induction on n. The case n = 1 was settled in Theorem 5. Now let T_1, ..., T_{n+1} be pairwise different proper V-topologies of K such that $T_i \nleq T$ for i = 1, ..., n + 1, T being a proper field topology of K. Put $T' = \sup(T,T_{n+1})$. Note that $T_i \nleq T'$ for i = 1, ..., n. Indeed, otherwise we should have $T_i \leq T'$ for some $1 \leq i \leq n$, implying that T_i and T' are dependent, by Theorem 2. On the other hand, the inductive assumption implies that T, T_i, T_{n+1} are independent, and in view of Theorem 4, T_i and $T' = \sup(T,T_{n+1})$ would be independent, a contradiction! Thus $T_i \nleq T'$ for i = 1, ..., n. Now, applying the inductive assumption to T', T_1, ..., T_n, we obtain independence of T', T_1, ..., T_n. Again, since $T' = \sup(T,T_{n+1})$, we see (by Theorem 4) that T, T_1, ..., T_n are independent.

The last statement of the theorem follows immediately from Lemma 1.

Theorem 6 shows that there are many independent field topologies on a field K.

Example 4 Let M be the topology defined on $\mathbb{Q}$ by Mutylin [3] (compare Example 11 of Chapter 1). M is a ring topology. Let M be defined by a sequence p_1, p_2, ... of prime numbers. Suppose that $T \in L_f(\mathbb{Q})$ is a proper topology satisfying $T \leq M$ (Chapter 1, Theorem 4). Mutylin [3, Th. 1] has proved that $T \nleq \sup_f\{T_p : p \in A\}$ for every nonempty subset A of $\mathbb{P}' = \mathbb{P} \cup \{\infty\}$ (where T_∞ is the ordinary order topology on $\mathbb{Q}$. If $p \in \mathbb{P}' \setminus \{p_1,p_2,\ldots\}$, then, by Theorem 6, the topologies T and T_p are independent.

As a corollary we obtain from Theorem 6 the following weaker form of the approximation theorem for valuations (see Chapter 4, Theorem 6).

THEOREM 7 (Approximation theorem for V-topologies; see Prestel and Ziegler [1, Th. (4.1)], and Stone [1, Th. 3.4]) Let $T_1, \ldots, T_n \in L_f(K)$ be proper, pairwise unequal V-topologies on a field K. Then $T_1, \ldots, T_n$ are independent.

Proof: Put $T = T_n$ and apply Theorem 6 to the topologies $T, T_1, \ldots, T_{n-1}$. We can apply Theorem 6 since $T_i \neq T_j$ if $i \neq j$, and, in particular, $T \neq T_k$ $(k = 1, \ldots, n - 1)$, i.e., $T \nleq T_k$, since every V-topology is minimal (see Chapter 2, Theorem 13, and Chapter 5, Theorem 8).

COROLLARY Let $T_i \in L_f(K)$ $(i = 1, \ldots, n)$ be proper, pairwise unequal V-topologies on a field K. Then for every proper topology $T \in L_f(K)$, $\sup(T,T_1,\ldots,T_n)$ is also a proper topology.

Proof: Since T and $T_i \neq \mathbf{0}_K$, hence also $S = \sup(T,T_1,\ldots,T_n) \neq \mathbf{0}_K$. Since $S \in L_f(K)$, it remains to prove that $S \neq \mathbb{1}_K$.

We proceed by induction on n. If $n = 1$, then $S = \sup(T,T_1) \neq \mathbb{1}_K$ if $T_1 \nleq T$, by Theorem 5, and $S = \sup(T,T_1) = T \neq \mathbb{1}_K$ otherwise.

Now let $T_1, \ldots, T_{n+1}$ be proper V-topologies on K, $T_i \neq T_j$ for every $i \neq j$. Let $S_1 = T_{n+1}$ and $S_2 = \sup(T,T_1,\ldots,T_n)$. The inductive assumption implies that $S_2 \neq \mathbb{1}_K$. If $S_1 \nleq S_2$, then S_1 and S_2 are independent, by Theorem 5, and in particular, $S = \sup(S_1,S_2) \neq \mathbb{1}_K$. If $S_1 \leq S_2$, then $S = \sup(S_1,S_2) = S_2 \neq \mathbb{1}_K$, by inductive assumption.

DEFINITION 2 A topology $T \in L_r(K)$ is called A-*linear* if K is the quotient field of an integral domain A and the A-submodules of K form a base of the neighborhoods of zero for T.

For $T \in L_r(K)$ put $\mathrm{supp}(T) = \{S \in L_r(K) : \mathbf{0}_K < S \leq T$ and S is of type V$\}$. It can happen that $\mathrm{supp}(T) = \emptyset$.

THEOREM 8 Let A be a Dedekind domain, K its quotient field, and P(A) the set of all prime ideals of A. If $T, S \in L_f(K)$ are two A-linear topologies, then the following conditions are equivalent.

(i) T and S are independent.

(ii) $\mathrm{supp}(T) \cap \mathrm{supp}(S) = \emptyset$.

(iii) $\inf(T,S) = \mathbf{0}_K$.

Proof: Since every A-linear field topology on K is the supremum of a family of p-adic topologies T_p ($p \in \mathbf{P}(A)$) (see Chapter 12, Theorem 1), hence $T = \sup\{T_p : p \in T\}$ and $S = \sup\{T_p : p \in S\}$, where $S \cup T \subset \mathbf{P}(A)$. Moreover, every finite family of p-adic topologies is independent.

(ii) $\Leftrightarrow$ (iii) Obvious.

(i) $\Rightarrow$ (iii) Suppose to the contrary that there exists a T_2 topology T_1 such that $0 < T_1 \leq \inf(T,S)$. Then there exist disjoint nonempty T_1-open sets on K, which contradicts the independence of T and S. Thus $\inf(T,S) = \mathbf{0}_K$.

(ii) $\Rightarrow$ (i) Since $\mathrm{supp}(T)$ and $\mathrm{supp}(S)$ are disjoint sets, every finite subset of $\mathrm{supp}(T) \cup \mathrm{supp}(S)$ is independent. Theorem 4 implies immediately that $T = \sup\{T_p : p \in \mathrm{supp}(T)\}$ and $S = \sup\{T_p : p \in \mathrm{supp}(S)\}$ are independent field topologies.

COROLLARY Let $T, S \in L_f(\mathbb{Q})$ be topologies having bases of the neighborhoods of zero consisting of additive subgroups of $\mathbb{Q}$. Then T and S are independent if and only if $\mathrm{supp}(T) \cap \mathrm{supp}(S) = \emptyset$.

Proof: Note the additive subgroups of $\mathbb{Q}$ are $\mathbb{Z}$-submodules of $\mathbb{Q}$ and then apply Theorem 8.

THEOREM 9 Let L be a finite field extension of a field K. If T_1, $T_2 \in L_f(K)$ are independent, then the product topologies S_1, $S_2 \in L_f(L)$ of T_1 and T_2 are independent, and conversely.

Proof: Let $w_1, w_2, \ldots, w_n$ be a base for extension L of K. It means that $L = Kw_1 + \cdots + Kw_n$. The sets $U_i = V_iw_1 + \cdots + V_iw_n$, where $V_i \in B(T_i)$ ($i = 1, 2$) form a base of zero neighborhoods for the Cartesian product topologies S_i. If $a, b \in L$, $a = a_1w_1 + \cdots + a_nw_n$ and $b = b_1w_1 + \cdots + b_nw_n$ ($a_j, b_j \in K$), then

$$(a + U_1) \cap (b + U_2) = ((a_1 + V_1)w_1 + \cdots + (a_n + V_1)w_n) \cap ((b_1 + V_2)w_1 + \cdots + (b_n + V_2)w_n) \quad (1)$$

Since $(a_i + V_1) \cap (b_i + V_2) \neq \emptyset$ for $i = 1, 2, \ldots, n$, let $c_i \in (a_i + V_1) \cap (b_i + V_2)$. Then $c = c_1w_1 + \cdots + c_nw_n \in (a + U_1) \cap (b + U_2)$.

The converse statement follows from formula (1).

We finish this section with the following:

THEOREM 10 Let K be a field and B a transcendence basis over the prime subfield of K. If $|K| = |B|$, then there are exactly $2^{2^{|K|}}$ pairs (T,S) of independent field topologies T, $S \in L_f(K)$.

Proof: It is sufficient to prove that there are at least $2^{2^{|K|}}$ pairs of independent field topologies T, S on K. Theorem 9 of Chapter 2 implies that K admits exactly $2^{|K|}$ pairwise independent (i.e., inequivalent) absolute values. Let A be a proper set of absolute values on K and let $w \notin A$ be arbitrary absolute values. Then, by Theorem 5, $T = \sup_f\{T_v : v \in A\}$ and $S = T_w$ are independent. Thus the cardinality of the set of all pairs (T,S) of independent field topologies on a field K equals $2^{2^{|K|}}$.

For stronger results in this direction, see Chapter 13.

3. TOPOLOGIES BOUNDED BY SUPREMA OF V-TOPOLOGIES

LEMMA 2 Let A be a ring with unity. Let $\mathbb{O}_A \neq T_1, T_2, T \in L_r(A)$ be such that $T_1 \leq T \leq \sup(T_1,T_2)$, where T_1 and T_2 are independent topologies. Then there exists $T_3 \in L_r(A)$, such that $T_3 \leq T_2$, $T = \sup(T_1,T_3)$ and T_1, T_3 are independent.

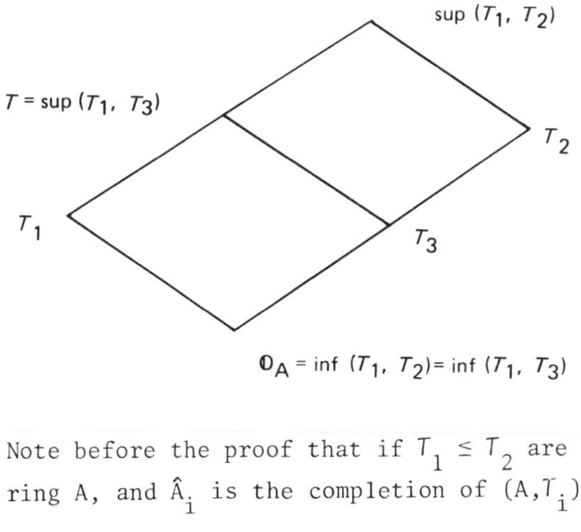

Note before the proof that if $T_1 \leq T_2$ are two ring topologies on a ring A, and $\hat{A}_i$ is the completion of (A,T_i) (i = 1, 2), then there exists a continuous ring homomorphism $f\colon \hat{A}_2 \to \hat{A}_1$ such that $f(a) = a$ holds for every $a \in A$. This follows immediately from the fact that $T_1 \le T_2$.

Proof: Let $\hat{A}$, $\hat{A}_1$, $\hat{A}_2$ be completions of (A,T), (A,T_1), (A,T_2), respectively. Theorem 3 implies that the completion of A at $\sup(T_1,T_2)$ equals $\hat{A}_1 \times \hat{A}_2$ (the direct sum of rings). Hence (see the remark before the proof) there exist continuous ring homomorphisms $f\colon \hat{A}_1 \times \hat{A}_2 \to \hat{A}$ and $g\colon \hat{A} \to \hat{A}_1$ satisfying $f(a,a) = a = g(a)$ for every $a \in A$. Define

$$D_1 = f(\hat{A}_1 \times \{0\}) \tag{1}$$

We shall prove that

$$f_1(x) = f(x,0) \qquad f_1\colon \hat{A}_1 \to D_1 \tag{2}$$

is a topological isomorphism. Put $g_1 = g|D_1$. We show that $g_1 \circ f_1 = \mathrm{id}_{\hat{A}_1}$. Let $x \in \hat{A}_1$. Since T_1 and T_2 are independent, there exists a net $(x_i)_{i\in I}$ (I is a directed set), $x_i \in A$ such that $((x_i,x_i))_{i\in I}$ is convergent to $(x,0)$ in $\hat{A}_1 \times \hat{A}_2$. Thus $x = \lim x_i$ in $\hat{A}_1$. Since $g \circ f$ is continuous and $g \circ f|A \times A = \mathrm{id}_A$, then $x = \lim(g \circ f)((x_i,x_i)) = (g \circ f)((x,0)) = (g_1 \circ f_1)(x)$. Let $e_1 = f(1,0)$, $e_2 = f(0,1)$. Then, for every $a \in A$,

$$e_1 + e_2 = 1 \qquad e_1 a = f_1(a) \qquad e_2 a = f(0,a) \tag{3}$$

In fact, $e_1 + e_2 = f((1,0) + (0,1)) = f(1,1) = 1$, and $e_1 a = f(a,0) = f_1(a)$, $e_2 a = f(0,1)f(a,a) = f((0,1)(a,a)) = f(0,a)$. Define $f_2\colon A \to \hat{A}$ by $f_2(x) = e_2 x$. Let T_3 be the weakest topology on A for which f_2 is continuous, i.e., open sets in (A,T_3) are elements of $\{f_2^{-1}(U) : U$ open in $\hat{A}\}$. Clearly, $T_3 \in L_r(A)$.

Let $(x_i)_{i\in I}$ be a net in A, $x \in A$. The following equivalences are easily seen.

$$x_i \to x \text{ in } (A,T_3) \Leftrightarrow e_2 x_i \to e_2 x \text{ in } \hat{A} \tag{4}$$

$$x_i \to x \text{ in } (A,T_1) \Leftrightarrow e_1 x_i \to e_1 x \text{ in } \hat{A} \tag{5}$$

$$x_i \to x \text{ in } (A, \sup(\mathcal{T}_1,\mathcal{T}_3)) \Leftrightarrow x_i \to x \text{ in } (A,\mathcal{T}) \tag{6}$$

$$x_i \to x \text{ in } (A,\mathcal{T}_2) \Rightarrow x_i \to x \text{ in } (A,\mathcal{T}_3) \tag{7}$$

Indeed, (4) follows from the definition of $\mathcal{T}_3$, and (5) from (3). To prove (6), note that $x_i \to x$ in $(A, \sup(\mathcal{T}_1,\mathcal{T}_3))$ if and only if $e_1x_i \to e_1x$ and $e_2x_i \to e_2x$ in A, by (4) and (5). Then, by (2) and (3), $e_1x_i + e_2x_i = x_i \to x = e_1x + e_2x$ in $(A,\mathcal{T})$. Conversely, let $x_i \to x$ in $(A,\mathcal{T})$. This implies, by continuity of multiplication, that $e_1x_i \to e_1x$ and $e_2x_i \to e_2x$ in $\hat{A}$. Finally, we prove (7). Let $x_i \to x$ in $(A,\mathcal{T}_2)$. Then $(0,x_i) \to (0,x)$ in $\hat{A}_1 \times \hat{A}_2$, i.e., $f(0,x_i) = e_2x_i \to e_2x = f(0,x)$ in $\hat{A}$. Thus $x_i \to x$ in $(A,\mathcal{T}_3)$, by (4). Consequently, $\mathcal{T}_3 \le \mathcal{T}_2$ by (7) and $\mathcal{T} = \sup(\mathcal{T}_1,\mathcal{T}_3)$, by (6). Since $\mathcal{T}_1$ and $\mathcal{T}_2$ are independent, then $\mathcal{T}_1$ and $\mathcal{T}_3$ are also independent (Lemma 1).

LEMMA 3 Let A be a ring with unity. Denote by $\mathbf{T}$ and $\mathbf{T}_0$ two sets of Hausdorff ring topologies on A. Let $\mathcal{T} \in \mathbf{T}$ and define

$$\mathcal{T}_1 = \sup\{\mathcal{T}'' \in \mathbf{T}_0 : \mathcal{T}'' \le \mathcal{T}\}$$

Suppose that:

(i) $\forall\, \mathcal{T}' \in \mathbf{T}\ \exists\, \mathcal{T}'' \in \mathbf{T}_0$, $\mathcal{T}'' \le \mathcal{T}'$.

(ii) Either $\mathcal{T} \le \mathcal{T}_1$ or there exists a Hausdorff ring topology $\mathcal{T}_2$ on A, such that $\mathcal{T} \le \sup(\mathcal{T}_1,\mathcal{T}_2)$, $\mathcal{T}_1$ and $\mathcal{T}_2$ being independent.

Then either $\mathcal{T} = \mathcal{T}_1$ or there exists a nontrivial $\mathcal{T}_3 \in \mathcal{L}_r(A)$ such that $\mathcal{T}_3 \notin \mathbf{T}$, $\mathcal{T} = \sup(\mathcal{T}_1,\mathcal{T}_3)$, and $\mathcal{T}_1$ and $\mathcal{T}_3$ are independent.

Proof: Let $\mathcal{T} \ne \mathcal{T}_1$. Since $\mathcal{T}_1 \le \mathcal{T}$, by definition, hence $\mathcal{T} \nleq \mathcal{T}_1$, and consequently, one can find $\mathcal{T}_2 \in \mathcal{L}_r(A)$ satisfying (ii). Lemma 2 guarantees existence of $\mathcal{T}_3 \in \mathcal{L}_r(A)$, with $\mathcal{T}_3 \le \mathcal{T}_2$ and $\mathcal{T} = \sup(\mathcal{T}_1,\mathcal{T}_3)$, where $\mathcal{T}_1$, $\mathcal{T}_3$ are independent. Since $\mathcal{T} \ne \mathcal{T}_1$, $\mathcal{T}_3$ is nontrivial. It remains to show that $\mathcal{T}_3 \notin \mathbf{T}$. Suppose to the contrary that $\mathcal{T}_3 \in \mathbf{T}$. Then there exists, by (i), $\mathcal{T}'' \in \mathbf{T}_0$, $\mathcal{T}'' \le \mathcal{T}_3$. Consequently, we should have $\mathcal{T}'' \le \mathcal{T}$, i.e., $\mathcal{T}'' \le \mathcal{T}_1$ by definition of $\mathcal{T}_1$. Finally, it would imply that $\mathcal{T}'' \le \inf(\mathcal{T}_1,\mathcal{T}_3)$. But $\inf(\mathcal{T}_1,\mathcal{T}_3) = \mathbf{0}_A$, by Lemma 1, as $\mathcal{T}_1$, $\mathcal{T}_3$ are independent. Thus $\mathcal{T}''$ would be non-Hausdorff, contradicting the assumption on $\mathbf{T}_0$. Thus $\mathcal{T}_3 \notin \mathbf{T}$.

THEOREM 11 Let A be a ring with unity and $S \subset L_r(A)$ be a set of Hausdorff topologies such that every finite subset of S is independent. Suppose that for every $T' \in L_r(A)$ satisfying $\mathbf{0}_A \neq T' \leq \sup S$ there exists $T'' \in S$ such that $T'' \leq T'$. Then for every $T \in L_r(A)$ satisfying $T \leq \sup S$, there exists a subset S_1 of S with $T = \sup S_1$.

Proof: Let T be a ring topology on A satisfying $T \leq \sup S$. If T is non-Hausdorff, then $T = \sup \emptyset$. For T Hausdorff put

$$\mathrm{T} = \{T' \in L_r(A) : T' \leq \sup S,\ T' \text{ a Hausdorff topology}\}$$

$$\mathrm{T}_0 = \{\sup S' : \emptyset \neq S' \subset S\}$$

$$S_1 = \{S \in S : S \leq T\} \qquad T_1 = \sup S_1$$

Then T and T_0 satisfy the assumptions of Lemma 3. Since for every nontrivial ring topology T_3 satisfying $T = \sup(T_1, T_3)$, $T_3 \in \mathrm{T}$, thus $T = T_1 = \sup S_1$, by Lemma 3.

THEOREM 12 Suppose that K is a field. Let $\mathbb{O}_K \neq T \in L_r(K)$ satisfy $T \leq \sup\{T_i : i \in I\}$, T_i a proper V-topology on K. Then there exists a subset J of I such that $\sup\{T_j : j \in J\} \leq T$.

Proof: Since T is a Hausdorff topology, there exists $U \in \mathcal{B}(T)$ satisfying $UU \subset K \setminus \{1\}$. By assumption, $T \leq \sup\{T_i : i \in I\}$, i.e., there exist V-topologies T_i and neighborhoods $U_i \in \mathcal{B}(T_i)$ $(i = 1, \ldots, n)$ such that $U_1 \cap U_2 \cap \cdots \cap U_n \subset U$. If $T_j \leq T$ for every $j = 1, \ldots, n$, then put $M_j = (K \setminus U_j)^{-1} \cup \{0\}$. Since T_j is a V-topology, then M_j are bounded. This implies, by Lemmas 16 and 17 of Chapter 1, that there exists $x \in U \setminus (M_1 \cup \cdots \cup M_n)$, i.e., $x^{-1} \in U_1 \cap \cdots \cap U_n \subset U$, and finally, $1 = xx^{-1} \in UU \subset K \setminus \{1\}$, a contradiction! Thus there exists T_j such that $T_j \leq T$. Let $J = \{j \in I : T_j \leq T\}$. Then $\sup\{T_j : j \in J\} \leq T$. Theorem 12 is proved.

THEOREM 13 Let K be a field and $T \in L_r(K)$. If there exist V-topologies T_i $(i \in I)$ such that $T \leq \sup\{T_i : i \in I\}$, there exists a subset $J \subset I$ with $T = \sup\{T_j : j \in J\}$.

Proof: Let V be the set of all V-topologies on a field K. If $T \leq \sup\{T_i : i \in I\}$ ($T_i \in V$ for every $i \in I$), there exists $T' \in V$

satisfying $\mathcal{T}' \leq \mathcal{T}$, by Theorem 12. Consequently, Theorem 11 implies that there exists a subset J of I, such that $\mathcal{T} = \sup\{\mathcal{T}_j : j \in J\}$.

NOTES

The results presented in this chapter are due mainly to Weber [2] and Więsław [10]. Results of Section 3 are taken from Weber [2]. Shanks' theorem (Theorem 5) appeared for the first time in Correl [1, Part 2, Ths. 1.2 and 1.7].

No important results on independent topologies are known in the case of division rings.

8

Fields Complete in Two Independent Topologies

In this chapter we consider fields K having the property that there are at least two independent field topologies T, $S \in L_f(K)$ such that both (K,T) and (K,S) are complete topological fields.

1. THE ABSOLUTE VALUE CASE

DEFINITION 1 A *field* K is said to be *multiply complete* if it is complete with respect to two independent field topologies.

Example 1 The field $\mathbb{C}$ of complex numbers is multiply complete with respect to two independent absolute values: the ordinary absolute value $\varphi(z) = |z|$ and the absolute value $\varphi_g(z) = |g(z)|$, where g is any discontinuous automorphism of $\mathbb{C}$, i.e., g is neither the identity nor the complex conjugate (compare Examples 1 and 2 of Chapter 1).

In this section only the topologies induced by absolute values will be considered.

THEOREM 1 Suppose that φ and ψ are nondiscrete and inequivalent absolute values on a field K. If (K,T_φ) is complete, the completion of (K,T_ψ) is an algebraically closed field.

THEOREM 2 An algebraically closed field K satisfying $|K|^{\aleph_0} = |K|$ is complete with respect to $2^{|K|}$ inequivalent absolute values of K.

With the help of Theorems 1 and 2 we will prove the fundamental

THEOREM 3 (Schmidt [1]) A field K is multiply complete with respect to two independent absolute values if and only if the following conditions are satisfied.

(i) K is an algebraically closed field.

(ii) $|K|^{\aleph_0} = |K|$.

Proof: (⇒) Since K is multiply complete with respect to two independent absolute values, Theorem 1 implies that K is an algebraically closed field. To prove (ii), note that K, being multiply complete, is complete with respect to a nondiscrete absolute value; i.e., every fundamental sequence in K has a limit in K. This shows that $|K|^{\aleph_0} \le |K|$. The converse inequality is obvious. Thus condition (ii) holds also.

(⇐) Follows easily from Theorem 2.

Our nearest aim is to prove Theorems 1 and 2. The proofs will be based on additional lemmas.

Let K be a field and φ be a nondiscrete absolute value on K.

DEFINITION 2 Define a norm in the ring K[X], putting for $g \in K[X]$, $g(X) = a_nX^n + \cdots + a_1X + a_0$, $\varphi(g) = \max\{\varphi(a_i) : 0 \le i \le n\}$.

DEFINITION 3 Let K be a field and $g,h \in K[X]$. The polynomials g and h have the same *decomposition type* over K if there exist p_i, $q_i \in K[X]$ $(i = 1, 2, \ldots, r)$ such that $g = p_1^{m_1}\cdots p_r^{m_r}$, $h = q_1^{m_1'}\cdots q_r^{m_r'}$ hold and the factors p_i, q_i satisfy $\deg(p_i) = \deg(q_i)$ and $m_i = m_i'$ for $i = 1, 2, \ldots, r$.

Hensel's lemma (Chapter 2, Theorem 6) will be useful in the following form:

LEMMA 1 Let g be a separable monic polynomial (i.e., the highest coefficient of g equals 1 and g has no multiple roots) over a field K complete with respect to a nontrivial absolute value φ. Then there exists $\delta > 0$ such that every element of $B_\delta(g) = \{f \in K[X] : \varphi(g - f) < \delta,\ f \text{ monic},\ \deg(f) = \deg(g)\}$ has the same decomposition type as g.

Hensel's lemma implies validity of Lemma 1 for non-Archimedean valued complete fields. Lemma 1 also remains true for complete fields with respect to Archimedean absolute values, namely, for $\mathbb{R}$

and $\mathbb{C}$, since in this case the roots of a polynomial are continuous functions of its coefficients.

DEFINITION 4 Let (K,φ) be an absolute-valued field. Let $b \in K^{\times}$ and $f(X) = a_nX^n + \cdots + a_0 \in K[X]$. Put $f_b(X) = b^n f(b^{-1}X) = a_nX^n + \cdots + a_1b^{n-1}X + a_0b^n$, and let $\Phi_b\colon K[X] \to K[X]$ be defined by $\Phi_b(f) = f_b$.

PROPOSITION 1 With the notation as defined above, we have:

(i) The mapping Φ_b does not change the decomposition type.

(ii) $\varphi(b) \geq 1$ implies that $\varphi(f) \leq \varphi(f_b)$ for every $f \in K[X]$.

(iii) Suppose that $\varphi(b) \geq 1$, $\delta > 0$, and that every element $f \in B_\delta(g)$ has the same decomposition type as g. Then every element $h \in B_\delta(g_b)$ has the same decomposition type as g.

Proof: Immediate from Definitions 3 and 4.

LEMMA 2 Let $(K,\mathcal{T}_\varphi)$, where φ is a nontrivial absolute value of K. Let $g \in K[X]$ be separable and put

$$M = \{g_b \in K[X] : 1 \leq \varphi(b)\}$$

Then every $h \in K[X]$, $\deg(h) = \deg(g)$, lying sufficiently near M has the same decomposition type as g.

Proof: Since g is separable, then by Lemma 1 one can find $\delta > 0$ such that for every $h \in K[X]$ satisfying $\varphi(g - h) < \delta$, h has the same decomposition as g. Applying Proposition 1(iii), we find that every $b \in K$, $\varphi(b) \geq 1$, $\varphi(g_b - h) < \delta$ implies that decomposition types of g_b and h are equal. Finally, every $h \in K[X]$ satisfying $\mathrm{dist}(h,M) < \delta$ has the decomposition type of g.

Now let K be a field provided with two inequivalent absolute values φ and φ_1. Let $\bar{K}$ be the completion of $(K,\mathcal{T}_\varphi)$ and $\bar{K}_1$ be the completion of $(K,\mathcal{T}_{\varphi_1})$. We have the following.

LEMMA 3 Let $\bar{g} \in \bar{K}[X]$ and $\bar{g}_1 \in \bar{K}_1[X]$ be separable with equal degrees. Then there exist $h \in K[X]$ having the same decomposition type over $\bar{K}$ as $\bar{g}$, and, moreover, the same decomposition type over $\bar{K}_1$ as $\bar{g}_1$.

Proof: (a) Both polynomials can be assumed to be monic. Indeed, the scalar multiplication changes neither the separability nor decomposition type of a polynomial.

(b) We can assume that $\bar{g}$, $\bar{g}_1 \in K[X]$ and are both monic. Indeed, it suffices to recall that K is dense both in $\bar{K}$ and $\bar{K}_1$, and then apply Hensel's lemma (Lemma 1). In fact, if $g \in K[X]$ satisfy $\varphi(g - \bar{g}) < \varepsilon$ for sufficiently small $\varepsilon > 0$, then g is separable and its decomposition type is the same as for $\bar{g}$.

(c) Since φ and φ_1 are nontrivial and inequivalent absolute values, there exists $c \in K$ with $\varphi(c) \geq 1$ and $\varphi_1(c) < 1$. Putting $b = c^m$ with $m \in \mathbb{N}$ sufficiently large, it can be assumed that $\varphi(b) \geq 1$ and $\varphi_1(b) < \varepsilon$, where ε is small enough.

Define $h \in K[X]$ as follows:

$$h(X) = X^n + a_{n-1}bX^{n-1} + \cdots + a_0b^n + a'_{n-1}b'X^{n-1} + \cdots + a'_0b'^n \qquad (1)$$

where

$$\bar{g}(X) = X^n + a_{n-1}X^{n-1} + \cdots + a_0 \in K[X]$$

and

$$\bar{g}_1(X) = X^n + a'_{n-1}X^{n-1} + \cdots + a'_0 \in K[X]$$

Equivalently, we can write h as

$$h(X) = \bar{g}_b(X) + a'_{n-1}b'X^{n-1} + \cdots + a'_0b'^n \qquad (2)$$

$$h(X) = (\bar{g}_1)_{b'}(X) + a_{n-1}bX^{n-1} + \cdots + a_0b^n \qquad (3)$$

where $b, b' \in K$ and $\varphi(b) \geq 1$, $\varphi_1(b) < 1$, $\varphi(b') < 1$, $\varphi_1(b') \geq 1$. If $\varphi(b')$ tends to zero, then dist(h,M) also tends to zero and h can be assumed to be near enough so that h and $\bar{g}$ could be of equal decomposition type. Keeping b' fixed, the same argument can be applied to guarantee equal decomposition types of h and $\bar{g}_1$ over $\bar{K}_1$.

LEMMA 4 A separably algebraically closed field K complete with respect to an absolute value φ is algebraically closed.

Proof: It is sufficient to assume that $\text{char}(K) = p \neq 0$ and to prove that for every $a \in K$ satisfying $\varphi(a) \leq 1$ there exists $b \in K$

with $b^p = a$, i.e., that $\sqrt[p]{a} \in K$. Indeed, if $\varphi(a) > 1$, then $\varphi(a^{-1}) < 1$ and $\sqrt[p]{a^{-1}} \in K$ implies that $\sqrt[p]{a} \in K$.

Let $g_n \in K[X]$ $(n \in \mathbb{N})$ satisfy

$$g_n(X) = X^p - c^n X - a \qquad c \in K,\ \varphi(c) < 1 \tag{4}$$

It is clear that $g_n \to g$, where $g(X) = X^p - a$, in the norm φ on $K[X]$ (see Definition 2). Moreover, each g_n is separable, being relatively prime to its derivative. Thus there are $\alpha_{in} \in K$ $(i = 1, 2, \ldots, p)$ such that $g_n(\alpha_{in}) = 0$. Since $\varphi(\alpha_{1n})\cdots\varphi(\alpha_{pn}) = \varphi(\alpha_{1n}\cdots\alpha_{pn}) = \varphi(a) \le 1$, there is an index i such that $\varphi(\alpha_{in}) \le 1$. For simplicity, put $\alpha_n = \alpha_{in}$ for the index i above. We have the following conditions:

$$\varphi(\alpha_n^p - a) = \varphi(c^n \alpha_n) \le \varphi(c)^n \to 0 \qquad \text{as } n \to \infty$$

Thus $\lim \alpha_n^p = a$. On the other hand, $\alpha_m^p - \alpha_n^p = (\alpha_m - \alpha_n)^p$, and $\varphi(\alpha_m^p - \alpha_n^p) = \varphi(\alpha_m - \alpha_n)^p$. This implies that

$$\begin{aligned}\varphi(\alpha_m - \alpha_n)^p &= \varphi((\alpha_m^p - a) - (\alpha_n^p - a)) \le \varphi(\alpha_m^p - a) + \varphi(\alpha_n^p - a)\\ &\le \varphi(c)^n + \varphi(c)^m \to 0 \qquad \text{as m and n tend to infinity}\end{aligned}$$

In this way we obtain that (α_n) is a Cauchy sequence in (K, T_φ), implying that there is $\alpha \in K$ such that $\alpha = \lim \alpha_n$, since (K, T_φ) is complete. Finally,

$$a = \lim \alpha_n^p = (\lim \alpha_n)^p = \alpha^p \qquad \alpha \in K$$

which proves that $\alpha = \sqrt[p]{a} \in K$.

Proof of Theorem 1: Let K be complete with respect to φ and let $\bar{K}_1$ be the completion of (K, T_ψ), $\psi \not\sim \varphi$. Let $\bar{g}_1 \in \bar{K}_1[X]$ be separable and irreducible, $\deg(\bar{g}_1) = n$, and let $\bar{g} \in K[X]$ have the form $\bar{g}(X) = (X - b_1)\cdots(X - b_n)$, $b_i \in K$. Then there is $h \in K[X]$, $\deg(h) = n$, which has decomposition type over K the same as $\bar{g}$, and over $\bar{K}_1$ the same as $\bar{g}_1$. Thus h is irreducible over $\bar{K}_1 \supset K$ and a fortiori over K. Since h is irreducible, being the product of linear factors, $n = 1$. Hence $\bar{K}_1$ is separably algebraically closed. We obtain the result, applying now Lemma 4.

LEMMA 5 If L is an algebraic extension of a field K, the trivial absolute value φ_0 on K has only a trivial extension to L.

Proof: Let φ be an extension of φ_0 to L. Suppose that φ is nontrivial. Then there is $b \in L$ satisfying $\varphi(b) > 1$. We can suppose that $\varphi(b) > 2$, taking, if necessary, a suitable power of b. Since b is algebraic over K, there exist $n \in \mathbb{N}$ and $a_i \in K$, not all zero, such that

$$b^n + a_{n-1}b^{n-1} + \cdots + a_0 = 0 \tag{5}$$

Then

$$0 = \varphi(b^n + a_{n-1}b^{n-1} + \cdots + a_0) \geq \varphi(b)^n - \sum_{i=0}^{n-1} \varphi(a_i)\varphi(b)^i$$

$$\geq \varphi(b)^n - \sum_{i=0}^{n-1} \varphi(b)^i = \varphi(b)^n - \frac{\varphi(b)^n - 1}{\varphi(b) - 1}$$

$$= \frac{\varphi(b)^n(\varphi(b) - 2) + 1}{\varphi(b) - 1} > 0$$

a contradiction. Thus φ must be trivial.

Remark Lemma 5 completes the proof of Theorem 16 of Chapter 2, in the case of trivial absolute value.

LEMMA 6 Let $K = P(B)$ where $B \neq \emptyset$ is a transcendence basis of K over its prime field P. If dim transc$[K:P] = m$ and $K \simeq K'$, then there exist on K' at least 2^m inequivalent absolute values.

Proof: Follows immediately from Theorem 9 of Chapter 2. Indeed, let $K' = P(B')$ with B' being a fixed transcendence basis of K'. If $f: B \to B'$ is any bijection, then f can be extended to an isomorphism $\bar{f}: K \to K'$. If $\alpha \in P(B)$,

$$\alpha = \frac{F(b_1,\ldots,b_n)}{G(b_1,\ldots,b_n)} \qquad \text{where } F,\ G \in P[X_1,\ldots,X_n]$$

then put

$$\bar{f}(\alpha) = \frac{F(f(b_1),\ldots,f(b_n))}{G(f(b_1),\ldots,f(b_n))} \tag{6}$$

It is easy to see that $\bar{f}$ maps K isomorphically onto K'. If φ is any absolute value on K, then $\bar{\varphi}$ defined by the formula

$$\bar{\varphi}(x) = \varphi(f(x)) \qquad x \in K \tag{7}$$

is an absolute value on K'. It is immediate that $\varphi \nsim \psi$ implies that $\bar{\varphi} \nsim \bar{\psi}$. Since K admits 2^m inequivalent absolute values, K' keeps this property, too.

LEMMA 7 Let K be an algebraically closed field having an infinite transcendence basis over its prime subfield. If $K' \simeq K$, then K' admits at least 2^m inequivalent absolute values, where $m =$ dim transc[K:P].

Proof: Write K = P(B)(A) and K' = P(B')(A'), where B and B' are the corresponding transcendence basis (the Steinitz decomposition of a field). Since A (resp. A') contains only elements algebraic over P(B) [resp. P(B')], then applying Theorem 19 of Chapter 2, and Lemma 6, the result follows straightforwardly.

Proof of Theorem 2: Let K be an algebraically closed field satisfying $|K|^{\aleph_0} = |K|$. Then $|K| > \aleph_0$, hence K admits a nontrivial absolute value, say φ (Chapter 2, Theorem 9). Let $\bar{K}$ denote the completion of (K, T_φ). Moreover, $|\bar{K}| \le |K|^{\aleph_0} = |K|$ and $|K| \le |\bar{K}|$ imply that $|\bar{K}| = |K|$. If P is the prime subfield of K, then $|\bar{K}| = |K|$ implies that dim transc$[\bar{K}:P]$ = dim transc[K:P]. Kürschak's theorem (Chapter 2, Theorem 7) shows that $\bar{K}$ is an algebraically closed field. Thus $K \simeq \bar{K}$, by the Steinitz decomposition theorem. Let $f\colon K \to \bar{K}$ be the isomorphism. Since $(\bar{K}, T_\varphi)$ is complete, it is easy to see that (K, T_w), $w = \varphi \circ f$, is also complete. Recall that K, being algebraically closed, possesses exactly $2^{|K|}$ automorphisms (see, e.g., Charnow [1]). Define

$$w_g(x) = w(g(x)) \qquad \text{for every } x \in K;\ g \in \mathrm{Aut}(K)$$

Clearly, (K, T_{w_g}) is complete, since (K, T_w) is complete. To finish the proof it remains to show that there exist $2^{|K|}$ discontinuous automorphisms of $(K, \mathcal{T}_w)$. Indeed, for every such automorphism g, $w_g \not\sim w$.

Let P be the prime subfield of K. Let B be a transcendence basis of K over P, and k = P(B). First note that $w|k$ is nontrivial. Indeed, otherwise w would be trivial on K, since K is algebraic over k (Lemma 5). Thus there exists $c \in k^{\times}$ satisfying $w(c) < 1$. Suppose that c is a rational function in variables $b_1, \ldots, b_n \in B$. Let $n(b) = 0$ if $b = b_i$ for some i, and otherwise let $n(b) \in \mathbb{Z}$ be taken in order to satisfy $w(c^{n(b)}b) = w(c)^{n(b)}w(b) > 1$. Let $B' = \{c^{n(b)}b : b \in B\}$. Since every element $b \in B$ is a rational function of elements of B' and conversely, every element $b' \in B'$ that can rationally be expressed by elements of B, B' is also a transcendence basis. Thus $k = P(B')$ and for every $b \in B' \setminus \{b_1, \ldots, b_n\}$, $w(b) > 1$.

Let $\emptyset \neq Y \subset B'$ be arbitrary. Define the function g_Y as follows:

$$g_Y(b) = \begin{cases} b^{-1} & \text{if } b \in Y \\ b & \text{if } b \in B' \setminus Y \end{cases}$$

Clearly, g_Y can be extended to an automorphism of k, by putting for every $f \in k$, $f = f(b_1', \ldots, b_s')$, $b_s' \in B'$:

$$g_Y(f) = f(g_Y(b_1'), \ldots, g_Y(b_s'))$$

and moreover, to an automorphism of K, since K is algebraically closed. Denote this automorphism again by g_Y. Then g_Y is discontinuous with respect to w, since for every $b \in Y$,

$$w_{g_Y}(b) = w(g_Y(b)) = w(b^{-1}) = \frac{1}{w(b)} < 1 < w(b)$$

This shows that the absolute values $w \circ g_Y$ and w are not equivalent. The same proof gives that if $Y \neq Z$, then $w \circ g_Y \not\sim w \circ g_Z$.

Thus K is complete with respect to at least $2^{|K|}$ nonequivalent absolute values $w \circ g_Y$, $\emptyset \neq Y \subset B'$.

Since every absolute value φ is an element of $\mathbb{R}_+^K$, there are at most

$$|\mathbb{R}_+^K| = |\mathbb{R}_+|^{|K|} = 2^{\aleph_0 |K|} = 2^{|K|}$$

i.e., exactly $2^{|K|}$, inequivalent absolute values on K.

2. THE VALUATION CASE

We start with the following.

DEFINITION 5 Let K be a field and v a valuation on K. The valued field (K,v) is said to be *maximally complete* and v is said to be maximal on K if (K,v) has no *immediate extensions*; i.e., every extension of (K,v) must change either the value group or the residue class field (compare Chapter 2, Section 5).

DEFINITION 6 A module M over a ring A is said to be *linearly compact* (in the discrete topology) if every family of cosets of M which has the finite intersection property has nonvoid intersection.

Equivalently, M is linearly compact if every finitely solvable system of congruences $x \equiv m_i \pmod{N_i}$ [N_i submodule of M $(i \in I)$] is solvable.

DEFINITION 7 A *valuation ring* A is called *maximal* if A is linearly compact as a module over A.

Schmidt's theorem (Theorem 3) has the following analog for valuations:

THEOREM 4 (Vámos [1], Th. A) Let K be a field. Then the following conditions are equivalent.

(i) K is algebraically closed and $|K|^{\aleph_0} = |K|$.

(ii) K is multiply complete with respect to two maximal valuations on K.

(iii) There exist independent valuations v and w on K such that v is maximal on K and (K,w) is Henselian.

Now let K be a field, $\Gamma \neq \{0\}$ a linearly ordered abelian group, and Γ_+ the positive cone of Γ. For a function $f: \Gamma \to K$ define $\operatorname{supp}(f) = \{\gamma \in \Gamma : f(\gamma) \neq 0\}$. Set

$$K[[\Gamma]] = \{f \in K^{\Gamma} : \text{supp}(f) \subset \Gamma_+ \text{ and supp}(f) \text{ is well ordered}\}$$
$$K((\Gamma)) = \{f \in K^{\Gamma} : \text{supp}(f) \text{ is well ordered}\} \tag{8}$$

Addition and multiplication are defined in $K((\Gamma))$ by

$$(f + g)(\alpha) = f(\alpha) + g(\alpha)$$
$$(fg)(\alpha) = \sum_{\beta+\gamma=\alpha} f(\beta)g(\gamma) \qquad f,g \in K((\Gamma)),\ \alpha,\beta,\gamma \in \Gamma \tag{9}$$

These operations are well defined and give $K((\Gamma))$ the structure of a field. The function $v: K((\Gamma)) \to \Gamma \cup \{\infty\}$, defined by $v(f) = \min \text{supp}(f)$ for $f \neq 0$, can be shown to be a maximal valuation on $K((\Gamma))$, with $K[[\Gamma]]$ as its valuation ring and K as its residue class field. Note that $K((\Gamma))$ can be identified with $K\{t^{\Gamma}\}$ (see Chapter 4, Example 2).

THEOREM 5 (Vámos [1], Th. B) (i) If K is a multiply complete field with respect to two maximal valuations, then $K \simeq K((\mathbb{Q}^+))$.
(ii) If K is an algebraically closed field and Γ a linearly ordered nonzero divisible group, then the field $K((\Gamma))$ is multiply complete with respect to two maximal valuations.
(iii) Let K be a multiply complete field with respect to two maximal valuations and v a maximal valuation on K with value group Γ. Then K is uniquely determined, up to a value-preserving isomorphism, by $\text{char}(K)$, $\bar{k}_v$, Γ and if $\text{char}(\bar{k}_v) = p \neq \text{char}(K)$, by $v(p)$. Moreover, if $\text{char}(K) = \text{char}(\bar{k}_v)$, there is a value-preserving isomorphism between K and $\bar{k}_v((\Gamma))$.

In connection with Theorem 3, I asked (Więsław [6], question 5, p. 137) whether a nonalgebraically closed field can be multiply complete with respect to two (independent) minimal field topologies. Berrondo [5] has answered this question positively.

Example 2 (Berrondo [5]) Let k be a field and E a linearly ordered set without any countable cofinal subset. Let $\Gamma = \mathbb{Z}^{(E)}$ be the linearly ordered group of maps from E to $\mathbb{Z}$ having finite support. For every $e \in E$ let X_e be the element of $k((\Gamma))$ defined by the characteristic function of $\{e\}$. If $M = \mathbb{N}^{(E)}$, then the set of mappings from Γ into k, having finite support contained in M, is the k-subalgebra $k[M] = k[X_i]_{i \in E}$ of $k((\Gamma))$. It can be shown (Berrondo [5, Th. 2]) that the quotient field $k(X_i)_{i \in E}$ of $k[X_i]_{i \in E}$ is closed in $k((\Gamma))$.

Now we are able to construct a nonalgebraically closed field multiply complete with respect to two independent valuations. Let $K = k(X_i, Y_j)_{i \in I, j \in E}$, where k is an arbitrary field; E is as above, and X_i, Y_j are all algebraically independent over k; $(i,j) \in E^2$. Let v_1 be the natural valuation on $k'((\Gamma))$ extending the trivial one from $k' = k(Y_j)_{j \in E}$. We shall consider v_1 on $K = k'(X_i)_{i \in E}$. Similarly, let v_2 be the natural valuation on $k''((\Gamma))$ extending the trivial one from $k'' = k(X_j)_{j \in E}$. Thus v_2 is a valuation on $K = k''(Y_i)_{i \in E}$. Since $(K, \mathcal{T}_{v_i})$, $i = 1, 2$, are closed in the suitable fields, K is multiply complete with respect to independent valuations v_1 and v_2. Indeed, since $X_i \to 0$ in $(K, \mathcal{T}_{v_1})$ and $X_i \not\to 0$ in $(K, \mathcal{T}_{v_2})$, then v_1 and v_2 are independent. Furthermore, K is not algebraically closed, as $Z^2 - X_i$ has no solutions in K.

Note, finally, that there exists a field K and proper field topologies $\mathcal{T}$ and $\mathcal{S}$ on K such that $(K, \mathcal{T})$ and $(K, \mathcal{S})$ are both complete and $\mathcal{T}$ is properly weaker than $\mathcal{S}$ (see Berrondo [6]).

NOTE

Let $(K, \mathcal{T}_v)$ be a valued, complete field satisfying the first axiom of countability. All the automorphisms of K are continuous if and only if K is not algebraically closed (Keller [1]).

9

Locally Bounded Ring Topologies on Fields

The notion of locally bounded field (or ring) topology was introduced in Section 5 of Chapter 1 and we will not repeat it here. Recall only that every locally bounded ring topology on a field defines an almost-order in a natural way and the class of equivalent almost-orders on a field can be identified with a suitable ring topology on the field. Moreover, as shown in Section 3 of Chapter 5, every locally bounded and minimal ring topology on a field is induced either by an absolute value or by a non-Archimedean valuation according to whether it has a nonzero topological nilpotent.

In this chapter we summarize successive generalizations of absolute values, valuations, and norms. The culmination will appear in Section 2, where every locally bounded ring topology will be described in terms of generalized valuations (Nakano's valuations).

1. SEVERAL GENERALIZATIONS OF VALUATIONS

Successive generalizations of the notion of valuation on a field lead to concepts of semivaluation and half-valuation. This has been done by Fuchs [1] and Jaffard [1].

DEFINITION 1 Let Γ be an ordered, directed, additively written commutative group. By a *half-valuation* we will mean any mapping v: $K^\times \to \Gamma$ (where K is a field) satisfying the following conditions:

(HV_1) $v(K^\times) = \Gamma$.

(HV_2) $v(ab) = v(a) + v(b)$.

(HV_3) $v(a) \geq v(c)$ and $v(b) \geq v(c)$ implies that $v(a - b) \geq v(c)$.

As in the case of valuations, we can adjoin to Γ the symbol ∞ and define a half-valuation as a function from K to $\Gamma \cup \{\infty\}$, where $v(0) = \infty$, $\infty + \gamma = \infty$, $\infty > \gamma$ for $\gamma \in \Gamma$ and $v(a) \neq \infty$ for every $a \in K^{\times}$.

DEFINITION 2 A partially ordered commutative group Γ is called a *lattice-ordered* group or *ℓ-group* if for every $\alpha, \beta \in \Gamma$, there exist $\sup(\alpha,\beta)$ and $\inf(a,b)$.

DEFINITION 3 A *semivaluation* s on a field K is a mapping $s\colon K^{\times} \to \Gamma$, where Γ is an ℓ-group, such that

(SV_1) $s(K^{\times}) = \Gamma$.

(SV_2) $s(ab) = s(a) + s(b)$.

(SV_3) $s(a + b) \geq \inf(s(a),s(b))$.

We can, moreover, define $s(0) = \infty$ (see the remark above).

PROPOSITION 1 (Jaffard [1]) Every ℓ-group is a semivaluation group of some field.

Proof: (sketch) Let Γ be an ℓ-group and K_0 a fixed field. Let A be the group algebra of Γ over K_0. Thus every element $x \in A$ has the form

$$x = a_1X^{\alpha_1} + \cdots + a_nX^{\alpha_n}$$

where $a_j \in K_0$, $\alpha_i \in \Gamma$, and $\alpha_i \neq \alpha_j$ for $i \neq j$. (Here X is an arbitrary element, transcendental over K_0.) Set

$$s(x) = \inf(\alpha_1,\ldots,\alpha_m)$$

Let K be the quotient field of A. Then for every $z \in K^{\times}$, $z = x/y$, where $x, y \in A$, put $s(z) = s(x) - s(y)$. It is not difficult to see that s is a semivaluation on K and Γ is its semivaluation group.

As for non-Archimedean valuations, we define a half-valuation ring (or a semivaluation ring) of s as the set $\{a \in K : s(a) \geq 0\}$. We have the following result of Yakabe [2]:

THEOREM 1 A proper subring R of a field K is a semivaluation ring of a suitable semivaluation on K if and only if K is the quotient field of R and for every two $a, b \in R^{\times}$ there exists the greatest common divisor (a,b) of a and b.

Another result of Yakabe [2, Th. 5] states:

THEOREM 2 For every semivaluation on a field, its semivaluation ring is the intersection of a family of valuation rings.

In other words, if s is a semivaluation on a field K, there exist valuations v_i $(i \in I)$ on K such that the semivaluation ring V_s of s has the form

$$V_s = \bigcap_{i \in I} V_{v_i} \tag{1}$$

We can now state:

THEOREM 3 Let s be a semivaluation on a field K and $\mathcal{T}_s$ the topology on K having the family of sets

$$B(\mathcal{T}_s) = \{aV_s : a \in K^\times\} \tag{2}$$

as a base of neighborhoods of zero, where V_s is the semivaluation ring of s. Then:

(i) $\mathcal{T}_s \in \mathcal{L}_r(K)$ is locally bounded and V_s is its order.

(ii) If the semivaluation ring of s is an intersection of only finitely many valuation rings in K, then $\mathcal{T}_s \in \mathcal{L}_f(K)$ is locally bounded and

$$\mathcal{T}_s = \sup\{\mathcal{T}_{v_i} : i \in I\} \qquad (|I| < \infty) \tag{3}$$

Proof: Theorem 3 follows immediately from Theorem 2, from Theorem 9 of Chapter 1, and from the fact that there is a one-to-one correspondence between the classes of almost-orders and locally bounded ring topologies on fields.

DEFINITION 4 Let Γ be an ℓ-group. We call a set Ω of positive elements in Γ an *upper class* of Γ if for every $\alpha \in \Omega$, $\alpha \leq \beta$ implies that $\beta \in \Omega$ and if, moreover, Ω is a sublattice of Γ. We shall call an upper class Ω of Γ an *ℓ-closed upper class* if it contains the greatest lower bound of every two-element subset of Ω. By the *order of a semivaluation* s we mean the number of maximal ℓ-closed upper classes of its value group. We shall say that a *semivaluation* is of the *first kind* if its order is finite; otherwise, we say that it is of the *second kind*.

In particular, a trivial semivaluation is of order 0, a non-trivial non-Archimedean valuation is a semivaluation of order 1.

Yakabe [4] is the author of the following extension theorem.

THEOREM 4 If k is a subfield of a field K, then any semivaluation s of the first kind and of order n on k can be extended to a semivaluation of the first kind of K and of the same order n. In particular, if K is a finite extension of k, then the number of distinct extensions of s in K, which are of order n, is finite and not greater than d^n, where d is the degree of separability of K over k.

Roughly speaking, a semivaluation can be treated as a generalization of a valuation in the same way as the norm is a generalization of an absolute value on a field. Moreover, s is a semivaluation of the first kind exactly when its semivaluation ring is the intersection of finitely many valuation rings (see Yakabe [2, proof of Th. 5]). Thus power-multiplicative norms have an analog for semivaluations, namely, semivaluations of the first kind. In both cases, the topology induced by the corresponding norm or semivaluation is a field topology and is the supremum of finitely many minimal and locally bounded field topologies (Theorem 8 of Chapter 3 and Theorem 3 of this chapter).

2. NAKANO'S VALUATIONS

The final step in a description of all locally bounded ring topologies on a field was made by Nakano in 1960 (see Nakano [1-3]). He introduced a notion of a generalized valuation which generalizes all known types of valuations, absolute values, semivaluations, norms, and so on. He treated in a unified fashion locally bounded ring topologies on fields, connecting almost-orders with generalized valuations. All locally bounded ring topologies on a field can be described in terms of generalized valuations.

First, some definitions.

DEFINITION 5 By a *value group of a generalized valuation* or, simply, by a *v-group* we shall mean a set Σ such that

(i) Σ is a multiplicative group and 1 is its unit element.

(ii) In Σ, there is defined an associative and commutative addition $(\alpha,\beta) \mapsto \alpha + \beta$.

(iii) $\alpha(\beta + \gamma) = \alpha\beta + \alpha\gamma$ and $(\alpha + \beta)\gamma = \alpha\gamma + \beta\gamma$ for every α, β, $\gamma \in \Sigma$.

LEMMA 1 Any v-group is a directed group.

Proof: Let Σ be a v-group. Note first that

$$\Sigma + \Sigma = \Sigma \tag{4}$$

This follows immediately from the identity

$$(\alpha + \beta)^{-1}\alpha + (\alpha + \beta)^{-1}\beta = 1 \tag{5}$$

which is true by Definition 5(iii). Put

$$\Omega = \{1\} \cup \{\lambda \in \Sigma : \exists\, \mu \in \Sigma,\ \lambda + \mu = 1\} \tag{6}$$

Then Ω is closed with respect to multiplication: $\Omega\Omega \subset \Omega$. Indeed, $\lambda_1 + \mu_1 = \lambda_2 + \mu_2 = 1$ implies that

$$(\lambda_1 + \mu_1)(\lambda_2 + \mu_2) = 1 = \lambda_1\lambda_2 + (\lambda_1\mu_2 + \mu_1\lambda_2 + \mu_1\mu_2)$$

$$\alpha\Omega = \Omega\alpha \qquad \text{for every } \alpha \in \Omega \tag{7}$$

If $\lambda + \mu = 1$, then $\alpha\lambda\alpha^{-1} + \alpha\mu\alpha^{-1} = 1$, i.e., $\alpha\lambda\alpha^{-1} \in \Omega$; thus $\alpha\Omega \subset \Omega\alpha$. The proof of the converse inclusion is similar.

$$\text{If } \lambda + \mu = 1, \text{ then } \lambda\mu = \mu\lambda. \tag{8}$$

In fact, set $\alpha = \lambda\mu^{-1}$. Then since $\alpha + 1 = \mu^{-1}$, we obtain $\lambda = \lambda\mu^{-1}\mu = (\lambda\mu^{-1})\mu = \alpha(\alpha + 1)^{-1}$. Hence $\mu^{-1}\lambda = (\alpha + 1)\alpha(\alpha + 1)^{-1} = (\alpha^2 + \alpha)(\alpha + 1)^{-1} = (\alpha + \alpha^2)(\alpha + 1)^{-1} = \alpha(\alpha + 1)(\alpha + 1)^{-1} = \alpha = \lambda\mu^{-1}$; thus (8) holds.

$$\lambda,\ \lambda^{-1} \in \Omega \Rightarrow \lambda = \lambda^{-1} = 1 \tag{9}$$

Let $\lambda + \mu = \lambda^{-1} + \nu = 1$. Then $\nu + \lambda^{-1}\mu = \nu(\lambda + \mu) + \lambda^{-1}\mu = \nu\lambda + (\lambda^{-1} + \nu)\mu = \nu\lambda + \mu = (\nu + \mu\lambda^{-1})\lambda$. Since $\nu + \lambda^{-1}\mu = \nu + \mu\lambda^{-1}$, then

$\nu + \lambda^{-1}\mu = (\nu + \lambda^{-1}\mu)\lambda$ and $\lambda = 1$ or $\lambda^{-1} = 1$. This allows us to define in Σ a partial order in the following way:

$$\alpha \geq \beta \Leftrightarrow \alpha^{-1}\beta \in \Omega \qquad \text{for } \alpha, \beta \in \Sigma \tag{10}$$

To prove that $(\Sigma,\leq)$ is a directed group, it remains to show that

$$\text{for every } \alpha, \beta \in \Sigma \text{ there exists } \gamma \in \Sigma \text{ with } \gamma \geq \alpha \text{ and } \gamma \geq \beta \tag{11}$$

Indeed, set $\gamma = \alpha + \beta$ and look at (5).

DEFINITION 6 Let K be a field and let Σ be a v-group. A function $\|\cdot\|: K \to 2^{\Sigma}$ assigning to every element $x \in K$ a subset $\|x\|$ of Σ is called a *generalized valuation* if the following conditions are satisfied for all $x, y \in K$.

(GV_1) $\|x\| = \Sigma \Leftrightarrow x = 0$.

(GV_2) $\|x\| + \|y\| \subset \|x + y\|$.

(GV_3) $\|x\|\cdot\|y\| \subset \|xy\|$.

Let $\sigma \in \Sigma$ be arbitrary. Set

$$U_\sigma = \{x \in K : \sigma \in \|x\| + \Sigma\} \tag{12}$$

THEOREM 5 The sets (12) form a base of neighborhoods of zero for a locally bounded ring topology on a field K. If, moreover, the generalized valuation satisfies the condition

(GV_4) For every $\sigma \in \Sigma$ there exists $x \in K^\times$ such that $\sigma \in \|x\|$.

then the topology induced by these sets is proper.

Proof: We will prove that the following conditions hold.

$$0 \in U_\alpha \tag{13}$$

$$U_\alpha \pm U_\beta \subset U_{\alpha+\beta} \tag{14}$$

$$U_\alpha U_\beta \subset U_{\alpha\beta} \tag{15}$$

$$\text{For every } \alpha, \beta \text{ there exists } \gamma \text{ with } U_\gamma \subset U_\alpha \cap U_\beta. \tag{16}$$

$$U_\alpha \text{ is bounded for every } \alpha \in \Sigma. \tag{17}$$

For every $\alpha \in \Sigma$ and $q \in U_\alpha$ there is $\varepsilon \in \Sigma$ such that

$$q + U_\varepsilon \subset U_\alpha. \tag{18}$$

Indeed, let $\|0\| = \Sigma$. Then $\alpha \in \|0\| + \Sigma = \Sigma$ for all α's. To prove (14), let $\alpha \in \|x\| + \Sigma$, $\beta \in \|y\| + \Sigma$. Since $\|x\| + \|y\| \subset \|x + y\|$, we have

$$\alpha + \beta \in (\|x\| + \Sigma) + (\|y\| + \Sigma) = (\|x\| + \|y\|) + \Sigma \subset \|x + y\| + \Sigma$$

i.e., $x + y \in U_{\alpha+\beta}$. Similarly, (15) holds, since Σ is a multiplicative group and

$$\begin{aligned}\alpha\beta \in (\|x\| + \Sigma)(\|x\| + \Sigma) &= \|x\|\cdot\|y\| + \|x\|\Sigma + \Sigma\|y\| + \Sigma \\ &= \|x\|\cdot\|y\| + \Sigma \subset \|xy\| + \Sigma\end{aligned}$$

that is, $xy \in U_{\alpha\beta}$.

Now let $\alpha, \beta \in \Sigma$ be arbitrary. Put $\gamma = (\alpha^{-1} + \beta^{-1})^{-1}$. Note first that $\gamma \leq \alpha$ and $\gamma \leq \beta$. In fact, $\alpha^{-1}\gamma \in \Omega$ if and only if $\alpha \geq \gamma$, by definition (10). But

$$\alpha^{-1}\gamma = \alpha^{-1}(\alpha^{-1} + \beta^{-1})^{-1} = ((\beta^{-1} + \alpha^{-1})\alpha)^{-1} = (\beta^{-1}\alpha + 1)^{-1}$$

and

$$\alpha^{-1}\gamma(\beta^{-1}\alpha + 1) = 1$$

i.e., $\alpha^{-1}\gamma + \alpha^{-1}\gamma\beta^{-1}\alpha = 1$. Thus $\alpha^{-1}\gamma \in \Omega$. Similarly, one can prove that $\gamma \leq \beta$.

Note that

$\sigma + \delta \geq \delta$ for every $\sigma, \delta \in \Sigma$, and if $\sigma > \delta$, then there exists $\varepsilon \in \Sigma$ such that $\sigma = \delta + \varepsilon$. (19)

The first statement of (19) follows from (5) and the second from the definition (10): If $\sigma > \delta$, then $\alpha^{-1}\delta + \eta = 1$ with $\eta \in \Omega$. Thus $\sigma = \delta + \sigma\eta$. Applying (19) to (14) we find that

$\sigma \leq \delta$ implies that $U_\sigma \subset U_\delta$ for every $\sigma, \delta \in \Sigma$. (20)

Now we are able to finish the proof of (16). Since $\gamma \leq \alpha$ and $\gamma \leq \beta$, $U_\gamma \subset U_\alpha$ and $U_\gamma \subset U_\beta$; i.e., (16) holds. To obtain (17), take arbitrary $\beta \in \Sigma$. Since $\beta = \alpha(\alpha^{-1}\beta)$, then (15) gives $U_\alpha U_{\alpha^{-1}\beta} \subset U_\beta$; thus U_α is bounded. Finally, let $q \in U_\alpha$. If $\beta \in \|q\| + \Sigma$, then by (19) there exists $\varepsilon \in \Sigma$ such that $\alpha = \beta + \varepsilon$. Applying (14), we find that $q + U_\varepsilon \subset U_\beta + U_\varepsilon \subset U_\alpha$; i.e., (18) holds. Conditions (13)-(18) imply that $\{U_\sigma : \alpha \in \Sigma\}$ is a base of the neighborhoods of zero for a ring topology $\mathcal{T}$ on K. The topology $\mathcal{T}$ is nontrivial by (18) and if (GV_4) is satisfied, $\mathcal{T}$ is also nondiscrete.

Example 1 The set $\mathbb{R}_+^\times$ of positive reals is a v-group under ordinary addition and multiplication.

Example 2 Every ℓ-group is a v-group if we define addition by $\alpha + \beta = \sup(\alpha,\beta)$. Therefore, $\mathbb{R}_+^\times$ is a v-group under ordinary multiplication and this addition.

Note that not every v-group is an ℓ-group. Indeed, the group $\Gamma = \mathbb{R}_+^\times \times \mathbb{R}_+^\times$ (with componentwise multiplication and addition) ordered by $(a,b) > (a',b')$ if and only if $a > a'$ and $b > b'$ is a v-group but not an ℓ-group, since $\inf\{(1,3),(3,1)\}$ does not exist in Γ.

Remark Conditions (GV_1) and (GV_2) imply that $\|x\| + \Sigma \subset \|x\|$; that is, $\alpha \in \|x\|$ and $\beta \geq \alpha$ imply that $\beta \in \|x\|$. We can therefore interpret the fact that $\alpha \in \|x\|$ as an indirect value of x, the value being smaller than α. But if there exists an infimum in $(\Sigma,\leq)$ of $\|x\|$ for every $x \in K^\times$, we may assign to x the value $\inf\|x\|$ without changing the topology defined by $\|\cdot\|$. Thus the concept of generalized valuation directly generalizes the notions of valuations, semivaluations, absolute values, and so on.

DEFINITION 7 Let v be a generalized valuation on a field K. Put $\mathcal{T}_v$ for the ring topology defined by a base of the neighborhoods of zero, as in (12).

THEOREM 6 (Nakano [1]) Let $\mathcal{T}$ be a proper ring topology on a field K. Then the following conditions are equivalent.

(i) $(K,\mathcal{T})$ is locally bounded.

(ii) $\mathcal{T}$ is induced by a generalized valuation; i.e., there exists a generalized valuation v such that $\mathcal{T} = \mathcal{T}_v$.

Proof: (ii) $\Rightarrow$ (i) Follows from Theorem 5.

(i) $\Rightarrow$ (ii) *Case 1.* $(K,\mathcal{T})$ has an open bounded subring. Let R be an open bounded subring of K. Then $\mathcal{B}(\mathcal{T}) = \{aR : a \in K^\times\}$ is a base of the neighborhoods of zero for $(K,\mathcal{T})$, by Theorem 8 of Chapter 1. Let $x \in K^{\times}$ be arbitrary. Then for every $V \in \mathcal{B}(\mathcal{T})$ there exists $a \in V^{\times}$ satisfying $x \in a^{-1}R$. Indeed, there is $W \in \mathcal{B}(\mathcal{T})$ such that $Wx \subset R$. Taking arbitrary $a \in W \cap V$, $a \neq 0$, we obtain $x \in a^{-1}R$.

Let $V \in \mathcal{B}(\mathcal{T})$ be arbitrary. We say that the subset $\{a_i : i \in I\}$ of V is a *gauge system* of $(K,\mathcal{T})$ if there exist a finite subset $F \subset I$ and integers α_i $(i \in F)$ such that

$$x \in \prod_{i \in F} a_i^{\alpha_i} R \tag{21}$$

The existence of a gauge system can easily be proved. Let Σ^* be the restricted direct product of the v-groups Σ_i $(i \in I)$, where $\Sigma_i = \mathbb{Z}$ with addition $\alpha + \beta = \min(\alpha,\beta)$ and multiplication defined as ordinary addition in $\mathbb{Z}$. Therefore, 0 is the identity. For $x \in K$ set

$$\|x\| = \{\alpha = (\alpha_i) \in \Sigma^* : \alpha_i = 0 \text{ for } i \notin F \text{ and (21) holds}\} \tag{22}$$

We will prove that conditions (GV_1)-(GV_4) are satisfied. Let

$$x \in \prod_{i \in F} a_i^{\alpha_i} R \qquad y \in \prod_{i \in F} a_i^{\beta_i} R$$

where some of the exponents may be equal to zero. Then

$$xy \in \left(\prod_{i \in F} a_i^{\alpha_i} R\right)\left(\prod_{i \in F} a_i^{\beta_i} R\right) \subset \prod_{i \in F} a_i^{\alpha_i + \beta_i} R$$

and

$$x + y \in \prod_{i \in F} a_i^{\gamma_i}(a_k^{\alpha_i - \gamma_i} R + a_i^{\beta_i - \gamma_i} R) \subset \prod_{i \in F} a_i^{\gamma_i} R$$

where $\gamma_i = \min(\alpha_i, \beta_i)$ for $i \in F$. Thus (GV_2) and (GV_3) are proved.

To prove (GV_1) note that $\|0\| = \Sigma^*$. Moreover, the set $\{x \in K : \|x\| = \Sigma^*\}$ is an ideal of K, and since this ideal is different from K, it must be zero. Hence (GV_1) holds. It is obvious that (GV_4) holds, by definition (22). Finally, since $R = U_1$, $\mathcal{T} = \mathcal{T}_{\|\cdot\|}$.

Case 2. $(K,\mathcal{T})$ has no bounded open subring. Let U be an almost-order defining $\mathcal{T}$ and $a \in U^\times$ its 2-addiator (see Chapter 1, Definition 9), i.e., $a(U \pm U) \subset U$. Let $R = \bigcap_{n\in\mathbb{Z}} a^n U$. It follows easily that $U \subset R$ and R is an open subring of K. Moreover, $\bigcap_{n\in\mathbb{N}} a^n U = \{0\}$. Otherwise, if $b \in \bigcap_{n\in\mathbb{N}} a^n U$ and $b \neq 0$, then $a^{-n}U \subset b^{-1}U$ for $n \in \mathbb{N}$, implying that $R \subset b^{-1}U$; therefore, R is bounded, a contradiction with our assumption.

For $x \in R^\times$ let

$$\nu(x) = n \Leftrightarrow x \in a^n U \setminus a^{n+1} U \tag{23}$$

and

$$W(x) = \begin{cases} 0 & x = 0 \\ 2^{\nu(x)} & x \neq 0 \end{cases} \tag{24}$$

By a standard computation it can be shown that $\nu(x + y) \geq \min(\nu(x), \nu(y)) - 1$ and $\nu(xy) \geq \nu(x) + \nu(y)$. This implies that $W(x + y) \leq 2 \max(W(x), W(y))$ and $W(xy) \leq W(x)W(y)$. Finally, we define

$$V(x) = \inf\left\{ \sum_{i=1}^{n} W(x_i) : x = \sum_{i=1}^{n} x_i, \text{ where } x_i \in R \text{ and } n \in \mathbb{N}\right\}$$

It can be verified that V is a norm on R satisfying $V(x) \leq W(x) \leq 2V(x)$ for all $x \in R$. Therefore, if $R = K$, the theorem is proved (see the proof of Theorem 5, Chapter 3, Section 3). Assume thus that $R \neq K$.

We select a gauge system $\{a_i : i \in I\}$ with $a_i \in U$ for every $i \in I$. Then $\nu(a_i) \geq 0$, i.e., $V(a_i) \leq W(a_i) \leq 1$. Let Σ be the restricted direct product of v-groups Σ_i $(i \in I)$, where $\Sigma_i = \mathbb{Z}$ with addition $\alpha + \beta = \min(\alpha,\beta)$ and multiplication defined by ordinary addition, and $\Sigma_0 = \mathbb{R}_+^\times$ with ordinary addition and multiplication. For $x \in K$ put

$$\|x\| = \left\{\bar{\alpha} = (\alpha_i) \in \Sigma : \alpha_i = 0 \text{ for } i \notin F,\ |F| < \infty \text{ and fixed}, \right.$$
$$\left. x = \prod_{i\in F} a_i^{\alpha_i} a_0,\ a_0 \in R,\ V(a_0) < \alpha\right\} \tag{25}$$

where $\bar{\alpha} = (\ldots,\alpha_i,\ldots,\alpha)$. If $y = \Pi_{i\in F} a_i^{\beta_i} b_0$, then $xy = \Pi_{i\in F} a_i^{\alpha_i+\beta_i} a_0 b_0$ and $V(a_0b_0) \le V(a_0)V(b_0) < \alpha\beta$; that is, $\|x\|\cdot\|y\| \subset \|xy\|$, i.e., (GV_3) holds. It can be proved similarly that (GV_2) is true. Moreover, (GV_1) is obvious, by definition (25). Finally, since $U_{(1+1)^{-1}} \subset U$ and $aU \subset U_1$, where 1 denotes the sequence $(0,\ldots,0,\ldots,1)$ of Σ, $U_{(1+1)^{-1}} \subset U \subset a^{-1}U_1$, which implies that $T = T_{\|\cdot\|}$. The proof is finished.

For further results dealing with generalized valuations, see Nakano [2,3].

We shall finish this section with the following:

Example 3 (Nakano [2]) Let $\Sigma^* = \Sigma \times \Sigma_0$, where Σ is the restricted direct product of $|I|$ copies $\Sigma_i = \mathbb{Z}$ $(i \in I)$ with $\alpha + \beta = \min(\alpha,\beta)$ as addition and the ordinary addition as multiplication in Σ_i, and $\Sigma_0 = \mathbb{R}_+^\times$ is taken with its natural operations. Let $K = \mathbb{Q}(X)$, $S = \mathbb{Q}[X]$, where $X = \{X_i : i \in I\}$ is a set of algebraically independent elements X_i over $\mathbb{Q}$. Put $\mathbb{P}$ for the set of all prime, pairwise non-associate elements of S. Suppose that I is infinite. Then $|I| = |\mathbb{P}| = |S|$. We shall use $\mathbb{P}$ as a gauge system with respect to S. Let ε_i be the element of Σ whose ith component is 1 and the others 0. Then every element ε of Σ can be written in the form

$$\varepsilon = \prod_{i\in F} \varepsilon_i^{n_i} \qquad \text{where } n_i \in \mathbb{Z}$$

and $F \subset I$ is a finite subset. For $x = ap_1^{n_1}\cdots p_s^{n_s}$, where $p_i \in \mathbf{P}$, $n_i \in \mathbf{Z}$, $a \in \mathbb{Q}$, define

$$\|x\| = \{\varepsilon = \varepsilon_1^{m_1}\cdots\varepsilon_s^{m_s} : m_i \ge n_i \text{ for } i = 1, 2, \ldots, s\} \tag{26}$$

It is not difficult to see that formula (26) defines a generalized valuation on K with the value group Σ.

Moreover, Σ^* can be realized as a value group of a suitable generalized valuation. To show this, for every $f \in S$, we denote by

$|f|$ the sum of the absolute values of all coefficients of f. Then for every f, g $\in$ S we have $|f \pm g| \le |f| + |g|$ and $|fg| \le |f| \cdot |g|$, i.e., $|\cdot|$ is a norm on S. Let

$$\mathbb{P}^* = \{p_i \in \mathbb{P} : |p_i| \le 1\}$$

Now we define a new generalized valuation $\|\cdot\|^*$ on K, putting for $x \in K$,

$$\|x\|^* = \left\{ \varepsilon = r \prod_{i \in F} \varepsilon_i^{n_i} : F \subset I, \text{ fixed and finite}, \right.$$

$$\left. x = a \prod_{i \in F} p_i^{n_i},\ a \in S,\ |a| < r,\ p_i \in \mathbb{P}^* \right\} \tag{27}$$

It is easy to see that the topology $\mathcal{T}$ induced on K by the generalized valuation $\|\cdot\|^*$ is nondiscrete. Moreover, for every $m \in \mathbb{N}$, $\|mx\|^* = m\|x\|^*$, thus $R^* = \{x \in K : 1 \in \|x\|^*\} \subset S$. If $f \in S$ and $r \in \|f\|^*$, then $|f| < r$. To show that the factor Σ_0 in Σ^* cannot be omitted, we assert that $(K,\mathcal{T})$ has no bounded subring. Suppose to the contrary that R is a bounded subring. Then there exists $t \in K$ such that $tR \subset R^*$. Let $f \in tR$ and let $m \in \mathbb{N}$ satisfy $m|f| \ge 1$. Since $mf \in tR \subset R^* \subset S$, then $1 \in \|mf\|^* = m\|f\|^*$, and $1/m \in \|f\|^*$, i.e., $|f| < 1/m$, a contradiction.

Thus Σ^* is the value group for the valuation $\|\cdot\|^*$.

Let Σ and Σ^* have the same meaning given in Example 3. Suppose that K is a field with a generalized valuation with Σ or Σ^* as the value group. If I is a finite set, we can find an element $a \in K$ satisfying $\varepsilon_i \in \|a\|$ for every $i \in I$. In this case a is a gauge element of the gauge set $R = \{x \in K : \varepsilon \in \|x\|\}$, which implies (by Cohn [1]) that there exists a non-Archimedean norm inducing the same topology on K as the generalized valuation. If Σ^* is the value group and I is finite, there is $a \in K$ such that 1/2 and ε_i $(i \in I)$ lie in $\|a\|^*$. An element a is a gauge element and $\{x \in K : 1 \in \|x\|^*\}$ is the gauge set. It implies that a suitable norm induces the topology on K.

3. LOCALLY BOUNDED RING TOPOLOGIES ON SOME CLASSES OF FIELDS

We have previously seen (Chapter 1, Theorem 9) that if $\mathcal{O}_i$ $(i \in I)$ are almost-orders of a field K and if for every $a_1, \ldots, a_n \in K^\times$, $a_1\mathcal{O}_1 \cap \cdots \cap a_n\mathcal{O}_n \neq \{0\}$, then $\mathcal{O}_I = \bigcap_{i\in I} \mathcal{O}_i$ is an almost-order of K and $\mathcal{O}_I$ defines a locally bounded (proper) ring topology $\mathcal{T}_I$ of K. Then one has $\mathcal{T}_I \geq \mathcal{T}_i$ for every $i \in I$, where $\mathcal{T}_i$ denotes the topology induced by $\mathcal{O}_i$. If, moreover, I is finite, then $\mathcal{T}_I = \sup_r\{\mathcal{T}_i : i \in I\}$. The question arises whether every locally bounded ring topology on a field has an almost-order of the form $\mathcal{O}_I$. This was conjectured by Endo [4]. Shanks and Warner [1,2] have proved that the answer is "yes" in the case $K = \mathbb{Q}$. In this case every locally bounded ring topology $\mathcal{T}$ on $\mathbb{Q}$ has an almost-order of the form $\bigcap_{v\in V} R_v$, where R_v is an almost-order for the topology induced by an absolute value v on $\mathbb{Q}$, i.e., $R_v = \{x \in \mathbb{Q} : v(x) \leq 1\}$ and V is a set of absolute values of $\mathbb{Q}$, which are given by the Ostrowski theorem (compare Chapter 2, Section 2). The idea of the proof lies in identifying the completion of $\mathbb{Q}$ at $\mathcal{T}$ (with an almost-order $\mathcal{O}(\mathcal{T}) = \bigcap \{R_v : v \in V\}$) with the local direct product of completions $\mathbb{Q}_v$ of $\mathbb{Q}$ at v, relative to open subrings $R_v = \{x \in \mathbb{Q} : v(x) \leq 1\}$, v being non-Archimedean. Thus the completion of $\mathbb{Q}$ at $\mathcal{T}$ equals the adèle ring. The crucial step is an application of two results of Mutylin [3].

(a) If $\mathcal{T} \in L_r(\mathbb{Q})$ is a proper locally bounded topology satisfying $\mathcal{T} \not\geq \mathcal{T}_\infty$ (the order topology on $\mathbb{Q}$), then for every $U \in \mathcal{B}(\mathcal{T})$ there exists an $a \in U$, $a \neq 0$, such that $\mathbb{Z}a \subset U$; see Mutylin [3, Lemma 10].

(b) If $\mathcal{T} \in L_r(\mathbb{Q})$ is locally bounded and satisfies $\mathcal{T}_\infty < \mathcal{T} < \mathbf{1}_\mathbb{Q}$, there exists a prime number p such that $\mathcal{T} < \mathcal{T}_I$, where $I = \mathbb{P}_1 \cup \{\infty\}$ and $\mathbf{P}_1$ is the set of all primes $q \neq p$; see Mutylin [3, Cor. 5 to Lemma 17].

Nichols and Cohen [1, Th. 3] extended this result to a class of quotient fields of Dedekind domains satisfying some conditions. This class contains quadratic imaginary fields and fields of algebraic functions of one variable over a finite field. The most general theorem was proved by Weber [3, Satz (3.3)], extending the result of Warner and Shanks to global fields. Moreover, Weber [4, Satz (3.3)] proved that the property formulated in this theorem characterizes global fields.

Another interesting class of fields is the class of algebraic function fields of one variable over a field k. Weber [4, Folgerung (4.4)] proved that for every proper locally bounded ring topology $\mathcal{T}$ on K for which k is bounded, an almost-order defining $\mathcal{T}$ is the intersection of a family of valuation rings of non-Archimedean absolute values of K, trivial on k. Other proofs of this theorem have been given by Cohen [2], Endo [5,6], and Stenzel [1].

It is not known how to describe locally bounded ring topologies on other fields, e.g., on $\mathbb{Q}(X)$, X transcendental over $\mathbb{Q}$ (see Chapter 15).

Let K be a field and S_0 the set of all pairwise nonequivalent absolute values of K (i.e., for every absolute value v on K there exists exactly one $w \in S$ satisfying $v \sim w$). Let S_∞ be the set of all Archimedean $v \in S_0$ such that $v|\mathbb{Q} = |\cdot|$ (the ordinary absolute value). Let $S \subset S_0$. For every $a \in K$ let

$$S(a) = \{v \in S : v(a) < 1\}$$

Let $\mathcal{T}_v$ denote the topology defined by $v \in S_0$. Let $\mathcal{T}(S) \in \mathcal{L}_r(K)$ be the locally bounded ring topology on K defined by the almost-order

$$\mathcal{O}(S) = \bigcap_{v \in S} V_v \qquad \text{where } V_v = \{x \in K : v(x) \le 1\},\ S \subset S_0$$

Put

$$\mathcal{T}^*(S) = \sup\{\mathcal{T} \in \mathcal{L}_f(K) : \mathcal{T} \le \mathcal{T}(S)\} \tag{28}$$

$$\hat{\mathcal{T}}(S) = \sup\{\mathcal{T}_v : v \in S\} \tag{29}$$

We shall prove that every proper locally bounded ring topology on a global field K has the form $\mathcal{T}(S)$ with a suitable $S \subsetneq S_0$. To prove it we shall need some technical results. The following conditions will play a special role in the proof.

Let $S \subset S_0$ be arbitrary.

(A.S) The approximation property for S: For every $S_1 \subset S$ the topologies $\mathcal{T}(S_1)$ and $\mathcal{T}(S \setminus S_1)$ are independent.

It is easy to see that (A.S) is equivalent to the following condition:

(A'.S) $\forall\, S_1 \subset S\ \forall\, a_1 \in K^\times\ \forall\, a_2 \in K^\times\ \exists\, x \in K^\times\ \forall\, v_1 \in S_1\ \forall\, v \in S \setminus S_1$
$v_1(x) \le v_1(a_1)$ and $v(x - 1) \le v(a_2)$

(FC.S) The finiteness condition for S: $\forall\, a \in K^\times,\ |S(a)| < \infty$.

Recall the following result (Chapter 2, Theorem 5):

LEMMA 2 (i) Let K be a global field, S_0 the set of all pairwise nonequivalent absolute values of K, and $S \subsetneqq S_0$ any nonempty subset. Then (A.S) and (FC.S) hold.
(ii) Let K be the quotient field of a Dedekind domain A and S the set of all non-Archimedean absolute values induced by maximal ideals of A. Then (A.S) and (FC.S) hold.

LEMMA 3 Let $S \subset S_0$ satisfy (A.S). Put $c = 1$ if $S_\infty = \emptyset$ and take $c \in K^\times$ satisfying $v(c) \le \min(1, v(1/2))$ for every $v \in S$ if $S_\infty \ne \emptyset$. Take $a \in K^\times$ and put $z = c/a$, $z' = 1/ac$. Then the following conditions are satisfied.

(i) $\forall\, S_1, S_2 \subset S,\ c\mathcal{O}(S_1 \cap S_2) \subset \mathcal{O}(S_1) + a\mathcal{O}(S_2)$.

(ii) $a\mathcal{O}(S(a)) \subset ((z + \mathcal{O}(S)^\times)^{-1}\mathcal{O}(S)$.

(iii) $((z + \mathcal{O}(S))^\times)^{-1} \subset \mathcal{O}(S) \Rightarrow S = S(a)$.

(iv) $S = S(a) \Rightarrow 0 \notin z' + \mathcal{O}(S)$ and $(z' + \mathcal{O}(S))^{-1} \subset \mathcal{O}(S)$.

Proof: (i) Let $x \in \mathcal{O}(S_1 \cap S_2)$. Then, by (A.S), there exists $y \in K^\times$ satisfying $v(y) \le \min(v(a), v(c))$ if $v \in S_2$ and $v(cx - y) \le 1$ if $v \in S_1 \setminus S_2$. If $v \in S_1 \cap S_2$, then $v(y) \le v(c)$, and consequently, $v(cx - y) \le v(c)v(x) + v(y) \le v(c)(v(x) + 1) \le 2v(c) \le 2\min(1, v(1/2)) \le 1$ if v is Archimedean, and $v(cx - y) \le \max(v(c)v(x), v(c)) = v(c)\max(v(x), 1) \le v(c) \le 1$ if v is non-Archimedean, implying that $cx - y \in \mathcal{O}(S_1)$. Moreover, $y \in a\mathcal{O}(S_2)$ and thus $cx = (cx - y) + y \in \mathcal{O}(S_1) + a\mathcal{O}(S_2)$.

(ii) Let $x \in \mathcal{O}(S(a))^{\times}$. There exists $y \in K$ such that for every $v \in S \setminus S(a)$, $0 < v(z + y) \le \min(v(c), v(1/ax))$, and for every $v \in S(a)$, $v(y) \le v(c)$ hold. Consequently, $v(y) \le 1$ if $v \in S(a)$. Let $v \in S \setminus S(a)$. Then $v(z + y) \le v(c)$ and $v(z) = v(c)/v(a) < v(c)$. This implies that

$$v(y) = v((z + y) - z) \le \max(v(z + y), v(z)) \le v(c) \le 1$$

if v is non-Archimedean. If v is Archimedean, one has $v(y) \le 2v(c) \le 2v(1/2) \le 1$, by definition of S_∞. Thus $y \in \mathcal{O}(S)$. To prove that $ax \in ((z + \mathcal{O}(S))^{\times})^{-1}\mathcal{O}(S)$, it suffices to show that $v(ax(z + y)) \le 1$ holds for every $v \in S$. If $v \in S \setminus S(a)$, then this follows from the choice of y. Otherwise, if $v \in S(a)$, then $v(y) \le v(c)$, $v(a) < 1$, and $v(c + ay) \le 1$, i.e., $v(ax(z + y)) = v(x)v(c + ay) \le 1$.

(iii) Since $((z + \mathcal{O}(S))^{\times})^{-1} \subset \mathcal{O}(S)$, (ii) implies that $a\mathcal{O}(S(a)) \subset \mathcal{O}(S)$, i.e., $\mathcal{O}(S(a)) = \mathcal{O}(S)$. Thus $S = S(a)$.

(iv) Let $x \in \mathcal{O}(S)$ and $v \in S$. Since $S = S(a)$, then $v(a) < 1$ and $v(z') > 1$. If $v \notin S_\infty$, then $v(z' + x) = v(z') > 1$. If $v \in S_\infty$, then $v(z' + x) \ge v(z') - v(x) \ge 2/v(a) - 1 > 1$, implying that $z' + x \ne 0$ and $(z' + x)^{-1} \in \mathcal{O}(S)$.

THEOREM 7 Suppose that $S \subset S_0$ and (A.S) is fulfilled. Then

(i) For every S_1, $S_2 \subset S$:

$$\sup\{\mathcal{T}(S_1), \mathcal{T}(S_2)\} = \mathcal{T}(S_1 \cup S_2)$$

$$\inf\{\mathcal{T}(S_1), \mathcal{T}(S_2)\} = \mathcal{T}(S_1 \cap S_2)$$

(ii) $\mathcal{T}(S) \in L_f(K) \Leftrightarrow \exists\, a \in K^{\times}$, $S = S(a)$.

(iii) $\mathcal{T}^*(S) = \sup\{\mathcal{T}(S(a)) : a \in K^{\times}\} = \sup\{\mathcal{T}(S(a)) : a \in \mathcal{O}(S)\}$.

Proof: Let c satisfy assumptions of Lemma 2.

(i) It is clear that $\sup\{\mathcal{T}(S_1), \mathcal{T}(S_2)\} \le \mathcal{T}(S_1 \cup S_2)$. Since $\mathcal{O}(S_1 \cup S_2) = \mathcal{O}(S_1) \cap \mathcal{O}(S_2)$, then $\sup\{\mathcal{T}(S_1), \mathcal{T}(S_2)\} \ge \mathcal{T}(S_1 \cup S_2)$.

Let $\mathcal{T}_0 = \inf\{\mathcal{T}(S_1), \mathcal{T}(S_2)\}$. In view of $\mathcal{T}(S_1 \cap S_2) \le \mathcal{T}_0$, we shall prove an opposite inequality. Let $U \in \mathcal{B}(\mathcal{T}_0)$. Take $V \in \mathcal{B}(\mathcal{T}_0)$ with $V + V \subset U$. Since $\mathcal{T}_0 \le \mathcal{T}(S_i)$ $(i = 1, 2)$ and $\mathcal{T}(S_i)$ is induced by the almost-order $\mathcal{O}(S_i)$, one can find a_1, $a_2 \in K^{\times}$ satisfying $a_1\mathcal{O}(S_1) \subset V$ and $a_1a_2\mathcal{O}(S_2) \subset V$. Consequently, by Lemma 3:

$$a_1 \in \mathcal{O}(S_1 \cap S_2) \subset a_1(\mathcal{O}(S_1) + a_2\mathcal{O}(S_2)) \subset V + V \subset U$$

i.e., U is a neighborhood of zero with respect to $\mathcal{T}(S_1 \cap S_2)$.

(ii) Follows from Lemma 3(ii) and (iii). Indeed, the implication ⇒ follows from Lemma 3(iii), and the converse implication from the same lemma [see Chapter 1, conditions (B_6) and (B_6')].

(iii) $\mathcal{T}(S(a)) \in L_f(K)$ for every $a \in K^\times$, by (ii). Put $\mathcal{T}_1 = \sup\{\mathcal{T}(S(a)) : a \in \mathcal{O}(S)^\times\}$. Since $\mathcal{T}(S(a)) \le \mathcal{T}(S)$ holds for every $a \in K^\times$,

$$\mathcal{T}_1 \le \sup\{\mathcal{T}(S(a)) : a \in K^\times\} \le \mathcal{T}^*(S)$$

We prove the opposite inequality, $\mathcal{T}^*(S) \le \mathcal{T}_1$. Indeed, take an arbitrary $U \in \mathcal{B}(\mathcal{T}^*(S))$. Since $\mathcal{T}^*(S) \in L_f(K)$, one can find $V \in (\mathcal{T}^*(S))$ satisfying $V(1 + v)^{-1} \subset U$. A topology $\mathcal{T}(S) \in L_r(K)$ is locally bounded and induced by the almost-order $\mathcal{O}(S)$. Thus there exists $b \in \mathcal{O}(S)^\times$ with $b\mathcal{O}(S) \subset V$. Put $a = bc \in \mathcal{O}(S)^\times$ and $z = c/a = 1/b$. Lemma 3(ii) implies that

$$a\mathcal{O}(S(a)) \subset ((z + \mathcal{O}(S))^\times)^{-1} \subset (1 + b\mathcal{O}(S))^{-1}b\mathcal{O}(S) \subset \frac{V}{1 + V} \subset U$$

i.e., U is a neighborhood of zero with respect to $\mathcal{T}_1$.

THEOREM 8 If $S \subset S_0$ satisfies (A.S) and (FC.S), then

(a) $\hat{\mathcal{T}}(S) = \mathcal{T}^*(S)$.

(b) The following conditions are equivalent.

(i) S is finite.

(ii) $\mathcal{T}(S) \in L_f(K)$, i.e., $\mathcal{T}(S)$ is a field topology.

(iii) $\mathcal{T}(S) = \hat{\mathcal{T}}(S)$.

(iv) $\hat{\mathcal{T}}(S)$ is locally bounded.

(c) $\mathcal{T}_v$, $v \in S$, are the only minimal topologies weaker than $\mathcal{T}(S)$, i.e., the only V-topologies weaker than $\mathcal{T}(S)$.

Proof: (a) $\hat{\mathcal{T}}(S) \le \mathcal{T}^*(S)$, by definition. Since every $a \in K^\times$, the set S(a) is finite, hence $\mathcal{T}(S(a)) = \hat{\mathcal{T}}(S(a)) \le \hat{\mathcal{T}}(S)$. Thus also $\mathcal{T}^*(S) \le \hat{\mathcal{T}}(S)$, by Theorem 7(iii).

(b) (i) ⇒ (ii) since for finite S, $\mathcal{T}(S) = \sup\{\mathcal{T}_v : v \in S\}$. (ii) ⇒ (iii) since $\mathcal{O}(S) \subset V_v$ for every $v \in S$; thus $\mathcal{T}_v \le \mathcal{T}(S)$, i.e.,

$\sup\{\mathcal{T}_v : v \in S\} = \hat{\mathcal{T}}(S) \le \mathcal{T}(S)$. But $\mathcal{T}(S) = \mathcal{T}^*(S)$, by assumption (ii). Since $\hat{\mathcal{T}}(S) = \mathcal{T}^*(S)$ by (a), then (iii) holds. (iii) $\Rightarrow$ (iv) is clear, by the assumption.

(iv) $\Rightarrow$ (i) Since $\hat{\mathcal{T}}(S)$ is locally bounded, S must be finite. Indeed, one can repeat arguments given in Chapter 1, Example 12.

We shall prove that if S is an infinite set of absolute values satisfying (A.S), then $\hat{\mathcal{T}}(S) = \sup\{\mathcal{T}_v : v \in S\}$ is not locally bounded. Let $U_v(b) = \{x \in K : v(x) < b\}$, $b \in \mathbb{R}_+^\times$, $v \in S$. Let $\mathcal{T} = \hat{\mathcal{T}}(S)$ and $U \in \mathcal{B}(\mathcal{T})$ be arbitrary. Then U can be written in the form $U = U_{v_1}(n_1) \cap \cdots \cap U_{v_s}(n_s)$. Let $v_0 \in S \setminus \{v_1,\ldots,v_s\}$ and put $V = U_{v_0}(1)$. Take an arbitrary $W \in \mathcal{B}(\mathcal{T})$. Thus $W = U_{w_1}(m_1) \cap \cdots \cap U_{w_t}(m_t)$. Since (A.S) is satisfied, one can find $q_1, \ldots, q_t, p_1, \ldots, p_s, a, p_0 \in K$ satisfying

$$w_j(q_j) < m_j, \quad w_j(q_i) \le 1 \ \text{ if } i \ne j; \quad v_0(q_j) \le 1, \ j = 1, \ldots, t$$

$$v_0(p_0) < 1, \quad v_0(p_i) \le 1, \quad v_i(p_0) \ge 1, \ i = 1, \ldots, s$$

$$v(p_0) \le 1 \ \text{ if } v \in \{v_1,\ldots,v_s,w_1,\ldots,w_t\}$$

$$v_i(p_i) < n_i; \quad v_i(p_j) \le 1 \ \text{ if } i \ne j; \quad w_j(a) \le 1, \quad i = 1, \ldots, t$$

Put $x = a^{-1}q_1\cdots q_t$, $y = p_0^{-N}p_1\cdots p_s$. Then $x \in W$, since $w_j(x) = w_j(a^{-1})\prod_{i=1}^t w_j(q_i) < w_j(q_j) < m_j$. Moreover, $y \in U$, since for every i,

$$v_i(p_0^{-N}p_1\cdots p_s) \le v_i(p_i) < n_i \qquad (i = 1, \ldots, s)$$

But $v_0(xy) = v_0(p_0^{-1})^N v_0(p_1\cdots p_s q_1\cdots q_t) > 1$ if N is large enough, since $v_0(p_1\cdots p_s q_1\cdots q_t)$ is a constant independent of N. Finally, this implies that $v_0(xy) > 1$, i.e., $xy \notin V$. This proves that $UW \not\subseteq V$, a contradiction!

Consequently, S is finite.

(c) Let $\mathcal{T} \le \mathcal{T}(S)$ be a minimal topology. Since $\mathcal{T}$ is a field topology, then $\mathcal{T} \le \hat{\mathcal{T}}(S)$. Theorem 12 of Chapter 7 implies the existence of a subset S_1 of S with $\sup\{\mathcal{T}_v : v \in S_1\} \le \mathcal{T}$. Minimality of $\mathcal{T}$ implies that $S_1 = \{v\}$, i.e., $\mathcal{T} = \mathcal{T}_v$ with a suitable $v \in S$.

LEMMA 4 Let v be an absolute value on a field K, $\mathcal{T} \in \mathcal{L}_r(K)$ a proper topology satisfying $\mathcal{T}_v \le \mathcal{T}$, and $U \subset K$ a bounded and multiplicatively closed (i.e., $UU \subset U$) neighborhood of zero in $(K,\mathcal{T})$. Then $U \subset V_v$.

Proof: There exists $a \in K^\times$ such that $a^{-1}U \subset V_v$. Thus for $u \in U$, $v(u) \le v(a)$. Since $u^n \in U$ for every $n \in \mathbb{N}$, then $v(u)^n \le v(a)$ holds for every $n \in \mathbb{N}$, implying that $v(u) \le 1$.

THEOREM 9 Let $S' \subset S \subset S_0$ and let (A.S) hold. Let $\mathcal{T}_v$ $(v \in S)$ be the only V-topologies weaker than $\mathcal{T}(S)$. Suppose that $\mathcal{T} \in \mathcal{L}_r(K)$ satisfies $\mathcal{T}(S') \le \mathcal{T} \le \mathcal{T}(S)$. Then there exists a subset S_1 of S such that $\mathcal{T} = \mathcal{T}(S_1)$.

Proof: If $\mathcal{T} = \mathbb{O}_K$, then $\mathcal{T} = \mathcal{T}(\emptyset)$. Let $\mathcal{T} \ne \mathbb{O}_K$. We shall apply Lemma 3 of Chapter 7, taking

$$T = \{\mathcal{T}' \in \mathcal{L}_r(K) : \mathbb{O}_K \ne \mathcal{T}' \le \mathcal{T}(S),\ \mathcal{T}' \text{ locally bounded}\}$$

$$T_0 = \{\mathcal{T}(S'') : \emptyset \ne S'' \subset S\} \qquad \mathcal{T}_1 = \sup\{\mathcal{T}'' \in T_0 : \mathcal{T}'' \le \mathcal{T}\}$$

We prove that $\mathcal{T}_1 = \mathcal{T}(S_1)$ with $S_1 = \{v \in S : \mathcal{T}_v \le \mathcal{T}\}$. First note that $\mathcal{T}(S_1) \le \mathcal{T}$, i.e., $\mathcal{O}(S_1)$ is a neighborhood of zero in $(K,\mathcal{T})$. Indeed, since $\mathcal{T}(S') \le \mathcal{T}$, then $\mathcal{O}(S')$ is a neighborhood of zero in $(K,\mathcal{T})$. Let U be a bounded neighborhood of zero in $(K,\mathcal{T})$ satisfying $UU \subset U$, contained in $\mathcal{O}(S')$. Lemma 4 implies that $U \subset \bigcap_{v \in S_1 \setminus S'} V_v$. Thus $U \subset \bigcap_{v \in S_1 \setminus S'} V_v \cap \mathcal{O}(S') = \mathcal{O}(S_1)$, and finally, $\mathcal{T}(S_1) \le \mathcal{T}$. This implies that $\mathcal{T}(S_1) \le \mathcal{T}_1 \le \mathcal{T}$. Note that $\mathcal{T}'' \le \mathcal{T}(S_1)$ holds for every $\mathcal{T}'' \in T_0$ satisfying $\mathcal{T}'' \le \mathcal{T}$. Indeed, $\mathcal{T}''$ has the form $\mathcal{T}'' = \mathcal{T}(S'')$ and is defined by an almost-order $\mathcal{O}(S'') = \bigcap_{v \in S''} V_v$ with $\mathcal{T}_v \le \mathcal{T}$. Thus $\mathcal{O}(S_1) = \bigcap_{v \in S_1} V_v \subset \mathcal{O}(S'')$. Consequently, $\mathcal{T}'' \le \mathcal{T}(S_1)$, and finally, $\mathcal{T}_1 = \mathcal{T}(S_1)$.

Now we prove that the conditions of Lemma 3 of Chapter 7 are satisfied.

To prove condition (i) of this lemma, note that for every $\mathcal{T}' \in T$ there exists a minimal locally bounded topology $\mathcal{T}'' \le \mathcal{T}'$. The topology $\mathcal{T}''$ is a V-topology (by Theorem 8 of Chapter 5), i.e., $\mathcal{T}'' = \mathcal{T}_v$ for some $v \in S$, by the assumptions of the theorem. To prove (ii), note that $\mathcal{T}(S) = \sup(\mathcal{T}(S_1), \mathcal{T}(S \setminus S_1)) \ge \mathcal{T}$ and $\mathcal{T}_1 = \mathcal{T}(S_1)$ and $\mathcal{T}_2 = \mathcal{T}(S \setminus S_1)$ are independent by (A.S). Thus we obtain (by Lemma 3 of

Chapter 7) that either $T = T_1$ or there exists a nontrivial $T_3 \in L_r(K)$ satisfying $T = \sup(T_1,T_3)$ and $T_3 \nleq T$, T_1 and T_3 being independent. The last condition is impossible, since T is locally bounded, but $\sup(T_1,T_3)$ is not locally bounded, T_3 being unbounded. Thus $T = T_1 = T(S_1)$, proving the theorem.

LEMMA 5 Let R be a ring with unity and $B_1 \subset R$. Define inductively $B_{n+1} = (B_n \pm B_n) \cup B_1B_n$, and $V_n \subset R$ with $V_{n+1} \pm V_{n+1} \subset V_n$, $V_1V_{n+1} \subset V_n$, $B_nV_{n+1} \subset V_n$ for every $n \in \mathbb{N}$. Then for every $r \in \mathbb{N}$ and every $a \in V_4$,

$$\left(\bigcup_{n=0}^{\infty} \sum_{s=0}^{n} B_r a^s\right) V_4^r \subset V_1$$

is satisfied (here $V_4^{r+1} = V_4^rV_4$ for every $r \in \mathbb{N}$).

Proof: Clearly, $0 \in V_{n+1} \subset V_n$, since $V_{n+1} - V_{n+1} \subset V_n$. First we prove by induction on m that

$$B_mV_4^m \subset V_2 \tag{30}$$

$m = 1$: $B_1V_4 \subset V_3 \subset V_2$. Let m be arbitrary. Suppose that $B_mV_4^m \subset V_2$ holds and take $x \in V_4^{m+1}$, $a \in B_{m+1}$. Then x and a have the form $x = yz$ with $z \in V_4$, $y \in V_4^m$ and either $a = b \pm c$ or $a = db$ with $b \in B_m$, $d \in B_1$. Thus, in the first case, if $a = b \pm c$, then one has

$$ax = (by)z \pm (cy)z \in V_2V_4 \pm V_3V_4 \subset V_3 \pm V_3 \subset V_2$$

and in the second case, if $a = bd$, then

$$ax = d(by)z \in B_1V_2V_4 \subset B_1V_3 \subset V_2$$

Let $r \in \mathbb{N}$, $a \in V_4$. Define inductively $M_0 = B_r$ and $M_n = M_{n-1}V_4 + M_0$ for every $n \in \mathbb{N}$. Then $\Sigma_{s=0}^n B_ra^s \subset M_n$, by an easy induction on n. Indeed, if $n = 0$, then $B_r \subset M_0$, by definition. If $\Sigma_{s=0}^n B_ra^s \subset M_n$, then

$$\sum_{s=0}^{n+1} B_ra^s = B_r + a\sum_{k=0}^{n} B_ra^k \subset M_0 + aM_n \subset M_0 + V_4M_n = M_{n+1}$$

Finally, we prove by induction that $M_nV_4^r \subset V_1$ holds for every nonnegative integer n. If n = 0, it follows from (30) that $M_0V_4^r = B_rV_4^r \subset V_2 \subset V_1$. Assume that $M_nV_4^r \subset V_1$. Since $M_0V_4^r \subset V_2$ and $M_{n+1} = M_nV_4 + M_0$, then $M_{n+1}V_4^r \subset M_nV_4^rV_4 + M_0V_4^r \subset V_1V_4 + V_2 \subset V_3 + V_2 \subset V_1$. Lemma 5 is proved.

LEMMA 6 Let K be a field, S a set of absolute values of K, and v_0 a fixed nontrivial absolute value of K. Let $a \in K$ satisfy $v_0(a) > 1$ and take $M \subset K$ satisfying $\mathcal{O}(S) \subset M + a\mathcal{O}(S)$. Assume that $c_0 = \sup v_0(M) < \infty$ and take $c > c_0(v_0(a) - 1)^{-1}$. Then for every $x \in \mathcal{O}(S)$ there exist $n \in \mathbb{N}$, $t_0, \ldots, t_{n-1} \in M$ and $t_n \in \mathcal{O}(S)$ such that

$$x = \sum_{s=0}^{n} t_s a^s \qquad \text{and} \qquad v_0(t_n) < c \tag{31}$$

Proof: Let $x \in \mathcal{O}(S)$. We shall construct sequences (x_n) and (r_n) such that conditions

$$\begin{aligned} &x_s \in \mathcal{O}(S) \qquad x = x_s a^s + \sum_{k=0}^{s-1} r_k a^k \qquad 0 \le s \le n \\ &x_{s+1} = \frac{x_s - r_s}{a} \qquad r_s \in M \qquad 0 \le s \le n \end{aligned} \tag{A_n}$$

hold for every nonnegative integer n. Finally, we shall prove that $v_0(x_n) < c$ holds for every $n \in \mathbb{N}$.

Put $x_0 = x$. Then (A_0) holds. Suppose that $x_0, \ldots, x_n$ and $r_0, \ldots, r_{n-1}$ satisfy (A_n). Since $x_n \in \mathcal{O}(S) \subset M + a\mathcal{O}(S)$, there exists $r_n \in M$ satisfying $x_n \in r_n + a\mathcal{O}(S)$, i.e., defining $x_{n+1} = (x_n - r_n)/a$, $x_{n+1} \in \mathcal{O}(S)$. Thus

$$\begin{aligned} x &= x_n a^n + \sum_{i=0}^{n-1} r_i a^i = (x_{n+1}a + r_n)a^n + \sum_{i=0}^{n-1} r_i a^i \\ &= x_{n+1}a^{n+1} + \sum_{i=0}^{n} r_i a^i \end{aligned}$$

and consequently, $x_0, \ldots, x_{n+1}$ and $r_0, \ldots, r_n$ satisfy (A_{n+1}).

Define recursively a sequence (b_n), putting $b_0 = v_0(x_0)$, $b_{n+1} = (b_n + c_0)/v_0(a)$. Inequalities

$$v_0(x_{n+1}) \le \frac{v_0(x_n) + c_0}{v_0(a)} \quad \text{and} \quad v_0(x_n) \le b_n$$

follow by an easy induction on $n \ge 0$. We shall prove that the sequence (b_n) is convergent and $\lim b_n < c$, showing that b_n is decreasing and bounded. Indeed, $b_n \ge b_{n+1}$ if and only if $b_n \ge c_0/[v_0(a) - 1]$ for every $n \ge 0$. If $b_n \ge b_{n+1}$, then

$$b_n - b_{n+1} = b_n - \frac{b_n + c_0}{v_0(a)} = \frac{b_n(v_0(a) - 1) - c_0}{v_0(a)} \ge 0$$

implies that $b_n \ge c_0(v_0(a) - 1)^{-1}$ since $v_0(a) > 1$. The converse implication can be proved similarly. The recursive formula for b_n implies that $b = \lim b_n$ satisfies

$$b = \frac{b + c_0}{v_0(a)} \qquad \text{i.e., } b = \frac{c_0}{v_0(a) - 1} < c$$

Now we formulate and prove the main result of this section.

THEOREM 10 (Weber [3, Satz (3.3)]) Let K be a global field, S_0 the set of all pairwise nonequivalent absolute values of K, and $\mathcal{T} \in L_r(K)$, $\mathcal{T} \ne \mathbb{1}_K$. Then the following conditions are equivalent.

(i) $(K,\mathcal{T})$ is locally bounded.

(ii) There exists a proper subset $S \subset S_0$ satisfying $\mathcal{T} = \mathcal{T}(S)$.

Proof: (ii) ⇒ (i) is obvious.

(i) ⇒ (ii) In view of Theorem 9 it suffices to prove that there exists a proper subset S of S_0 for which $\mathcal{T} \le \mathcal{T}(S)$ holds. Since $\mathcal{T} \ne \mathbb{1}_K$, there exists $v_0 \in S_0$, such that $\mathcal{T}_{v_0} \nleq \mathcal{T}$. Otherwise, $\mathcal{T}(S_0) \le \mathcal{T}$, implying that $\mathcal{O}(S_0)$ is a neighborhood of zero in $(K,\mathcal{T})$. Since $\mathcal{O}(S_0)$ is an almost-order of $(K,\mathcal{T})$, then $\mathcal{O}(S_0)\mathcal{O}(S_0)^{\times})^{-1} = K$. But by the product formula (Chapter 2, Theorem 12) holding for K, $\prod_{v\in S_0} v(a) = 1$ for every $a \in K^{\times}$. Let $a \in \mathcal{O}(S_0)$. Then, in view of the product formula, $v(a) = 1$ holds for every $v \in S_0$. Since every $x \in K^{\times}$ has the form $x = a_1 a_2^{-1}$ ($a_1, a_2 \in \mathcal{O}(S_0)^{\times}$), then $v(x) = 1$ holds for every $x \in K^{\times}$ and every $v \in S_0$; i.e., every $v \in S_0$ is trivial on K, a contradiction! Thus there exists $v_0 \in S_0$ satisfying $T_{v_0} \nleq T$.

If char(K) = 0, then put $S'_{\infty} = S_{\infty}$. If char(K) = p, then there exists a finite field k and an element $X \in K$, transcendental over k, satisfying $[K:k(X)] < \infty$. Then put $S'_{\infty} = \{v \in S_0 : v|k(X) = v_{\infty}\}$ (for definition of v_{∞}, see Chapter 2, Section 2). Moreover, let $S_1 = S_0 \setminus S'_{\infty}$ and $S = S_0 \setminus \{v_0\}$. $\mathcal{O}(S_1)$, being a Noetherian ring, is finitely generated, say, by elements $a_1, \ldots, a_m$. Since (A.S_1) holds, there exists $a_0 \in K^{\times}$ satisfying $v(a_0^{-1}) \leq 1$ for every $v \in S_1$ and $v_0(a_0^{-1}) < 1$ if $v_0 \in S_1$. Then $\mathcal{O}(S)$ is contained in the subring of K generated by $B_1 = \{a_0, a_1, \ldots, a_m\}$. Define B_n recursively as in Lemma 5. Then for every finite subset F of $\mathcal{O}(S)$ there exists $r \in \mathbb{N}$ with $F \subset B_r$. Since (K,T) is locally bounded, there exists a base of bounded neighborhoods V_n of zero satisfying the conditions of Lemma 5. Since $T_{v_0} \nleq T$, there exists $a \in V_4$ such that $v_0(a) > 1$. It is possible to find a finite subset $M \subset \mathcal{O}(S)$ and a real number $c > 0$ such that the assumptions of Lemma 6 are fulfilled. Let $N = \{x \in \mathcal{O}(S) : v_0(x) \leq c\}$. The set N is finite, by the product formula. Consequently, $M \cup N$, being finite, is contained in some B_r. Lemma 6 implies that $\mathcal{O}(S) \subset \bigcup_{n=0}^{\infty} \Sigma_{s=0}^{n} B_r a^s$, and consequently $\mathcal{O}(S) a^r \subset V_1$. Finally, this implies that $T \leq T(S)$. To finish the proof we apply Theorem 9.

COROLLARY 1 Every locally bounded field topology T of a global field has the form

$$T = \sup\{T_1, \ldots, T_n\}$$

where the T_i are induced by nontrivial absolute values.

COROLLARY 2 Every nontrivial norm p of a global field K is equivalent to a norm of the form $\sup(v_1, \ldots, v_n)$, where v_i are nontrivial absolute values of K.

Proof: Since every nontrivial norm induces a proper locally bounded field topology, it is sufficient to apply Corollary 1.

Remark Corollary 2, due to Mahler [2-4] and Kiyek [6], was proved directly in Chapter 3.

THEOREM 11 Let K be a global field and $\mathcal{T} \in \mathcal{L}_r(K)$ a proper locally bounded topology. Then the following conditions are equivalent.

(i) $\mathcal{T} \in \mathcal{L}_f(K)$.

(ii) $\mathcal{T}$ is induced by a nontrivial norm.

(iii) $N_0(\mathcal{T}) \neq \{0\}$.

(iv) There exist nontrivial absolute values $w_1, \ldots, w_n$ on K satisfying

$$\mathcal{T} = \sup\{\mathcal{T}_{w_1}, \ldots, \mathcal{T}_{w_n}\}$$

Proof: It is easy to see that the following implications hold: (iv) ⇒ (iii) ⇔ (ii) ⇒ (i). Indeed, (ii) ⇔ (iii) by Cohn's theorem (Chapter 3, Theorem 4). (ii) ⇒ (i) is clear, since every norm topology is a field topology (Chapter 3, Theorem 1). (iv) ⇒ (iii) since every element $a \in K^\times$ satisfying $\max(w_1(a), \ldots, w_n(a)) < 1$ lies in $N_0(\mathcal{T})$. (i) ⇒ (iv) by Corollary 1.

We finish this section with some remarks on locally bounded topologies on finite extensions of the rational function field k(X) for which k is bounded.

THEOREM 12 (Weber [4, Folgerung (4.4)]) Let K be an algebraic function field of one variable over a field k. If $\mathcal{T}$ is a proper locally bounded ring topology on K such that k is a bounded set in $(K, \mathcal{T})$, there exists a set S of nontrivial (non-Archimedean and discrete on k) absolute values on K such that $\mathcal{O}(S)$ is an almost-order inducing $\mathcal{T}$, i.e., $\mathcal{T} = \mathcal{T}(S)$.

To describe locally bounded ring topologies on algebraic function fields of one variable, we need a generalization of Theorem 10.

LEMMA 7 Suppose that $\mathcal{T} \in \mathcal{L}_r(K)$ is locally bounded. Let S_0 be the set of all pairwise nonequivalent absolute values of K. Let $v_0 \in S_0$ satisfy $\mathcal{T}_{v_0} \nleq \mathcal{T}$. Put $S = S_0 \setminus \{v_0\}$ and suppose that

(i) $\mathcal{O}(S)(\mathcal{O}(S)^\times)^{-1} = K$.

(ii) For every $c > 0$ the set $M(v_0,c) = \{x \in \mathcal{O}(S) : v_0(x) \le c\}$ is bounded in $(K,\mathcal{T})$.
(iii) For every $a \in \mathcal{O}(S)^{\times}$ there exists $c > 0$ satisfying $\mathcal{O}(S) \subset M(v_0,c) + a\mathcal{O}(S)$.
Then $\mathcal{T} \le \mathcal{T}(S)$.

Proof: Let U be an almost-order defining $\mathcal{T}$, and let $z \in U^{\times}$ satisfy $z(U \pm U) \subset U$. Since $\mathcal{T}_{v_0} \nleq \mathcal{T}$, there exists $a \in zU$ satisfying $v_0(a) > 2$. There exist a_1, $a_2 \in \mathcal{O}(S)$ and a real number $c_0 > 0$ such that $a = a_1a_2^{-1}$ and $\mathcal{O}(S) \subset M(v_0,c_0) + a_1\mathcal{O}(S) \subset M(v_0,c_0) + a\mathcal{O}(S)$. Since $M(v_0,c_0)$ is bounded in $(K,\mathcal{T})$, there exists $b \in K^{\times}$ satisfying $M(v_0,c_0) \subset bU$. Lemma 6 implies that $\mathcal{O}(S) \subset \bigcup_{n=1}^{\infty} \Sigma_{s=0}^{n} bUa^s$. Since $a(U \pm U) \subset U$, then, iterating, we obtain $\mathcal{O}(S) \subset (b/a)U$; i.e., U is a neighborhood of zero in $(K,\mathcal{T}(S))$.

LEMMA 8 Suppose that $\mathcal{O}(S_0)(\mathcal{O}(S_0)^{\times})^{-1} \ne K$ and let $\mathcal{T} \in \mathcal{L}_r(K)$, $\mathcal{T} \ne \mathbb{1}_K$ be locally bounded. Then there exists $v_0 \in S_0$ such that $\mathcal{T}_{v_0} \nleq \mathcal{T}$.

Proof: Assume that for every $v \in S_0$, $\mathcal{T}_v \le \mathcal{T}$. Let U be an almost-order inducing $\mathcal{T}$. Then $U \subset V_v$, by Lemma 4, implying that $U \subset \mathcal{O}(S_0)$ and $K = U(U^{\times})^{-1} = \mathcal{O}(S_0)(\mathcal{O}(S_0)^{\times})^{-1}$, a contradiction!

Proof of Theorem 12: Let S_0 denote the set of all pairwise nonequivalent absolute values of K trivial on k. It can be shown that $\mathcal{O}(S_0)(\mathcal{O}(S_0)^{\times})^{-1} \ne K$. Let $v_0 \in S_0$ be such that $\mathcal{T}_{v_0} \nleq \mathcal{T}$ (Lemma 8). Put $S = S_0 \setminus \{v_0\}$. Then (A.S) and (FC.S) hold. Consequently, $K = \mathcal{O}(S)(\mathcal{O}(S)^{\times})^{-1}$. Since $a\mathcal{O}(S)$ is a vector space over k and the quotient vector space $\mathcal{O}(S)/a\mathcal{O}(S)$ is finite-dimensional over k (Duering [2, p. 23]), there exist $a_1, \ldots, a_n \in \mathcal{O}(S)$ satisfying

$$\mathcal{O}(S) = \sum_{s=1}^{n} ka_s + a\mathcal{O}(S)$$

Put $c = \max(v_0(a_1),\ldots,v_0(a_n))$. Then $\mathcal{O}(S) \subset M(v_0,c) + a\mathcal{O}(S)$. Since $M(v_0,c)$ is a finite-dimensional vector space over k (Duering [2]), then $M(v_0,c)$ is bounded with respect to every ring topology on K for which k is bounded. Now applying Lemmas 7 and 8 and Theorem 9, the result follows.

4. TOPOLOGICAL CHARACTERIZATIONS OF $\mathbb{R}$ AND $\mathbb{C}$

It is well known (Chapter 6, Theorem 3) that the real and the complex number fields are the only locally compact and connected topological fields (Pontrjagin [1]). Moreover, $\mathbb{R}$ and $\mathbb{C}$ are the only normed field extensions of the reals, by the Gelfand-Mazur theorem.

In this section we present another characterization of these fields. For other characterizations, see Chapter 10.

We shall prove the following.

THEOREM 13 (Mutylin [3, Th. 5]) Real and complex number fields are the only locally bounded field extensions of the reals.

Note that $\mathbb{R}$ and $\mathbb{C}$ are the only finite-dimensional field extensions of $\mathbb{R}$, since $\mathbb{C}$ is algebraically closed. Furthermore, since $\mathbb{R}$ is a complete absolute-valued field, the only field topology on $\mathbb{C}$ extending the topology on $\mathbb{R}$ is the Cartesian product topology, i.e., the topology induced by the absolute value.

Proof: Suppose to the contrary that $(K,\mathcal{T})$ is an infinite-dimensional locally bounded field extension of the reals. Set $L = K$ if $i = \sqrt{-1} \in K$ and $L = K(i)$ otherwise. In the last case we equip L with the Cartesian product topology, being again a locally bounded field topology by Theorem 17 of Chapter 2. The definition of L implies that $\mathbb{C}$ is a topological subfield of $(L,\mathcal{T})$. Let R be an almost-order inducing topology $\mathcal{T}$ on L. Since $\mathbb{R} \subset L$ topologically, L contains nonzero topological nilpotents. Thus the set N_0 of topological nilpotents in $(L,\mathcal{T})$ is open (see Chapter 3, Lemma 3). Take $U \in \mathcal{B}(\mathcal{T})$ such that $U + U \subset R$. There exists nonzero $h \in U \cap N_0$ satisfying $hR \subset U$. It follows that $h(R \pm R) \subset R$; i.e., h is a topological nilpotent 2-addiator for $(L,\mathcal{T})$. We can suppose that $h \in N_0 \cap \mathbb{R}$, that is, $0 < h < 1$. Thus every element of hR is a topological nilpotent, as R is bounded. Let $x \in L^\times$. Then there exist $U, V \in \mathcal{B}(\mathcal{T})$ satisfying $xU \subset hR$ and $x^{-1}V \subset hR$. Thus one can find positive constants $\alpha, \beta \in \mathbb{R}$ with $\alpha x, (\beta x)^{-1} \in N_0$; i.e., αx is a topological nilpotent but βx is not a topological nilpotent. This allows us to define

$$p(x) = (\sup\{\alpha \in \mathbb{R}_+^\times : \alpha x \in N_0\})^{-1} \tag{32}$$

It is clear that $p(0) = 0$ and $p(x) \neq 0$ for $x \in L^\times$; moreover, $p(x)$ is finite. Take $x \in \mathbb{C}^\times$. Then $\alpha x \in N_0 \cap \mathbb{C} = \{z \in \mathbb{C} : |z| < 1\}$, $\alpha < 1/|x|$. It follows that $p(x) = (\sup\{\alpha \in \mathbb{R}_+^\times : \alpha < |x|^{-1}\})^{-1} = |x|$.

$$p(x) = |x| \text{ for } x \in \mathbb{C} \quad \text{and} \quad p(xy) \leq p(x)p(y) \quad \text{if } x,y \in L \tag{33}$$

Indeed, the first statement of (33) was shown earlier. To prove the second, take arbitrary $x, y \in L^\times$. Let $A = \{\alpha \in \mathbb{R}_+^\times ; \alpha x \in N_0\}$, $B = \{\beta \in \mathbb{R}_+^\times : \beta y \in N_0\}$, $C = \{\gamma \in \mathbb{R}_+^\times : \gamma xy \in N_0\}$. Then $AB \subset C$, since $N_0 N_0 \subset N_0$. Thus

$$0 < \sup A \cdot \sup B = \sup AB \leq \sup C$$

and finally, $p(xy) = (\sup C)^{-1} \leq (\sup A \cdot \sup B)^{-1} = p(x)p(y)$. Hence we get, applying (33), that

$$p(xy) = |x|p(y) \qquad \text{for } x \in \mathbb{C} \text{ and } y \in L \tag{34}$$

It follows immediately that

$$p(x^N) = p(x)^N \qquad \text{for every } x \in L \text{ and } N \in \mathbb{N} \tag{35}$$

Let $B = \{\beta \in \mathbb{R}_+^\times : \beta x^N \in N_0\}$ and $A = \{\gamma \in \mathbb{R}_+^\times : \gamma x \in N_0\}$. If $x^N \in N_0$, then $x \in N_0$ (compare the proof of Lemma 3 of Chapter 3). Thus

$$B = \{\alpha^N \in \mathbb{R}_+^\times : (\alpha x)^N \in N_0\} \subset \{\gamma^N \in \mathbb{R}_+^\times : \gamma x \in N_0\} \subset A^N$$

(here $A^N = \{\alpha^N : \alpha \in A\}$); hence

$$0 < \sup B \leq \sup A^N = (\sup A)^N$$

and

$$p(x)^N = (\sup A)^{-N} \leq (\sup B)^{-1} = p(x^N)$$

Furthermore, we have

$$p(x + y) \leq p(x) + p(y) \qquad \text{for every } x, y \in L \tag{36}$$

It is sufficient to show that (36) holds for $x, y \in L$ satisfying $p(x) < 1$ and $p(y) < 1$. Indeed, if $x, y \in L$ are arbitrary, then $p(\alpha x) < 1$ and $p(\alpha y) < 1$ hold for every $\alpha \in \mathbb{R}_+^\times$ with $\alpha < \min(p(x)^{-1}, p(y)^{-1})$.

Hence suppose that $p(x) < 1$ and $p(y) < 1$. There exists $N_1 \in \mathbb{N}$ such that $x^n, y^n \in R$ for every $n \geq N_1$. Furthermore, there exists $N_2 \in \mathbb{N}$ such that for every k, $0 \leq k < N_1$, we have $x^k y^n, x^n y^k \in R$ for all $n \geq N_2$. Take $n \geq N_2$. By a simple induction we obtain $(x + y)^n \in h^{-n}R$, since $R + R \subset h^{-1}R$. It means that $h^{2n}(x + y)^n \in h^n R$; that is, $(h^2(x + y))^n$ tends to zero in $(L,\mathcal{T})$. It proves that $p(h^2(x + y)) < 1$ and $p(x + y) < 1/h^2$, by (33). We have

$$\begin{aligned} p(x + y) &= 2 \max(p(x),p(y)) \\ &\quad \times p\left(\frac{x}{2 \max(p(x),p(y))} + \frac{y}{2 \max(p(x),p(y))}\right) \\ &< \frac{2}{h^2} \max(p(x),p(y)) \end{aligned}$$

Setting $M = 2h^{-2}$ and $m = 2^n - 1$, we obtain, applying (35), that

$$\begin{aligned} p(x + y)^m &= p((x + y)^m) \\ &= p\left(\sum_{k=0}^{m} \binom{m}{k} x^k y^{m-k}\right) \leq M^n \max\left\{p(\binom{m}{k} x^k y^{m-k}) \ : \ 0 \leq k \leq m\right\} \\ &\leq M^n \max\left\{\binom{m}{k} p(x)^k p(y)^{m-k} \ : \ 0 \leq k \leq m\right\} \\ &\leq M^n \sum_{k=0}^{m} \binom{m}{k} p(x)^k p(y)^{m-k} \\ &= M^n (p(x) + p(y))^m \end{aligned}$$

Thus $p(x + y) \leq M^{n/m}(p(x) + p(y))$. Since n/m tends to zero for n tending to infinity, (36) holds.

It follows from (33)-(36) that L is a normed overfield of $\mathbb{C}$. The Gelfand-Mazur theorem implies that L is finite-dimensional over the reals, contrary to our assumption. This proves Theorem 13.

Remark The Gelfand-Mazur theorem is not necessary for the proof. Since p is power-multiplicative [see (35)], $p = \sup\{w_1,\ldots,w_N\}$, w_i being nonequivalent absolute values of L (by Theorem 8 of Chapter 3). Since $p(x) = |x|_\infty$ for every $x \in \mathbb{C}$, then $w_i \sim |\cdot|_\infty$ for $i = 1, \ldots, N$, and thus every w_i is Archimedean. Consequently, there exists an embedding of $(L,\mathcal{T}_{w_i})$ into $(\mathbb{C},\mathcal{T}_\infty)$ by Ostrowski's theorem (Chapter 3, Corollary 2 to Theorem 2). The last theorem can be proved directly, without application of the Gelfand-Mazur theorem (see Ostrowski [3, p. 284]).

NOTES

First we shall present the notion of Chion's valuation (Chion [2]). Let A be a ring, not necessarily commutative, and Γ a totally ordered multiplicative semigroup with zero such that $\gamma 0 = 0\gamma = 0$ and $\gamma \leq 0$ for every $\gamma \in \Gamma$, having the property that for every α, β, $\gamma \in \Gamma$, each condition $\alpha\gamma = \beta\gamma \neq 0$ and $\gamma\alpha = \gamma\beta \neq 0$ implies that $\alpha = \beta$. The mapping $w: A \to \Gamma$ from A onto Γ is called a *Hahn valuation* if the conditions

(H_1) $w(a) = 0 \Leftrightarrow a = 0$
(H_2) $w(ab) = w(a)w(b)$
(H_3) $w(a - b) \leq \max(w(a),w(b))$

hold for every a, b $\in$ A.

We shall call Γ the value semigroup of the Hahn valuation w. It is easy to see that every totally ordered semigroup Γ is the value semigroup of a Hahn valuation of a suitably chosen ring (Chion [2]). Viswanathan [2] continued Chion's study of Hahn valuations, considering completeness and compactness of Hahn valued rings.

Fukawa [2] has considered Krull valuations for rings. An *F-valuation* v of a ring A is a mapping $v: A \to \Gamma \cup \{\infty\}$ of A into a totally ordered abelian group Γ (additively written) plus an infinity element satisfying the following conditions.

(F_1) $v(1) = 0$, $v(0) = \infty$
(F_2) $v(xy) = v(x) + v(y)$
(F_3) $v(x + y) \geq \min(v(x),v(y))$

Fukawa proved several extensions of approximation theorems for fields with valuations. Here an essential role is played by the assumption that Γ is a group.

Bandyopadhyay [1,3] generalized a notion of the Hahn valuation. A multiplicative semigroup Γ which is also an upper semilattice is called a U-*semigroup* if there exists $0 \in \Gamma$ such that $0\gamma = \gamma 0 = 0$ and $0 \le \gamma$ for all $\gamma \in \Gamma$. Let A be a ring and Γ a U-semigroup. The mapping $v\colon A \to \Gamma$ of the ring A into the U-semigroup Γ is called a B-*valuation* if conditions

(B_1) $v(0) = 0$

(B_2) $v(ab) = v(a)v(b)$

(B_3) $v(a - b) \le \sup(v(a),v(b))$

are satisfied for every a, b $\in$ A. It can be shown that in the case of suitably chosen classes of Γ, every B-valuation can be decomposed into a product of Hahn valuations (Bandyopadhyay [1,3]).

Other generalizations of valuations can be found in Adams [1,2], Mathiak [1-3] (see the Notes for Chapter 4), Nagata [2], Rado [1], and Varouchakis [1]. Generalized valuations on rings have been considered by Nakano [3].

There exists no description of locally bounded ring topologies on a given division ring, e.g., on the division ring of rational quaternions.

Theorem 13 also holds in the noncommutative case (Warner [7]). Let K be a division ring equipped with a proper ring topology $\mathcal{T}$. Then $(K,\mathcal{T})$ is topologically and algebraically isomorphic to either $\mathbb{R}$, $\mathbb{C}$, or $\mathbb{H}$, if and only if the following conditions are fulfilled: $(K,\mathcal{T})$ is complete, locally bounded, additively generated, and the commutator subgroup $[K^\times,K^\times]$ of $K^\times$ is left bounded.

Shu-Bun Ng and Warner [1] proved that every homomorphism from a complete metrizable topological ring into $\mathbb{R}$ is continuous.

Recently, Kallman [1] obtained the following characterization of the reals $\mathbb{R}$ among other topological rings.

THEOREM Let $\mathbb{R}$ be the ring of real numbers, topologized as usual, let R be a topological ring whose topology is given by a complete separable metric, and let $f\colon R \to \mathbb{R}$ be a ring isomorphism. Then f is a topological isomorphism.

10

Connected Fields of Arbitrary Characteristic

DEFINITION 1 A topological field $(K,\mathcal{T})$ is connected if the underlying topological space $(K,\mathcal{T})$ is connected.

THEOREM 1 Every topological field is either connected or totally disconnected.

Proof: Suppose that $(K,\mathcal{T})$ is a proper topological field and that H is the connected component of zero. If $H = \{0\}$, then $(K,\mathcal{T})$ is totally disconnected. If $H \neq \{0\}$, then let $a \in H$ be nonzero. Then $a^{-1}H$ is a connected subset of K and $1 \in a^{-1}H$. Similarly, for arbitrary $b \in K^{\times}$, $ba^{-1}H$ is connected and it contains b. Thus

$$K = \bigcup_{b \in K} ba^{-1}H$$

is the connected component of zero or, equivalently, $(K,\mathcal{T})$ is connected.

THEOREM 2 (Pontrjagin [1]) The only locally compact and connected topological fields are the real and the complex number fields.

Proof: This follows as a corollary from Theorem 3 of Chapter 6.

Now we shall prove a result on locally bounded additively generated fields.

THEOREM 3 (Kowalsky [1, Satz 11]) Let $(K,\mathcal{T})$ be a proper locally bounded and additively generated topological field. Then the set $N_0(\mathcal{T})$ of topological nilpotents of $(K,\mathcal{T})$ is nonzero.

Proof: Let $U \in \mathcal{B}(\mathcal{T})$ be an arbitrary and bounded neighborhood. Let $n \cdot U = U + \cdots + U$ (n times). Then $\bigcup_{n\in\mathbb{N}} n \cdot U$ is an open subgroup of the additive group K^+ of K. Thus $K = \bigcup_{n\in\mathbb{N}} n \cdot U$, by our assumption. Since $(K,\mathcal{T})$ is locally bounded, there is an $x \in K^\times$ such that $xU + xU \subset U$ or $2 \cdot U \subset x^{-1}U$. By a simple induction we obtain $2^k \cdot U \subset x^{-k}U$ for $k \in \mathbb{N}$. Since $n \cdot U \subset m \cdot U$ for $n \le m$, we have

$$K = \bigcup_{k=0}^{\infty} 2^k \cdot U = \bigcup_{k=0}^{\infty} x^{-k}U$$

We will show that $x \in N_0(\mathcal{T})$. This is equivalent to

$$\bigcap_{n\in\mathbb{N}} x^nU = \{0\}$$

Indeed, for otherwise there would exist a $y \in K^\times$ such that $y \in x^nU$ for every $n \in \mathbb{N}$ and this would imply that $x^{-n} \in y^{-1}U$, i.e.,

$$K = \bigcup_{n=0}^{\infty} x^{-n}U = y^{-1}U^2 \qquad K = U^2$$

which is impossible, since U^2 is bounded and every bounded topological field is discrete, contradicting the assumption that $\mathcal{T}$ is a proper topology on K (Chapter 1, Lemma 15).

COROLLARY Let $(K,\mathcal{T})$ be a proper locally bounded and connected topological field. Then there exists a nondiscrete norm on K inducing the topology $\mathcal{T}$.

Proof: Apply Theorem 3 of this chapter and Theorem 4 of Chapter 3.

THEOREM 4 (Baer and Hasse [1]) The complex number field $\mathbb{C}$ is the only two-dimensional (in a topological sense) subfield of $\mathbb{C}$.

In contrast to Theorem 4, there exists a one-dimensional subfield $L \neq \mathbb{R}$ of $\mathbb{C}$ (see Kapuano [1]).

Similarly, there exists a connected subfield M of $\mathbb{C}$ (equipped with the ordinary topology) different from $\mathbb{R}$ and $\mathbb{C}$. We present this construction next.

THEOREM 5 (Dieudonné [2]) There exist subfields M of the complex number field $\mathbb{C}$ (equipped with the ordinary topology) which are connected, locally connected, dense in $\mathbb{C}$, algebraically isomorphic with $\mathbb{C}$, and satisfy $\mathbb{R} \neq M \neq \mathbb{C}$.

The construction is based on the following:

LEMMA 1 If a subset A of $\mathbb{R}^2$ has a nonempty intersection with every nonempty perfect subset of $\mathbb{R}^2$, then A is connected.

Proof: Since A intersects every perfect set, A is everywhere dense in $\mathbb{R}^2$. Suppose to the contrary that A is disconnected. Thus there are nonempty and open subsets V and W of $\mathbb{R}^2$ with $A = A \cap V \cup A \cap W$ and $A \cap V \cap W = \emptyset$. This implies that $V \cap W = \emptyset$ since A is everywhere dense. Since $W \subset \mathbb{R}^2 \setminus V$ there is a perfect set $B \subset Fr(V)$. But $Fr(V) \cap V = \emptyset$ and $Fr(V) \cap W = \emptyset$, implying that $B \cap A = \emptyset$, a contradiction.

Proof of Theorem 5: First, we order the perfect sets in $\mathbb{R}^2$ in a transfinite sequence $\{P_\alpha\}$ such that for every α, $|\{\beta : \beta \leq \alpha\}| < c$. Define now the transfinite sequence $\{u_\alpha\}$ of elements $u_\alpha \in \mathbb{C}$ such that the field $K_\alpha = \mathbb{Q}(\{u_\beta : \beta \leq \alpha\})$ is a pure transcendental extension of $\mathbb{Q}$ and $K_\alpha \cap P_\beta \neq \emptyset$ for every $\beta \leq \alpha$. Suppose that this sequence is defined for all $\beta < \alpha$. Put $K'_\alpha = \bigcup_{\beta<\alpha} K_\beta$. It follows immediately that K'_α is a pure transcendental extension of $\mathbb{Q}$. Indeed, for every finite subset $F \subset \{u_\beta : \beta < \alpha\}$, there exists $\gamma < \alpha$ satisfying $F \subset K_\gamma$; that is, F is algebraically independent over $\mathbb{Q}$, by our inductive assumption. Since $|\{u_\beta : \beta < \alpha\}| < c$ and $\mathbb{Q}$ is countable, $|K'_\alpha| < c$. But $|P_\alpha| = c$, which implies that there is $u_\alpha \in P_\alpha$ transcendental over K'_α. Let $K_\alpha = K'_\alpha(u_\alpha)$. The field K_α is purely transcendental over $\mathbb{Q}$ and it has the desired properties. Put $K = \bigcup_\alpha K_\alpha$. The subfield K of $\mathbb{C}$ is purely transcendental over $\mathbb{Q}$. Moreover, since $K \cap P_\alpha \neq \emptyset$ for every perfect set $P_\alpha \subset \mathbb{R}^2$, by applying Lemma 1 we obtain that K is connected and dense in $\mathbb{C}$. Obviously, $\mathbb{R} \neq K \neq \mathbb{C}$, since $\mathbb{C} \setminus \mathbb{Q}$ contains algebraic numbers.

If we take a ball $\{x \in \mathbb{R}^2 : |x| = \sqrt{x_1^2 + x_2^2} < r\}$ instead of $\mathbb{R}^2$ in Lemma 1, where $r > 0$, then applying this modified lemma to K we obtain the local connectedness of K.

Let B be a transcendence basis of the reals over $\mathbb{Q}$. Thus, by the Steinitz theorem, $\mathbb{R} = \mathbb{Q}(B)(A) = K_0(A)$, where $K_0 = \mathbb{Q}(B)$ and every element of A is algebraic over K_0. Since $|B| = |\{u_\alpha : u_\alpha \in K\}| = \mathfrak{c}$, the fields K_0 and K are (algebraically) isomorphic. Let L be the algebraic closure of K in $\mathbb{C}$. Since L is isomorphic with $\mathbb{C}$, there is a subfield M of L, $K \subset M$ such that $M \simeq \mathbb{R}$ algebraically. The field M just constructed has the desired properties.

In a natural way, a problem arises as to whether there are any connected fields of nonzero characteristic. Such examples were given independently by Mutylin [1,4] and Waterman and Bergman [1]. They have shown that every discrete field can be embedded in a connected topological field equipped with a suitably chosen topology. In particular, there exists a connected field of arbitrary nonzero characteristic. Mutylin [4] has shown, moreover, that every discrete field can be embedded in a complete and connected topological field. This shows that the local boundedness in Theorem 7 is an essential assumption.

We present here the construction due to Waterman and Bergman [1]. Their result can be formulated precisely, in the following way:

THEOREM 6 Every discrete field k can be embedded as a subfield in an arcwise connected topological field K.

COROLLARY There exist connected topological fields of arbitrary characteristic.

Sketch of the proof: Let k be a discrete field. Put

$$R = k[\{T_\alpha\}_{\alpha \in (0,1)}] \tag{1}$$

where T are indeterminates and define, moreover, $T_0 = 0$, $T_1 = 1$. Let

$$U_\varepsilon = \bigcup_{\substack{\alpha,\beta \in [0,1]^n \\ \Sigma|\alpha_i - \beta_i| < \varepsilon}} (T_{\alpha_1} - T_{\beta_1}, \ldots, T_{\alpha_n} - T_{\beta_n}) \tag{2}$$

where $\alpha = (\alpha_1,\ldots,\alpha_n)$, $\beta = (\beta_1,\ldots,\beta_n)$, and $(a_1,\ldots,a_n) = Ra_1 + \cdots + Ra_n$ is an ideal in R generated by $a_1, \ldots, a_n$. For every $f \in R$, define

$$\|f\| = \inf\{\varepsilon : f \in U_\varepsilon\} \tag{3}$$

Since for every $f \in R$, $f \in R = (T_1 - T_0)$, then $\|f\| \leq 1$. We say that f contains T_γ $(\gamma \neq 0, 1)$ if $f \notin k[\{T_\alpha\}_{\alpha\neq\gamma}]$. Moreover, f contains T_0 and T_1, by definition.

LEMMA 2 The function $f \to \|f\|$, mapping R into $\mathbb{R}_+$, has the following properties.

(i) $\|f\| \geq 0$; $\|f\| = 0 \Leftrightarrow f = 0$.

(ii) $\|f + g\| \leq \|f\| + \|g\|$.

(iii) $\|fg\| \leq \min(\|f\|, \|g\|)$.

In particular, R is a topological ring with the topology induced by the metric $d(f,g) = \|f - g\|$.

Proof of Lemma 2: Let $\|f\| = \inf\{\varepsilon : f \in U_\varepsilon\} = 0$. Suppose that $f \neq 0$ and depends on the variables $T_{\alpha_1}, \ldots, T_{\alpha_n}$ with $\alpha_1 < \cdots < \alpha_n$. Thus $f \in U_\varepsilon \cap k[T_{\alpha_1},\ldots,T_{\alpha_n}]$. Take $\varepsilon < \min\{|\alpha_i - \alpha_j| : i \neq j\}$. Since for such ε, $U_\varepsilon \cap k[T_{\alpha_1},\ldots,T_{\alpha_n}] = k$, hence $f \in k$, a contradiction that proves (i).

To prove (ii) note that $U_\varepsilon + U_\eta \subset U_{\varepsilon+\eta}$ holds for every ε and η, which implies that

$$\begin{aligned}\|f + g\| &= \inf\{\varepsilon : f + g \in U_\varepsilon\} = \inf\{\varepsilon + \eta : f + g \in U_{\varepsilon+\eta}\} \\ &\leq \inf\{\varepsilon + \eta : f \in U_\varepsilon, g \in U_\varepsilon\} = \|f\| + \|g\|\end{aligned}$$

Every U_ε is a union of prime ideals in R. This implies that for every f, $g \in R$, $fg \in U_\varepsilon$ implies that $f \in U_\varepsilon$ or $g \in U_\varepsilon$. Thus (iii) holds.

Finally, we note that the mapping of $I = [0,1]$ into R defined by $\alpha \mapsto T_\alpha$ is an isometry; i.e., $|\alpha - \beta| = \|T_\alpha - T_\beta\|$ for every α, $\beta \in I$. Hence 0 and 1 are connected by an arc in R.

Let K be the quotient field of R. We sketch below, following Waterman and Bergman [1], how to define on K a field topology extension that is defined on R by the norm (3).

Let

$$X = \{\varepsilon : (0,1] \xrightarrow{\varepsilon} (0,1], \text{ e is a monotone increasing function}\} \tag{4}$$

and put

$$Y = \{\varepsilon = (\varepsilon_1, \varepsilon_2, \ldots) : \varepsilon_i \in X, \forall x \in (0,1] \forall i, \quad \varepsilon_{i+1}(x) \le \varepsilon_i(x)\} \tag{5}$$

For a fixed $\varepsilon \in Y$ let

$$V_\varepsilon = \left\{a = \sum_{i=1}^{n} \frac{f_i}{g_i} \in K : \forall i, \|f_i\| \le \varepsilon_i(\|g_i\|)\right\} \tag{6}$$

It can be shown (Waterman and Bergman [1]) that the following lemma holds.

LEMMA 3 The family of sets $\mathcal{B} = \{V_\varepsilon : \varepsilon \in Y\}$ defines a field topology on $K = k(\{T_\alpha\}_{\alpha\in[0,1]})$. An embedding of R [provided with the topology induced by the norm (3)] in $(K,\mathcal{T})$ is homeomorphic.

Theorem 6 follows immediately from Lemma 3. Indeed, let $f: I = [0,1] \to R \subset K$ be an arc defined by the isometry $f(\alpha) = T_\alpha$. Let $a, b \in K^\times$ be arbitrary. Then $f_{ab}: I \to K$, defined by the formula

$$f_{ab}(x) = \begin{cases} af\left(2\left(\frac{1}{2} - x\right)\right) & \text{for } 0 \le x \le \frac{1}{2} \\ bf\left(2\left(x - \frac{1}{2}\right)\right) & \text{for } \frac{1}{2} \le x \le 1 \end{cases}$$

is an arc from a to b in $(K,\mathcal{T})$. Since elements $a, b \in K$ were taken arbitrarily, $(K,\mathcal{T})$ is an arcwise-connected topological field, containing a field k as a discrete subfield.

The sketch of the proof of Theorem 6 is finished.

THEOREM 7 (Mutylin [4, Th. 1]; Warner [7, Th. 1]) Let $(K,\mathcal{T})$ be a proper topological field. Then $(K,\mathcal{T})$ is topologically isomorphic either to $\mathbb{R}$ or $\mathbb{C}$ if and only if the following conditions are satisfied.

(i) $(K,\mathcal{T})$ is complete.

(ii) $(K,\mathcal{T})$ is locally bounded.

(iii) $(K,\mathcal{T})$ is additively generated.

Proof: It is obvious that $\mathbb{R}$ and $\mathbb{C}$ satisfy (i)-(iii). Indeed, $\mathbb{R}$ and $\mathbb{C}$ being connected, are additively generated.

Conversely, suppose that conditions (i)-(iii) are satisfied. We shall follow the proof of Warner [7], whose proof is simpler than that of Mutylin [4].

Theorem 3 implies that $(K,\mathcal{T})$ contains a nonzero topological nilpotent. Moreover, since $(K,\mathcal{T})$ is locally bounded, the topology $\mathcal{T}$ is induced by a nontrivial norm, say p (by Theorem 4 of Chapter 3), i.e., $\mathcal{T} = \mathcal{T}_p$. Let p_{sp} be the spectral norm on K induced by p. Since p_{sp} is power-multiplicative, it can be represented in the form

$$p_{sp} = \sup\{f_a : a \in K^\times\}$$

where for every $a \in K^\times$, f_a is an absolute value satisfying $f_a \le p_{sp}$ and $f_a(a) = p_{sp}(a)$ (Chapter 3, Theorem 8, especially its proof). For every $a \in K^\times$, $\{x \in K : f_a(x) \le 1\}$ is a neighborhood of zero in $(K,\mathcal{T}_p)$, since the topology induced on K by f_a is not stronger than $\mathcal{T}_p$, and therefore is not an additive subgroup of K^+. Thus f_a is an Archimedean absolute value. It means that char(K) = 0 and there exists an embedding of $\mathbb{Q}$ into K, $g: \mathbb{Q} \to K$, satisfying $f_a(g(x)) = |x|^{r_a}$, where $0 < r_a \le 1$, and $x \in \mathbb{Q}$, since the usual absolute value is the only Archimedean absolute value of $\mathbb{Q}$ (see Chapter 2, Theorems 2 and 10). Let $a \in \mathbb{Q}^\times$. Since $|a|^{r_a} = f_a(a) = p_{sp}(a)$, then $|a| < 1$ if and only if $p_{sp}(a) < 1$. Since $p_{sp}(a) < 1 \Leftrightarrow a \in N_0(\mathcal{T}_p)$, consequently $|a| < 1$ if and only if $a \in N_0(\mathcal{T}_p)$. This implies that $\mathcal{T}|\mathbb{Q} \ne \mathbb{I}_{\mathbb{Q}}, \mathcal{T}_q$ (q a prime number). Indeed, $\mathcal{T}|\mathbb{Q}$ is not the q-adic topology $\mathcal{T}_q$, since $N_0(\mathcal{T}_q)$ is a nonzero subgroup of $\mathbb{Q}^+$, hence an open subgroup in $(\mathbb{Q},\mathcal{T}_p)$. Therefore, $\mathcal{T}|\mathbb{Q}$ is the ordinary topology on $\mathbb{Q}$ induced by the Archimedean absolute value. This implies, by Theorem 14 of Chapter 3, that the closure of $\mathbb{Q}$ in K equals $\mathbb{R}$. Thus K is a complete and locally bounded extension of the reals $\mathbb{R}$. Finally, Theorem 13 of Chapter 9 implies that either $K \simeq \mathbb{R}$ or $K \simeq \mathbb{C}$, both isomorphisms being topological.

COROLLARY The real and complex number fields are the only locally bounded, complete, and connected topological fields.

Proof: Since every connected topological field is additively generated, we can apply Theorem 7.

DEFINITION 2 A topological field $(K,\mathcal{T})$ is called *locally convex* if there exists a base $\mathcal{B}(\mathcal{T})$ of neighborhoods of zero for $(K,\mathcal{T})$ such that $n \cdot U = nU$ holds for all $n \in \mathbb{N}$ and $U \in \mathcal{B}(\mathcal{T})$. Here $nU = \{na : a \in U\}$ and $n \cdot U = U + \cdots + U$ (n times).

THEOREM 8 The only locally convex, complete, and additively generated topological fields are the real and complex number fields.

Proof: Let $(K,\mathcal{T})$ be a proper locally convex, complete, and additively generated topological field. Then K has zero characteristic. In fact, if $\text{char}(K) = p \neq 0$, then $pU = \{0\} \neq pU$, since $(K,\mathcal{T})$ is nondiscrete. Thus $\mathbb{Q}$ can be identified as a subfield of K. Since K is additively generated, every neighborhood of zero in $(K,\mathcal{T})$ generates the group K^+. In particular, if $\mathcal{B}(\mathcal{T})$ is a base of the neighborhoods of zero consisting of locally convex sets and $U \in \mathcal{B}(\mathcal{T})$ is arbitrary, then $K = \bigcup_{n\in\mathbb{N}} n \cdot U$, since the underlying additive group of $(K,\mathcal{T})$ is open. Let $V \in \mathcal{B}(\mathcal{T})$ be a fixed neighborhood. Then for every $U \in \mathcal{B}(\mathcal{T})$ there exists $n \in \mathbb{N}$ such that $V \subset n \cdot U$. But $n \cdot U = nU$, thus $(1/n)V \subset U$, $1/n \in K^{\times}$. It means (Chapter 1, Lemma 12) that V is a bounded neighborhood of 0. Since V was arbitrary, $\mathcal{B}(\mathcal{T})$ consists of bounded sets; that is, $(K,\mathcal{T})$ is a locally bounded topological field. Now the result follows from Theorem 7.

NOTES

Recall first the following theorem (Więsław [8, Th. 3]): Let $(K,\mathcal{T})$ be a topological division ring, complete and connected with respect to a topology $\mathcal{T}$ induced by a Krull-Schilling valuation. Then $(K,\mathcal{T})$ is topologically and algebraically isomorphic to $\mathbb{R}$, $\mathbb{C}$, or $\mathbb{H}$. The following theorem can be deduced from the theorem above by a slight modification of its proof. Every nonlinear (i.e., having no base of the neighborhoods of zero consisting of additive subgroups) V-topological division ring is topologically and algebraically isomorphic to a subfield of the division ring of real quaternions (for a proof, see, e.g., Szambien [1]). Both results are special cases of Theorem 7. For another proof of Theorem 8 see Matzinger [1].

Recently, Šahmatov [1, Th. 8; 2, Th. 4] has proved the following result.

THEOREM There exist connected topological fields (resp. division rings) of arbitrary characteristic such that the underlying topological space is not normal.

This solves problem 1.3.6 of Arnautov, Vodinčar, and Mihalev [1]. Šahmatov has also proved the existence of a topological field whose underlying space is normal but not hereditary normal (Šahmatov [2, Th. 5]). We mention another interesting result due to Šahmatov [1, Cor. 5 to Th. 9]. Namely, a compact space is metrizable if and only if it can be embedded as a subspace in a topological field. Uspenskiǐ [1] has proved that every metrizable topological space is homeomorphic to a subspace of a metrizable topological field.

Theorem 6 can be extended to rings. Arnautov and Ursul [1] proved that every topological ring, not necessarily associative, can be embedded as a closed subring in an arcwise-connected topological ring.

If V is a nonzero compact vector space over a topological division ring $(K,\mathcal{T})$, then $\mathcal{T} = \mathbb{1}_K$. Moreover, a nonzero compact vector space over a discrete division ring K is connected if and only if $\mathrm{char}(K) = 0$ (Warner [2, Ths. 5 and 9]).

Recently, Shell [4] proved that the real number field is the only proper arcwise-connected subfield of the complex number field.

Indeed, Kallmann and Simmons [1, Prop. (1.3)] proved that if K is a continuum in the plane which does not lie on a line, then the set of differences $K - K = \{a - b : a,b \in K\}$ contains an open set. Let $\mathbb{R} \neq R \subset \mathbb{C}$ be an arcwise connected subfield of $\mathbb{C}$. Let K be the union of arcs connecting any three noncollinear points in R. Then $K - K$, hence $R - R$, contains an open set $U \subseteq \mathbb{C}$. Consequently, $R = \mathbb{C}$.

A detailed study of connected and disconnected fields can be found in Shell [4].

11

Ring Topologies on the Rational Number Field

In this chapter we summarize the fundamental results about ring topologies on the rational number field $\mathbb{Q}$. Recall first that locally bounded minimal ring topologies on $\mathbb{Q}$ are induced by nontrivial absolute values; that is, these topologies are the p-adic topologies T_p and the ordinary order topology T_∞ on $\mathbb{Q}$ as follows from the Ostrowski theorem (Chapter 2, Theorem 10). The result of Shanks and Warner [1,2] (see Chapter 9, Theorem 10, the case $K = \mathbb{Q}$) implies that every nondiscrete locally bounded field topology T on $\mathbb{Q}$ has the form

$$T = \sup\{T_i : i \in I\} \tag{1}$$

where I is a finite subset of $\mathbb{P}' = \mathbb{P} \cup \{\infty\}$. Furthermore, a result of Correl [3] states that every field topology T on $\mathbb{Q}$ having a base of zero neighborhoods consisting of additive subgroups of $\mathbb{Q}$ has the form (1), where I is a subset of $\mathbb{P}$. We shall prove a more general result in Chapter 12.

DEFINITION 1 A ring topology T on a field (ring) K is called *linear* if there exists a base of the neighborhoods of zero for T consisting of additive subgroups of K.

Note that the locally bounded topology $T_{\mathbb{Z}} \in L_r(\mathbb{Q})$ which has the integers as an order is the unique strongest linear topology on $\mathbb{Q}$. Every additive subgroup of $\mathbb{Q}$ is open in $(\mathbb{Q}, T_{\mathbb{Z}})$. Furthermore, $T_{\mathbb{Z}}$ is also a maximal locally bounded ring topology on $\mathbb{Q}$, as follows immediately from Theorem 10 of Chapter 9. We shall prove, following Shell [3], that $T_{\mathbb{Z}}$ is nonmaximal in $L_r(\mathbb{Q})$.

LEMMA 1 (Shell [3, Lemma 2.1]) If $\mathcal{B}$ is a base of the neighborhoods of zero of a ring topology on $\mathbb{Z}$, then $\{mV : m \in \mathbb{N}, V \in \mathcal{B}\}$ is a subbase of neighborhoods of zero of a topology $T \in L_r(\mathbb{Q})$ satisfying $T_{\mathbb{Z}} \leq T$.

Proof: If $U_i - U_i \subset V_i$ for $i = 1, 2, \ldots, n$, then $\bigcap_{i=1}^n m_i U_i - \bigcap_{i=1}^n m_i U_i \subset \bigcap_{i=1}^n m_i V_i$. If $aU_i \subset V_i$ for $i = 1, 2, \ldots, n$, then

$$\frac{a}{b} \bigcap_{i=1}^{n} bm_i U_i \subset \bigcap_{i=1}^{n} m_i V_i \qquad \text{where } a, b, m_i \in \mathbb{N}$$

Since $\mathbb{Z} \in \mathcal{B}$, then $\mathbb{Z} \in \mathcal{B}(T)$, thus $n\mathbb{Z} \in \mathcal{B}(T)$ for every $n \in \mathbb{N}$. This implies that $a\mathbb{Z} \in \mathcal{B}(T)$ for every $a \in \mathbb{Q}^{\times}$; therefore, $T_{\mathbb{Z}} \leq T$.

THEOREM 1 (Shell [3, Th. 2]) Suppose that $a_n \in \mathbb{N}$ $(n \in \mathbb{N})$ satisfy the following conditions:

(i) $a_n > 2$ and $n! \mid a_n$.

(ii) $a_n \mid a_{n+1}$.

(iii) $a_{n+1} > (2na_n + 1)^{2^{n+1}}$.

If $V_{rn} = \{x \in \mathbb{Z}: x \equiv a \pmod{2a_n}$ for some $a \in \mathbb{Z}$ and $|a| < (2a_n)^{1/2^r} - 1\}$, $V_r = \bigcap_{n \in \mathbb{N}} V_{rn}$, then $\mathcal{C} = \{mV_r : m, r \in \mathbb{N}\}$ is a subbase of the neighborhoods of zero of a topology $T \in L_r(\mathbb{Q})$ satisfying $T_{\mathbb{Z}} < T < \mathbb{1}_{\mathbb{Q}}$. This means that $T_{\mathbb{Z}}$ is a nonmaximal ring topology on $\mathbb{Q}$.

Proof: We shall first prove that $\{V_r : r \in \mathbb{N}\}$ is a base of the neighborhoods of zero for a ring topology on $\mathbb{Z}$. We have

$$V_{r+1} + V_{r+1} \subset V_r \qquad V_{r+1}V_{r+1} \subset V_r \quad \text{for } r \in \mathbb{N} \tag{2}$$

Indeed, set $c = (2a_n)^{1/2^r}$. Suppose that $x, y \in V_{r+1}$ and

$$x \equiv a \pmod{2a_n} \qquad |a| < \sqrt{c} - 1$$

$$y \equiv b \pmod{2a_n} \qquad |b| < \sqrt{c} - 1$$

If $c < 4$, then $a = b = 0$, which implies that $|a + b| < c - 1$. If $c \geq 4$, then $|a + b| < 2\sqrt{c} - 2 < 2\sqrt{c} - 1 \leq c - 1$. Thus $x + y \in V_r$. Furthermore, $|ab| < (\sqrt{c} - 1)^2 = c - (2\sqrt{c} - 1) < c - 1$, implying (2). Moreover,

$$kV_{r+|k|} \subset V_{r+|k|} + \cdots + V_{r+|k|} \subset V_r$$
$$(2^{|k|} \text{ times})$$

Applying Lemma 1, we obtain that the topology T satisfies $T_{\mathbb{Z}} \leq T$.

Now we shall show that $T \neq T_{\mathbb{Z}}$ by showing that only the zero ideal of $\mathbb{Z}$ is contained in V_1. Since $a_n \geq 2$, $\sqrt{2a_n} \leq a_n$. As $\pm a_n$ are the representatives of a_n modulo $2a_n$ with the smallest absolute value, $a_n \notin V_{1n}$, hence $a_n \notin V_1$. But since $n!|a_n$, the sequence a_n is eventually in every nonzero ideal of $\mathbb{Z}$.

Finally, T is nondiscrete. Indeed, if $N \geq \max(r, m_1 \ldots m_t)$, then $x = 2a_n m_1 \cdots m_t \in m_1 V_r \cap \cdots \cap m_t V_r$. Since a_n divides x/m_i for $n \leq N$, $x/m_i \in V_{rn}$ for $n \leq N$. Otherwise, for $n > N$, we have

$$0 \leq \frac{x}{m_i} \leq 2Na_N \leq a_{N+1}^{1/2^{N+1}} - 1 \leq a_n^{1/2^r}$$

DEFINITION 2 Let (K,T) be a topological field. We say that T is a *full topology* on K if the completion $\hat{K}$ of (K,T) is a field.

It can be shown that the following lemma holds.

LEMMA 2 (Kowalsky [2, Satz, p. 164]; Mutylin [6, Th. 3']) Every minimal field topology is full.

Remark The converse statement is not true: There exist nonminimal field topologies with the property that the completions with respect to them are fields. See Chapter 3, Section 6, where suitable examples are presented.

Now we shall discuss some examples of locally unbounded ring topologies on $\mathbb{Q}$. Such examples were given in Chapter 1, Examples 11 and 12. It is easy to see that $T = \sup_f\{T_{p_i} : i \in \mathbb{N}\}$ in Example 12, using Lemma 4 of Chapter 1. Another example is the following:

Example 1 (Mutylin [4]) Let

$$U_n = \left\{ \sum_{r=n}^{\infty} \frac{k_r p_r}{\ell_r p_r + m_r} : k_r, \ell_r, m_r \in \mathbb{Z};\ |k_r| \leq a_{rn}, \right.$$
$$\left. |\ell_r| \leq a_{rn},\ |m_r| \leq b_{rn} \right\} \qquad (3)$$

where $k_r = 0$ for all but a finite number of r's; here p_r $(r \in \mathbb{N})$ is a sequence of prime numbers. Mutylin has found conditions ensuring that the sets of the form (3) form a base of zero neighborhoods for a ring topology $\mathcal{T}$ on $\mathbb{Q}$. The topology $\mathcal{T}$ can be weakened to a field topology $\mathcal{S}$. It can be shown (Mutylin [4, Cor. 4 to Lemma 8]) that $\mathcal{S}$ is a full topology on $\mathbb{Q}$. Furthermore, since $p_n/(1 + p_n) \to 0$ as n tends to infinity in $(\mathbb{Q},\mathcal{S})$, then $\mathcal{S} \neq \mathcal{T}_p$ for every $p \in \mathbb{P}' = \mathbb{P} \cup \{\infty\}$. This implies that there exists a complete and locally unbounded topological field of zero characteristic having no proper closed subfields.

Further discussions around Example 1 led Mutylin to the following:

THEOREM 2 (Mutylin [4, Th. 2]) There exist uncountably many full and locally unbounded field topologies on the rational number field.

It is not known if these topologies are minimal.

In connection with Theorem 2 it is worthwhile to notice that every countable field admits $2^{2^{\aleph_0}}$ field topologies which fail to be the supremum of locally bounded ring topologies (Podewski [1]).

Let A_i be subsets of a ring A. We write $\Sigma'^{}_{i=n} A_i$ for the set of all sums $a_n + \cdots + a_{n+k}$, where $a_i \in A_i$ and $k = 0, 1, 2, \ldots,$.

DEFINITION 3 (Zobel [1, Defs. 2.1 and 3.1]) Let A be a commutative ring. We say that $\mathcal{T}$ is the direct ring topology on A if there exist a base $\mathcal{B} = \{U_n : n \in \mathbb{N}\}$ of neighborhoods of zero for $(A,\mathcal{T})$ and sets M_{ik} $(i, k \in \mathbb{N};\ i \geq k)$ such that

$$U_n = \sum_{s=n}' M_{sn} \qquad \text{for every } n \in \mathbb{N} \tag{4}$$

and the system of sets $\mathcal{G} = \{M_{ik} : i, k \in \mathbb{N};\ i \geq k\}$ has the following properties:

(a) $0 \in M_{nn}$ for every $n \in \mathbb{N}$.

(b) $M = -M$ for $M \in \mathcal{G}$.

(c) $M_{ik} + M_{ik} \subset M_{i,k-1}$ for all $i, k \in \mathbb{N}$ with $2 \leq k \leq i$.

(d) $(\Sigma^r_{s=1} M_{s1}) \cap M_{r+1,1} = \{0\}$ for every $r \in \mathbb{N}$.

(e) $M_{k,n+1} \Sigma_{s=n+1}^{k-1} M_{s,n+1} + M_{k,n+1} M_{k,n+1} \subset M_{k,n}$ holds for every k, $n \in \mathbb{N}$ satisfying $k \geq n + 1$.

(f) For every $a \in A$ and every $n \in \mathbb{N}$ there exists $k \in \mathbb{N}$ such that for every $i \geq n + k$,

$$aM_{i,n+k} \subset M_{in}$$

We can now formulate the following.

THEOREM 3 (Zobel [1, Satz 4.6]) Every ring topology on $\mathbb{Q}$ is the supremum of infima of direct ring topologies.

The direct ring topologies in Theorem 3 can be effectively described, but we shall omit the details.

The topology $\mathcal{T}$ in Mutylin's example (see Chapter 1, Example 11) is a direct ring topology on $\mathbb{Q}$ and it has the property that the completion of $(\mathbb{Q},\mathcal{T})$ is a proper integral domain (Zobel [1, Satz 3.6]) in which the only invertible elements are those of $\mathbb{Q}^{\times}$.

It is not known whether it can be true that the completion of a field with respect to a field topology is a proper integral domain. For other problems of this kind, see Chapter 15. A good summary of results dealing with ring topologies on $\mathbb{Z}$ and $\mathbb{Q}$ can be found in Weber [1].

12

Linear Field and Ring Topologies on the Quotient Field of Arithmetical Rings

Let A be a Dedekind domain, $\mathbb{P}$ the set of prime ideals of A, and K the quotient field of A. Let $\mathcal{T}_p$ be a p-adic topology on K, where $p \in \mathbb{P}$ (see Chapter 1, Example 9). It is clear that the sets

$$\{x \in K : v_p(x) \geq n\} \qquad n \in \mathbb{N} \tag{1}$$

are A-submodules of K and form a base of neighborhoods of zero for $(K, \mathcal{T}_p)$.

The following result holds:

THEOREM 1 (Jebli [1]) Let K be the quotient field of a Dedekind domain A, and let $(K, \mathcal{T})$ be a proper topological field. Then the following conditions are equivalent.

(i) $(K, \mathcal{T})$ has a base of the neighborhoods of zero consisting of A-submodules of K.

(ii) $\mathcal{T}$ is the supremum of a family of p-adic topologies $\mathcal{T}_p$.

Proof: (ii) ⇒ (i) Let $\mathcal{T} = \sup_{p \in I} \mathcal{T}_p$, $I \subset \mathbb{P}$. Since every member U of $\mathcal{B}(\mathcal{T})$ is a finite intersection of A-submodules of the form (1), where $n \in \mathbb{N}$, $p \in I$ (by Lemma 4 of Chapter 1), $\mathcal{B}(\mathcal{T})$ consists of A-submodules of K.

(i) ⇒ (ii) The proof will be based on some properties of submodules of the quotient fields of Dedekind domains. Let $G \neq (0)$ be an A-submodule of K.

If a, b, c are ideals of A such that $a \subset G$, $abc^{-1} \subset G$ and $b + c = A$, then ac^{-1} is an A-submodule of G. (2)

Indeed, $ac^{-1} = (b + c)ac^{-1} = bac^{-1} + cac^{-1} = abc^{-1} + aA \subset G + G \subset G$.

If a, b, c are ideals of A such that $ab^{-1} \subset G$, $ac^{-1} \subset G$, and $b + c = A$, then $a(bc)^{-1} \subset G$ is an A-submodule. (3)

Let $p \in \mathbb{P}$ be arbitrary. If p divides bc, then p divides b or c and p does not divide b and c simultaneously. Hence p does not divide $b + c$. It implies that $bc + (b + c) = A$. As

$$a(b + c)(bc)^{-1} = ac^{-1} + ab^{-1} \subset G$$

then applying (2), we have $a(bc)^{-1} \subset G$.

By an easy induction we can prove

If a, a_1, ..., a_n are ideals of A satisfying the following conditions; $aa_i^{-1} \subset G$ for $i = 1, 2, \ldots, n$ and $a_i + a_j = A$ if $i \neq j$, then $a(a_1 \cdots a_n)^{-1} \subset G$. (4)

Let $G \neq (0)$ be an A-submodule of K. Set $a_G = A \cap G$. If $p \in \mathbf{P}$ is arbitrary, we define

$$h_G(p) = \begin{cases} +\infty & \text{if } a_G p^{-n} \subset G \text{ for every } n \in \mathbb{N} \\ n & \text{if } a_G p^{-n} \subset G \text{ and } a_G p^{-(n+1)} \not\subset G \end{cases} \tag{5}$$

Note that $h_G(p) = 0$ if p exactly divides a_G [i.e., if $v_p(a_G) = 1$]. Furthermore, if $m \geq 0$, then $a_G p^{-m} \subset G$ if and only if $m \leq h_G(p)$.

LEMMA 1 Let $G \neq (0)$ be an A-submodule of K and $x \in K$. Then $x \in G$ if and only if there exists an ideal b of A and a set of nonnegative integers $(k_p)_{p\in\mathbf{P}}$ such that $0 \leq k_p \leq h_G(p)$, $k_p = 0$ for all but a finite number of p's, and

$$Ax = ba_G \prod_{p\in\mathbf{P}} p^{-k_p} \tag{6}$$

Proof of Lemma 1: ($\Leftarrow$) Let $p_0 \in \mathbf{P}$ be such that $k_{p_0} \neq 0$. Then, by assumption,

$$ba_G p_0^{-k_{p_0}} = Ax \prod_{p \neq p_0} p^{k_p}$$

i.e., $Ax \prod_{p \neq p_0} p^{k_p} \subset A \cap G$. Let $a_i = p_i^{k_{p_i}}$ for those p_i for which $k_{p_i} \neq 0$. Then $a_i + a_j = A$ for $i \neq j$. Setting $a = ba_G$, we have $aa_i^{-1} = ba_G p_i^{-k_{p_i}} \subset G$. It implies, by (4), that

$$Ax = ba_G \prod_p p^{k_p} = a(a_1 \cdots a_n)^{-1} \subset G$$

Thus $x \in G$.

($\Rightarrow$) Let $Ax = cd^{-1}$, where c and d are ideals of A satisfying $c + d = A$. Then $c = dAx \subset a_G = G \cap A$, since $x \in G$. This implies that there exists an ideal b of A having the property $c = ba_G$. Note that $c + d = A$. Indeed, $A = c + d \subset b + d = A$. Furthermore, $a_G bd^{-1} = Ax \subset G$. Thus $a_G d^{-1} \subset G$, by (2). Let $q \in \mathbb{P}$ be arbitrary. Then

$$a_G q^{-v_q(d)} = a_G d^{-1} \prod_{p \neq q} p^{v_p(d)} \subset G$$

showing that $v_p(d) \leq h_G(p)$, and that

$$Ax = ba_G \prod_{p \in \mathbb{P}} p^{-v_p(d)}$$

LEMMA 2 Let a be an ideal of A and $h\colon \mathbb{P} \to \mathbb{N} \cup \{\infty\}$ any function satisfying $h(p) = 0$ for every $p \in \mathbb{P}$ dividing a. Then there exists exactly one A-submodule G of K such that $a = a_G$ and $h = h_G$.

Proof of Lemma 2: Let G be the union of all A-submodules $ba \prod_p p^{-k_p}$, where b is an ideal of A, $0 \leq k_p \leq h(p)$ holds for every $p \in \mathbb{P}$ and $k_p = 0$ for all but a finite number of p's. It is sufficient to prove, by Lemma 1, that G is an A-submodule of K and $a = a_G$.

Let $x \in A$. Since xb is an ideal of A, $xba\,\Pi_p\, p^{-k_p} \subset G$. Furthermore, we have

$$ba \prod_p p^{-k_p} + ca \prod_p p^{-\ell_p} = bab_1 \prod_p p^{-\max(k_p,\ell_p)} + cac_1 \prod_p p^{-\max(k_p,\ell_p)}$$
$$= a(bb_1 + cc_1) \prod_p p^{-\max(k_p,\ell_p)} \subset G$$

since $bb_1 + cc_1$ is an ideal of A and $0 \le \max(k_p,\ell_p) \le h(p)$.

Let $a' = ba\,\Pi_p\, p^{-k_p} \subset a_G$, where b is an ideal of A and $0 \le k_p \le h(p)$ holds for every $p \in \mathbb{P}$, almost all k_p being zero. Thus $ba = a'\,\Pi_p\, p^{k_p}$. Since $k_p > 0$ implies that $p \nmid a$, then p^{k_p} divides b. This shows that $\Pi_p\, p^{k_p}$ divides b or a divides a'.

Since $a \subset G$, by definition, and a is an ideal of A, so $a \subset a_G = G \cap A$. To prove the converse, take arbitrary $x \in a_G$. Let $Ax = q_1^{s_1} \cdots q_t^{s_t}$, where $q_i \in \mathbb{P}$. Furthermore, there exists an ideal b of A and nonnegative integers k_p, $k_p = 0$ for almost all $p \in \mathbb{P}$, $0 \le k_p \le h(p)$, such that $Ax \subset ba\,\Pi_p\, p^{-k_p}$; i.e., $ba\,\Pi_p\, p^{-k_p}$ divides Ax. This implies that $ba \,|\, q_1^{s_1} \cdots q_t^{s_t}\, \Pi_p\, p^{k_p}$. Since $p \nmid a$ for $k_p > 0$, the previous remark gives that $a \,|\, q_1^{s_1} \cdots q_t^{s_t}$, or $Ax = q_1^{s_1} \cdots q_t^{s_t} \subset a$. In particular, $x \in a$. This proves that $a_G \subset a$. Finally, $a_G = a$.

> Let $\mathcal{B} = \mathcal{B}(T)$ be a base of neighborhoods of zero for the field (K,T) consisting of A-submodules of K. If there exists $G \in \mathcal{B}$ with $h_G(q) < \infty$, then $T_q \le T$. (7)

Indeed, let $n \in \mathbb{N}$ be arbitrary. Then there exists $H \in \mathcal{B}$ such that

$$Hq^{-h_G(q)-n} \subset G \tag{8}$$

The assumption implies that $a_G q^{-h_G(q)} \subset G$, and consequently,

$(a_G q^n)q^{-h_G(q)-n} \subset G$. Since $a_G \in \mathcal{B}$, there is $H \in \mathcal{B}$ satisfying (8). Thus $H \subset q^{h_G(q)+n} G$ and $a_H \subset a_G q^{h_G(q)+n}$, implying that $a_G q^{h_G(q)+n} | a_H$. This shows that $q^n | a_H$, i.e., $H \subset \{x \in K : v_q(x) \geq n\}$, since $h_H(q) = 0$, and (7) follows.

If $G \in \mathcal{B}$, then $h_G(p) = \infty$ for all but a finite number of $p \in \mathbf{P}$. (9)

In fact, the continuity of $x \mapsto x^{-1}$ implies that there exists $H \in \mathcal{B}$ with $(H + 1)^{-1} \subset G + 1$. Set

$$S = \{p \in \mathbf{P} : p \nmid a_H\} \tag{10}$$

Then for every $n \in \mathbb{N}$ and $p \in S$,

$$p^n + a_H = A \tag{11}$$

We shall prove that

$$a_G p^{-n} \subset G \tag{12}$$

Let $x \in a_G p^{-n}$. If $x = ab$, where $a \in a_G$, $b \in p^{-n}$, then $bp^n \subset A$. Take $b' \in p^n$ satisfying $bb' = 1$. Then $a_H = -1 - b' \in a_H$, by (11); i.e., $(a_H + 1)^{-1} = -1/b' \in G + 1$. We have for $x = ab$:

$$x = ab = a/b' = (-a)/(-b') \in (-a)(G + 1) \subset a_G(G + 1)$$
$$\subset Ga_G + a_G \subset G + A \cap G \subset G$$

as G is an A-submodule of K.

If $x = a_1b_1 + \cdots + a_Nb_N$, where $a_i \in a_G$, $b_i \in p^{-n}$, we repeat our arguments for each pair a_i, b_i. Since $n \in \mathbb{N}$ was arbitrarily chosen, $h_G(p) = \infty$ for every $p \in S$.

Proof of Theorem 1, continued: Now we are able to prove that (i) ⇒ (ii). Let $\mathcal{T}' = \sup_{p \in I} \mathcal{T}_p$, where I is the set of $p \in \mathbf{P}$ such that there exists $G \in \mathcal{B}$ with the property $h_G(p) < \infty$. Then $\mathcal{T}' \leq \mathcal{T}$ by (7). It follows from (9) that the set

$$T = \{p \in \mathbf{P} : h_G(p) < \infty\} \tag{13}$$

is finite. Let

$$V = \{x \in K : v_p(x) \le v_p(a_G) \text{ for every } p \in T\} \tag{14}$$

Obviously, $V \in \mathcal{B}(\mathcal{T}')$. We shall prove that $V \subset G$. Let $x \in V$ be any nonzero element. If $Ax = \Pi_p\, p^{v_p(x)}$, then set

$$n_p = \begin{cases} v_p(x) - v_p(a_G) & \text{for } p \in T \\ \max(v_p(x), 0) & \text{for } p \notin T \end{cases}$$

If $b = \Pi_p\, p^{n_p}$, then $Ax = ba_G\, \Pi_p\, p^{-k_p}$, where $k_p = 0$ for $p \in T$, and $k_p = n_p - v_p(x) + v_p(a_G) \ge 0$ for $p \notin T$. Lemma 1 implies that $x \in G$. Thus $G \in \mathcal{B}(\mathcal{T}')$. The proof is finished.

COROLLARY 1 Suppose that K is the quotient field of a Dedekind domain A equipped with an A-linear field topology $\mathcal{T}$. Then $(K,\mathcal{T})$ is a locally bounded topological field if and only if there exists a finite set $I \subset \mathbb{P}$ satisfying $\mathcal{T} = \sup_{p \in I} \mathcal{T}_p$.

Proof: ($\Leftarrow$) Obvious.

($\Rightarrow$) We have $\mathcal{T} = \sup_{p \in I} \mathcal{T}_p$, by Theorem 1. If I were infinite, it could be shown (Chapter 1, Example 12) that $\mathcal{T}$ is locally unbounded, contrary to our assumption (Chapter 9.3, proof of Th. 8).

COROLLARY 2 Let K be the quotient field of a principal ideal domain A. If $\mathcal{T} \in \mathcal{L}_f(K)$, then $\mathcal{T}$ has a base of the neighborhoods of zero consisting of A-submodules of K if and only if $\mathcal{T}$ is the supremum of a family of p-adic topologies, where p is a prime element of A.

In the case of ring topologies, we have the following.

THEOREM 2 (Heine and Warner [1, Th. 2]) Let K be the quotient field of a Dedekind domain A and $B \ne 0$ be an A-submodule of K. Of all the ring topologies on K for which B is open, let $\mathcal{T}_B$ be the weakest. If $\mathcal{B}$ is a base of neighborhoods of zero for a ring topology $\mathcal{T}$ on K, consisting of A-submodules of K, then

$$\mathcal{T} = \sup_{B \in \mathcal{B}} \mathcal{T}_B$$

Now we shall prove

THEOREM 3 (Heine and Warner [1, Th. 12]) Let K be the quotient field of a Dedekind domain A. If $T \in L_r(K)$ is a proper topology having a base of the neighborhoods of zero consisting of A-submodules of K, the following conditions are equivalent.

(i) $T = \sup_{p \in F} T_p$, where $F \subset \mathbb{P}$ is a finite subset.

(ii) There is a norm p on K, such that $T = T_p$.

(iii) $T \in L_f(K)$ and it is locally bounded.

Proof: (i) ⇒ (ii) If $T = \sup_{p \in F} T_p$, F finite, and if $v(x) = \max_{p \in F} v_p(x)$ for $x \neq 0$, then $T = T_p$, where $p(x) = e^{-v(x)}$, is easily seen to be a norm on K.

(ii) ⇒ (iii) Indeed, a nondiscrete topology defined on a field by a norm is a locally bounded field topology (see Chapter 3, Theorem 1).

(iii) ⇒ (i) by Corollary 1 to Theorem 1.

It is interesting to note that the property stated in Theorem 1 is characteristic of Dedekind domains. Namely, we have the following:

THEOREM 4 (Jebli [5, Prop. A.1.12]) Let A be a Krull domain, K its quotient field, and $\mathbb{P}_1$ the set of prime ideals of height 1. Then the following conditions are equivalent.

(i) Every proper topology $T \in L_f(K)$ with an A-submodule base of the neighborhoods of zero has the form

$$T = \sup_{p \in S} T_p \qquad \text{where } S \subset \mathbb{P}_1$$

(ii) A is a Dedekind domain.

THEOREM 5 (Jebli [5, Th. B.1.5]) Let A be any noetherian domain and K its quotient field. Then the following conditions are equivalent.

(i) Every proper topology $T \in L_f(K)$ with an A-submodule base of the neighborhoods of zero has the form

$$T = \sup_{p \in S} T_p \qquad \text{where } S \subset \mathbb{P}$$

(ii) A has dimension one and all its localizations are analytically irreducible.

DEFINITION 1 Let A be an integral domain and T a ring topology on A. We say that T is a *valuation topology* if there exists a valuation v on the quotient field K of A such that $T = T_v | A$.

DEFINITION 2 (Abouabdillah [2]) An integral domain A is called *topologically Prüferian* if for every prime ideal p of A the p-adic topology is a valuation topology.

It can be shown that the following result holds.

THEOREM 6 (Abouabdillah [2, Th. (2.12)]) Let A be any noetherian integral domain and K its quotient field. Then the following conditions are equivalent.

(i) Every proper A-linear topology $T \in L_f(K)$ has the form $T = \sup_{p \in T} T_p$, where T is a subset of the set $\mathbf{P}(A)$ of prime ideals of A.

(ii) A is topologically Prüferian.

For other variations on this theme, especially for other results dealing with suprema of p-adic topologies on the quotient fields of arithmetical domains, see Abouabdillah [1-3], Jebli [1-6], and Tabaâ [1].

We finish this chapter with the following general result.

THEOREM 7 (Cohen [8, Th. 1]) Let D be a Dedekind domain which is not a field, let K be the quotient field of D, and let $\mathbf{P}$ be the set of nonzero proper prime ideals of D. Suppose the class number of D is finite. If $T \in L_f(K)$ is proper, for which D is a bounded set, then there exists a nonempty subset $\mathbf{P}_1$ of $\mathbf{P}$ such that $T = \sup\{T_p : p \in \mathbf{P}_1\}$.

13

Locally Unbounded Topologies on Fields

This chapter can be treated as a continuation of Chapter 1, Section 5. Recall first that a ring topology T on a field K is called *locally unbounded* if it is not locally bounded. The first example of such a topology was given by Kowalsky [2] (see Chapter 12, Example 12) and Williamson [1] (see Chapter 1, Example 8). In order to see immediately that the topology defined by him is locally unbounded, one should apply Mutylin's result (Chapter 9, Theorem 13). Another example was provided by Gould [1] (Chapter 1, Example 10). Generalizing Gould's construction, Endo [2] gave an example of a locally unbounded field topology on the rational function field K(X), where K is provided with a locally bounded, proper field topology. The topologies in examples of Mutylin [3,4] (see Chapter 1, Example 11, and Chapter 11, Example 1) are also locally unbounded. Heine [2] has introduced the following.

DEFINITION 1 Let K be a field and $T \in L_r(K)$. The *topology* T is called *of type* N if T fails to be the supremum of locally bounded ring topologies on K.

We are going to show that every infinite field admits a ring topology of type N.

DEFINITION 2 A *field* K is called *algebraic* if it is a subfield of the algebraic closure of a finite field.

In other words, a field K is algebraic if char(K) = p $\neq$ 0 and K is algebraic over $\mathbb{F}_p$.

First we shall prove

THEOREM 1 (Kiltinen [1, Th. 6.1]) Let K be a field. Then the following conditions are equivalent.

(i) K is not algebraic.

(ii) K admits a nontrivial absolute value.

(iii) K admits a proper locally bounded ring topology.

Proof: (i) ⇔ (ii) by Theorem 8 of Chapter 2 (see also the proof of Theorem 6 of Chapter 1). (ii) ⇒ (iii) is obvious, since for every absolute value v on K, $\mathcal{T}_v$ is a locally bounded field (hence ring) topology (see Chapter 2, Theorem 13).

It remains to show that (iii) ⇒ (i). Suppose that K is algebraic and $\mathcal{T} \in \mathcal{L}_r(K)$ is nondiscrete and locally bounded. Let $U \in \mathcal{B} = \mathcal{B}(\mathcal{T})$ be arbitrary. Since $(K,\mathcal{T})$ is locally bounded, there exists $V \in \mathcal{B}$ such that $V \subset U$ and $VV \subset V$. Indeed, if $U_0 \subset U$, $U_0 \in \mathcal{B}$ is bounded, then $V = \{x \in U_0 : xU_0 \subset U_0\}$ has the desired property (see Chapter 1, Theorem 8). Thus $V^n \subset V$ holds for every $n \in \mathbb{N}$. Since K is algebraic, for every $x \in K^\times$ there exists $n \in \mathbb{N}$ satisfying $x^{-1} = x^n$. Thus, for $x \in V$, $x \neq 0$, we have $x^{-1} = x^n \in V^n \subset V$, or $(V^\times)^{-1} \subset V$.

Let $a \in K$ be arbitrary. Then there exists $W \in \mathcal{B}$ such that $aW \subset V$, by definition (see Chapter 1, Theorem 1). As $\mathcal{T}$ is nondiscrete, there is $b \in V \cap W$, $b \neq 0$, Then $ab \in V$, and finally, $a \in b^{-1}V \subset (V^\times)^{-1}V \subset V \subset U$. Thus $K \subset V \subset U$, so $\mathcal{T} = \mathbf{0}_K$; i.e., $\mathcal{T}$ is trivial, a contradiction.

Now we can prove

THEOREM 2 Every algebraic field K admits a ring topology of type N.

Proof: K admits no locally bounded ring topology by Theorem 1. Thus every ring topology on K is of type N. Since K is infinite, there is a proper field topology $\mathcal{T}$ on K (by Theorem 6 of Chapter 1) which is of type N.

THEOREM 3 (Heine [2, Th. 3.1]) Let K be a field. Then the following conditions are equivalent.

(i) K is infinite.

(ii) K admits an unbounded field topology.

Proof: (ii) ⇒ (i) since if K is finite, then $\mathbf{0}_K$ and $\mathbf{1}_K$ are the only ring topologies on K, both being locally bounded.

(i) ⇒ (ii) Let P be the prime subfield of K and B be a transcendence basis of K over P. If $B = \emptyset$ and $\operatorname{char}(K) \neq 0$, K is an infinite algebraic field, admitting a proper field topology T (by Theorem 6 of Chapter 1). Consequently, (K,T) is locally unbounded, by Theorem 1. Therefore, let K be nonalgebraic. If $B = \emptyset$ and $\operatorname{char}(K) = 0$, then let $\mathbb{P}$ denote the set of all primes. If $B \neq \emptyset$, then let $x \in B$ and let $\mathbf{P}$ be the set of all monic irreducible polynomials $p \in P(B \setminus \{x\})[x]$.

In both cases we have $|\mathbf{P}| = |K| \geq \aleph_0$. For every $p \in \mathbb{P}$ let T_p denote the p-adic topology on P(B) and T_p^* any extension of T_p to K. If $p, q \in \mathbf{P}$, $p \neq q$, then $T_p^* \neq T_q^*$. Let $Y \subset \mathbf{P}$ be an infinite subset. Then

$$F_Y = \sup\{T_p^* : p \in Y\} \tag{1}$$

is an unbounded field topology on K (see Chapter 9, Theorem 8).

As a corollary we obtain

THEOREM 4 (Heine [2, Th. 3.2]) Every nonalgebraic field K possesses at least $2^{|K|}$ unbounded field topologies.

Proof: Let $Y, Z \subset \mathbb{P}$ be infinite subsets (see the proof of Theorem 3). Assume that $p \in Y \setminus Z$. Then the set $\{a \in K : v_p(a) \geq 1\}$ (v_p the p-adic exponential valuation on K) is a neighborhood of zero in (K,F_Y) but not in (K,F_Z). Consequently, $F_Y \neq F_Z$. Thus K admits at least $2^{|\mathbb{P}|} = 2^{|K|}$ unbounded field topologies.

Now we shall prove some quantitive results on the topologies of type N.

Theorem 2 follows immediately from the following.

THEOREM 5 (Podewski [1]) Every countable field admits $2^{2^{\aleph_0}}$ field topologies of type N.

For some fields the number of all ring topologies with prescribed properties can be given. Namely, the following results hold.

THEOREM 6 (Heine [1, Sätze 2.1, 2.2, 3.2]) Let K be a nonalgebraic field. Then:

(i) There are exactly $2^{|K|}$ locally bounded ring topologies on K.

(ii) There are exactly $2^{2^{|K|}}$ ring topologies on K.

(iii) There are exactly $2^{2^{|K|}}$ ring topologies of type N on K.

Proof: We shall prove only statements (i) and (ii). The proof of (iii) is of the same type but more technical in its nature. Let $\mathbb{P}$ have the same meaning as in the proof of Theorem 3.

(i) Recall that $\mathcal{T}(A)$ denotes the locally bounded ring topology on K defined by the almost-order $\mathcal{O}(A) = \bigcap_{p \in A} V_p$, where V_p is the valuation ring of the p-adic valuation of K. Let B, C $\subset \mathbb{P}$, and B $\neq$ C. Assume that $p \in B \setminus C$. Then $\mathcal{T}(B) \neq \mathcal{T}(C)$. Indeed, since $\mathcal{O}(B) \subset \mathcal{O}(C)$, thus $\mathcal{T}(C) \leq \mathcal{T}(B)$. It remains to show that $\mathcal{T}(B) \not\leq \mathcal{T}(C)$. Assume the contrary, i.e., $\mathcal{T}(B) \leq \mathcal{T}(C)$. Then there exists an $x \in K^\times$ satisfying $x\mathcal{O}(C) \subset \mathcal{O}(B)$. Note that $p^{-1} \in \mathcal{O}(C)$, since for every q-adic exponential valuation ($q \in C$), $v_q(p^{-1}) = 0$. Consequently, $xp^{-n} \in \mathcal{O}(B)$ holds for every $n \in \mathbf{N}$. Thus

$$x \in \bigcap_{n \in \mathbf{N}} p^{-n}\mathcal{O}(B) \subset \bigcap_{n \in \mathbf{N}} p^n V_p = \{0\}$$

since $\{p^n V_p : n \in N\}$ is a base of the neighborhoods of zero in the p-adic topology $\mathcal{T}_p$, a contradiction! Thus K admits at least $2^{|\mathbb{P}|} = 2^{|K|}$ bounded ring topologies. On the other hand, every such topology is induced by an almost-order, being a subset of K. Consequently, there exist at most $2^{|K|}$ bounded ring topologies on K.

(ii) Let $X \neq \emptyset$ be any set and $\emptyset \neq \mathcal{C} \subset 2^X$. Recall that the family $\mathcal{C}$ of subsets of X is called a dual filter on X if the conditions

(DF_1) $A, B \in \mathcal{C} \Rightarrow A \cup B \in \mathcal{C}$

(DF_2) $A \in \mathcal{C}, B \subset A \Rightarrow B \in \mathcal{C}$

hold for every A, B. This means that the set $\mathcal{F} = \{C' = X \setminus C : C \in \mathcal{C}\}$ is a filter on X. It is easy to see that there exist exactly $2^{2^{|X|}}$ dual filters on a set X, i.e., as many filters as on the set X. Put

$$S(\mathcal{E}) = \sup\{\mathcal{T}(E) : E \in \mathcal{E}\} \tag{2}$$

where $\mathcal{E}$ is a dual filter on $\mathbf{P}$. Let $\mathcal{C}$, $\mathcal{D}$ be dual filters on $\mathbf{P}$, $\mathcal{C} \neq \mathcal{D}$. Assume that $C_0 \in \mathcal{C} \setminus \mathcal{D}$. We shall prove that

$$S(\mathcal{C}) \neq S(\mathcal{D}) \qquad \text{if } \mathcal{C} \neq \mathcal{D} \tag{3}$$

Suppose that $S(\mathcal{C}) = S(\mathcal{D})$. Then $\mathcal{T}(C_0) \leq \sup\{\mathcal{T}(D) : D \in \mathcal{D}\}$, i.e., there exist $x_i \in K^\times$ and $D_i \in \mathcal{D}$ satisfying

$$x_1\mathcal{O}(D_1) \cap \cdots \cap x_n\mathcal{O}(D_n) \subset \mathcal{O}(C_0) \tag{4}$$

Since $D_1 \cup \cdots \cup D_n \in \mathcal{D}$, then $C_0 \not\subset D_1 \cup \cdots \cup D_n$. Take $a \in C_0 \setminus (D_1 \cup \cdots \cup D_n)$ and $t \in x_1\mathcal{O}(D_1) \cap \cdots \cap x_n\mathcal{O}(D_n)$. Thus $t = x_ig_i$ with $g_i \in \mathcal{O}(D_i)$, $i = 1, \ldots, n$. Then $g_ia^{-m} \in \mathcal{O}(D_i)$ if $i = 1, \ldots, n$ and $m \in \mathbb{N}$. Consequently, $a^{-m}t = x_ig_ia^{-m} \in x_i\mathcal{O}(D_i)$ for $i = 1, \ldots, n$ and $m \in \mathbb{N}$, i.e., $a^{-m}t \in x_1\mathcal{O}(D_1) \cap \cdots \cap x_n\mathcal{O}(D_n) \subset \mathcal{O}(C_0)$, a contradiction, since $a^{-m}t \notin \mathcal{O}(C_0)$ for sufficiently large $m \in \mathbb{N}$. This implies that $S(\mathcal{C}) \neq S(\mathcal{C})$. The proof is finished.

Similar results were obtained by Kiltinen [4]. He has remarked that if a field K has infinite transcendence degree over its prime subfield, it admits $2^{2^{|K|}}$ field topologies of type N (Kiltinen [4, p. 35]). For similar results see also J.-A. Cohen [6].

We finish this chapter with the following result.

THEOREM 7 (Berrondo [6]) For every field k there exists a topological field $(L,\mathcal{T})$ such that

(i) k is a discrete subfield of L; i.e., $\mathcal{T}|k = \mathbf{1}_k$.

(ii) $(L,\mathcal{T})$ is linear, locally unbounded, and complete.

Sketch of the proof: Let k be an arbitrary field and K a pure transcendental field extension of k, admitting a nonmetrizable additive Krull valuation v (i.e., such that the value group $\Gamma = v(K^\times)$ has no cofinal countable subset). We take as K the field defined in Example 2 of Chapter 8 (see Berrondo [5]). Let A be the valuation ring of v. Since v is nonmetrizable, it can be shown that K((X)) is the quotient field of A[[X]]. Let $\mathcal{T}$ be a topology on K((X)) defined by the set of all nonzero ideals of A[[X]]. It can be shown that $(K((X)),\mathcal{T})$ is a complete topological field. The valuation v can be extended to K(X) by putting

$$\bar{v}\left(\sum_{k=0}^{N} a_k X^k\right) = \inf\{v(a_k) : k = 0, 1, \ldots, N\} \tag{5}$$

if not all a_i are equal to zero. It turns out that if $(K(X),\mathcal{T}_{\bar{v}})$ is complete, then $(K(X),\mathcal{T})$ is also complete.

Now let $a \in K$ satisfy $v(a) = 0$. Put $f_a(X) = X + a$, $f_a|K = id_K$. The function f_a can be extended to the automorphism f_a of K(X). If $\mathcal{T}_a = f_a(\mathcal{T})$, the topological fields $(K(X),\mathcal{T})$ and $(K(X),\mathcal{T}_a)$ are topologically and algebraically isomorphic. Now putting

$$T = \sup\{\mathcal{T}_a : a \in K, v(a) = 0\}$$

we obtain that the topological field (K(X),T) is complete, linear, and locally unbounded. The field K(X) contains k as a discrete subfield.

14

An Extension Problem for Topological Fields

In this chapter the following problem will be discussed: Let $(K,\mathcal{T})$ be a proper topological field and let L be a field extension of K. Can the topology $\mathcal{T}$ always be extended to a field topology on L? If L is a finite extension, it is always possible by Theorem 16 of Chapter 2. In particular, if L is a finite field extension of a field K endowed with a field topology $\mathcal{T}_p$, induced by a norm p, then the Cartesian product topology on L is also induced by a norm, as follows from Theorem 6 of Chapter 3. The case when $\mathcal{T}$ is induced by a valuation was considered in Chapter 4, Section 4.

Let $(K,\mathcal{T})$ be a locally bounded topological field and let X be an arbitrary topological space. Denote by C(X,K) the ring of all continuous functions from X into $(K,\mathcal{T})$. In the case of $K = \mathbb{R}$ ($\mathcal{T} = \mathcal{T}_\infty$), Henriksen (see Hinrichs [1]) posed the following question:

Can C(X,K) be topologized as an algebra over $(K,\mathcal{T})$
in such a way that every maximal ideal is closed? (1)

Hinrichs [1] has remarked that the following equivalence is true in the real case.

LEMMA 1 Let X be an arbitrary topological space. Then the following conditions are equivalent.

(i) C(X,K) admits an algebra topology over $(K,\mathcal{T})$ having all maximal ideals closed.

(ii) For every maximal ideal m of C(X,K), the residue class field C(X,K)/m admits a Hausdorff algebra topology.

The proof is similar to that of the real case. Thus the extension problem for topological fields arises in a natural way. On the other hand, the Steinitz decomposition theorem for fields reduces the general extension problem to the case of a purely transcendental and algebraic field extension. We shall discuss these two cases separately.

DEFINITION 1 Let X be a topological space and let $(K,\mathcal{T})$ be a topological field. We say that a *function* $f \in C(X,K)$ *is bounded* if $f(X)$ is a bounded set in $(K,\mathcal{T})$. Let $C^*(X,K)$ be the algebra of all bounded functions from $C(X,K)$.

Obviously, $C^*(X,K)$ is a K-subalgebra of $C(X,K)$. It is well known that if K is a locally compact and connected topological field (i.e., if K is topologically isomorphic either to $\mathbb{R}$ or to $\mathbb{C}$, by Pontrjagin's theorem (Chapter 10, Theorem 2), then for every maximal ideal m of $C^*(X,K)$, $C^*(X,K)/m \simeq K$ topologically, where X is an arbitrary topological space. This fact is known as Stone's theorem.

It motivates the following.

DEFINITION 2 Let X be a topological space and $(K,\mathcal{T})$ a topological field. If

$$C^*(X,K)/m \simeq K \quad \text{(topologically)} \tag{2}$$

holds for every maximal ideal m of $C^*(X,K)$, we say that Stone's theorem holds for $C^*(X,K)$.

DEFINITION 3 Let $(K,\mathcal{T})$ be a topological field. We say that $(K,\mathcal{T})$ is a *Stone's field* if Stone's theorem holds for $C^*(X,K)$ for every topological space X.

Correl and Henriksen [1] and Goldhaber and Wolk [1] obtained several results on Stone's theorem and Stone's fields. The most general result was given by Kowalsky [3]:

THEOREM 1 (Kowalsky [3, Sätze 4 and 5]) Every locally compact topological field is a Stone's field. Conversely, every Stone's field of type V is locally compact.

This theorem gives another characterization of the locally compact topological fields from the point of view of functional analysis.

We present next some partial solutions of the extension problem. We start with the following.

THEOREM 2 (Podewski [2]) Let $(K,\mathcal{T})$ be a proper topological field with countable K. If X is a transcendental element over K, $\mathcal{T}$ can be extended to a field topology on K(X).

If K is a countable field and L a purely transcendental field extension of K, having at most countable transcendence degree over K, every field topology can be extended from K to L. In the uncountable case, little is known.

Generalizing a construction of Williamson [1] (see Chapter 1, Example 8), Hinrichs obtained the following:

THEOREM 3 (Hinrichs [1, Th. 2.1]) Let $(K,\mathcal{T})$ be a topological subfield of a locally compact field and let L be a purely transcendental field extension of K. Then $\mathcal{T}$ admits an extension to a field topology on L.

A proof of this theorem is similar to that of Example 8 of Chapter 1, where one should take Haar's measure instead of Lebesgue's measure.

Mutylin [4] modified Williamson's example and proved that there exists a locally unbounded and complete field extension of $(\mathbb{C},\mathcal{T}_\infty)$ (Mutylin [4, Th. 4]). This shows that the extension problem may have solutions not so "regular" as the properties of an extended field topology. A detailed account of the known results is presented in Chapter 15.

Now it is necessary to say something on solutions of the extension problem for algebraic extensions of fields. Hinrichs [1, Th. 3.4] proved that every locally bounded field topology can be extended to a ring topology from a field to its algebraic extension of at most countable degree. Arnautov and Vizitiu extended this result to arbitrary algebraic field extensions. That is, they proved the following:

THEOREM 4 (Arnautov and Vizitiu [2, Ths. 1 and 2]) Let L be an algebraic field extension of a field K. If $T \in L_i(K)$ (i = r, f) is a locally bounded topology, there is a topology $S \in L_i(L)$ (i = r, f) such that $S|K = T$.

However, the topology in their construction cannot be locally bounded (except in the case of finite-dimensional extensions).

DEFINITION 4 Let m be a cardinal number. A topological space X is called an *m-space* if for every $n < m$ the intersection of n arbitrary open sets is an open set. By an *m-topology* we shall mean a topology of an m-space. A *topology* T is called *m-minimal* if the only m-topology on X, properly weaker than T, is the trivial topology.

The following theorem holds.

THEOREM 5 (Vizitiu [1]) Let (K,T) be a proper topological field and L an algebraic field extension of K. If T is an m-topology and the dimension of L over K is less than m, there exists an m-topology $S \in L_f(L)$, extending T. Furthermore, if T is a locally bounded field topology, then S also is a locally bounded field topology.

However, in the most interesting case of $m = \aleph_0$, Theorem 5 gives nothing new, since this case was considered earlier (Chapter 2, Theorem 17). Following Weber [4], we shall prove the following.

THEOREM 6 (Weber [4, Satz (1.6)]) Every proper locally bounded ring topology can be extended from a field to a proper locally bounded ring topology of its arbitrary algebraic extension.

We shall need the following:

LEMMA 2 Let L be a field extension of a field K. Suppose that $T \in L_r(L)$, $T \neq \mathbb{1}_L, \mathbb{0}_L$. Let $U \in B(T)$ be a bounded set. Suppose that $A \subset U$ has the property that for every $a \in L$ there exists a nonzero polynomial f with coefficients in A such that $f(a) = 0$. Then $L = U(A^\times)^{-1}$.

Proof: Note at first that if $z \in L^\times$ satisfies $z(U + U) \subset U$, then $Uz + Uz^2 + \cdots + Uz^n \subset U$ for every $n \in \mathbb{N}$. We shall apply induction. For $n = 1$ the assumption is true. Suppose that $Uz + Uz^2 + \cdots + Uz^n \subset U$. Then

$$\begin{aligned} Uz + Uz^2 + \cdots + Uz^n + Uz^{n+1} &= Uz + z(Uz + \cdots + Uz^n) \\ &\subset Uz + Uz \subset U \end{aligned}$$

Now let $a \in L^\times$ be arbitrary and put $b = -1/a$. There exists $z_1 \in U^\times$ such that $z_1(U + U) \subset U$ and $z_1 b(U + U) \subset U$. Indeed, since U is bounded, $U + U$ also is bounded. Thus there is $V \in \mathcal{B}(T)$ satisfying $V(U + U) \subset U$. Moreover, there exists $V_1 \in \mathcal{B}(T)$ with $bV_1 \subset V$, by (B_5) (Chapter 1, Theorem 1). Let $V_2 = V_1 \cap bV_1$. Obviously, $V_2 \in \mathcal{B}(T)$ and V_2 satisfies $V_2(U + U) \subset U$. Taking arbitrary $z_1 \in V_2^\times$, we have $z_1(U + U) \subset U$ and $bz_1(U + U) \subset U$. Put $z_2 = bz_1$. By assumption, there exist $a_0, a_1, \ldots, a_n \in A$, $a_0 \neq 0$, such that

$$\sum_{k=0}^{n} a_k z_2^k = 0$$

Let $c = z_1 \Sigma_{k=1}^n a_k z_2^{k-1}$. Then

$$\begin{aligned} c &= z_1\left(a_1 + \sum_{k=1}^{n-1} a_{k+1} z_2^k\right) \in z_1\left(A + \sum_{k=1}^{n-1} A z_2^k\right) \subset z_1\left(U + \sum_{k=1}^{n-1} U z_2^k\right) \\ &\subset z_1(U + U) \subset U \end{aligned}$$

Since

$$a_0 + bc = a_0 + bz_1\left(a_1 + \sum_{k=1}^{n-1} a_{k+1} z_2^k\right) = a_0 + a_1 z_2 + \cdots + a_n z_2^n = 0$$

then

$$a = -\frac{1}{b} = \frac{c}{a_0} \in U(A^\times)^{-1}$$

LEMMA 3 Suppose that $T \in L_r(L)$ is a proper locally bounded topology on a field L. If $U \in \mathcal{B}(T)$ is a bounded neighborhood of zero and K is a subfield of L, $V = U \cap K$, then

(i) $T|K = \mathbb{1}_K$ or $\mathcal{B}(T|K) = \{aV : a \in K^\times\}$.

(ii) $T|K \neq \mathbb{1}_K \Leftrightarrow L = U(V^\times)^{-1}$.

(iii) If L is algebraic over K and $K = V(V^\times)^{-1}$, then $\mathcal{B}(T|K) = \{aV : a \in K^\times\}$.

Proof: Suppose that $T|K \neq \mathbf{1}_K$. Then $\{aU : a \in K^\times\}$ is a base of the neighborhoods of zero for T. Thus the sets

$$aV = a(U \cap K) = aU \cap K \qquad a \in K^\times$$

form a base of the neighborhoods of zero for $T|K$. Thus (i) holds.

(ii) If $T|K \neq \mathbb{1}_K$, then for every $a \in L$ there exists $v \in V^\times$ such that $av \in U$, i.e., $a \in U(V^\times)^{-1}$, or $L = U(V^\times)^{-1}$. Conversely, if $L = U(V^\times)^{-1}$, then for every $a \in L^\times$ there exist $u \in U$, $v \in V^\times a^{-1}$, $a^{-1} = uv^{-1}$. Thus $v = ua \in Ua \cap V^\times = Ua \cap U \cap K^\times \subset Ua \cap K^\times$. Thus $V \neq \{0\}$ and $T|K \neq \mathbf{1}_K$.

To prove (iii), note that $L = U(K^\times)^{-1}$, by Lemma 2. Since $K = V(V^\times)^{-1}$, we have

$$L = U(K^\times)^{-1} = U(V^\times(V^\times)^{-1})^{-1} = UV^\times(V^\times)^{-1} \subset U(V^\times)^{-1}$$

since $V \subset U$ and $L = U(V^\times)^{-1}$. Condition (ii) implies that $T|K \neq \mathbf{1}_K$. Finally, $\mathcal{B}(T|K) = \{aV : a \in K^\times\}$, by (i).

Proof of Theorem 6: Let K be a field and T a proper locally bounded ring topology on K. Let $V \in \mathcal{B}(T)$ be an almost-order inducing T and $z \in V^\times$ an 2-addiator of V, i.e., $z(V + V) \subset V$. Suppose that L is an algebraic extension of K. To prove the theorem it is sufficient to prove existence of an almost-order U in L satisfying $U \cap K = V$. Then the topology S induced in L by U is locally bounded and it extends T. We shall apply the Kuratowski-Zorn lemma. Let the set

$$\mathcal{X} = \{W \subset L : W \cap K = V,\ z(W + W) \subset W,\ WW \subset W,\ W(W^\times)^{-1} = W(V^\times)^{-1}\}$$

be partially ordered by inclusion. Let U be a maximal element of $\mathcal{X}$. It is easy to see that $F = U(U^\times)^{-1}$ is a field. Indeed, if $x \in F^\times$, then $x^{-1} \in F^\times$. Let $x = ab^{-1}$, $y = cd^{-1}$, with $a, c \in U$; $b, d \in U$. Thus

$$xy = ac(bd)^{-1} \in U^2((U^\times)^2)^{-1} \subset U(U^\times)^{-1} = F^\times$$

$$x + y = \frac{ad + bc}{bd} \in (U + U)(U^{\times})^{-1} = z(U + U)(zU^{\times})^{-1} \subset U(U^{\times})^{-1} = F$$

since $z \in V \subset U$, and $UU \subset U$. Moreover, K is a subfield of F. In fact, $U \cap K = V$ and $K = V(V^{\times})^{-1}$, which implies that $K \subset F$.

We shall prove that $F = L$. Suppose to the contrary that $F \neq L$ and take $d \in L \setminus F$. Since d is algebraic over K, then d is algebraic over F. Let

$$f(X) = a_0 + a_1X + \cdots + a_nX^n \in F[X]$$

be the minimal polynomial of d over F. We can suppose that $a_j \in U$ for $j = 0, 1, \ldots, n$, since every a_j has a form c_j/d_j with c_j, $d_j \in U$. Now put

$$B = U + dU + \cdots + d^{n-1}U \quad \text{and} \quad W = \{x \in F(d) : xB \subset B\}$$

It is not difficult to see that $UB \subset B$, $z(B + B) \subset B$, $U = B \cap F$, which leads to $W \cap F = U$, and this implies that W has the following properties:

$$WW \subset W \qquad z(W + W) \subset W \qquad W \cap K = (W \cap F) \cap K = U \cap K = V$$

We shall prove that

$$W(W^{\times})^{-1} = W(U^{\times})^{-1}$$

The inclusion $\supset$ is obvious since $U \subset W$. Note first that $za_nd \in W$. Indeed, $za_nd^n = -z(a_0 + a_1d + \cdots + a_{n-1}d^{n-1}) \in zB$, as $a_j \in U$. Hence

$$\begin{aligned} za_ndB &= za_n(dU + \cdots + d^{n-1}U) + za_nd^nU \\ &= za_n(dU + \cdots + d^{n-1}U) + z(-a_0 - a_1d - \cdots - a_{n-1}d^{n-1})U \\ &\subset z(dU + \cdots + d^{n-1}U) + z(U + dU + \cdots + d^{n-1}U) \\ &\subset zB + zB \subset B \end{aligned}$$

Since $a_n \neq 0$ (f is the minimal polynomial of d), we obtain that $d \in W(U^{\times})^{-1}$. Now let $\alpha, \beta \in W$ be arbitrary. Then, as $\alpha/\beta \in F(d)$, it has the form

$$\gamma = c_0 + c_1 d + \cdots + c_{n-1} d^{n-1} \quad \text{with } c_j \in F$$

Since $d \in W(U^\times)^{-1}$, also $d^2 \in W(U^\times)^{-1}$, and similarly $d^k \in W(U^\times)^{-1}$ for $k = 0, 1, \ldots, n-1$ (we recall that $UU \subset U$ and $WW \subset W$). Note that $\gamma \in W(U^\times)^{-1}$. Indeed,

$$\gamma \in F + F \cdot W(U^\times)^{-1} + \cdots + F \cdot W(U^\times)^{-1}$$

But $W(U^\times)^{-1} + W(U^\times)^{-1} \subset W(U^\times)^{-1}$, and by induction,

$$\underbrace{W(U^\times)^{-1} + \cdots + W(U^\times)^{-1}}_{(n \text{ times})} \subset W(U^\times)^{-1}$$

Let $a, a' \in W$; $b, b' \in U^\times$. Then

$$\frac{a}{b} + \frac{a'}{b'} = \frac{ab' + a'b}{bb'} = \frac{z(ab' + a'b)}{zbb'} \in z(WU + WU)(z(U^\times)^2)^{-1}$$

$$\subset z(W + W)(zU^\times)^{-1} \subset W(U^\times)^{-1}$$

since $z \in V \subset U$. In this way we have proved that $W(W^\times)^{-1} = W(U^\times)^{-1}$. Thus $W \in X$, $U \subsetneq W$, a contradiction! This shows that F, defined as $U(U^\times)^{-1}$, equals L. Finally, the locally bounded ring topology of L defined by U has the desired properties.

Some new results in this direction have been obtained by Arnautov [9] and Kiltinen [6].

THEOREM 7 (Arnautov [9]) If L is a simple transcendental field extension of a field K, then every locally bounded metrizable ring topology on K can be extended to a ring topology on L.

THEOREM 8 (Kiltinen [6, Th. 11]) If L is any countable field and K is an arbitrary subfield of L, then every locally bounded metrizable field topology on K can be extended to a ring topology on L.

Sketch of the proof of Theorem 7: Let $L = K(x)$ be a simple transcendental field extension of K, and $\mathcal{T} \in \mathcal{L}_r(K)$. The topological field $(K,\mathcal{T})$ satisfies, by assumption, the first axiom of countability. Then there exists a sequence $a_n \in K^\times$ such that $a_n \to 0$ in $(K,\mathcal{T})$. Let $V_0 \in \mathcal{B}(\mathcal{T})$ be an almost-order defining $\mathcal{T}$. In particular, $V_0V_0 \subset V_0$. Moreover, $a_n^{-1}V_0$ are bounded sets and $K = \bigcup_{n=1}^{\infty} a_n^{-1}V_0$. Since the join, product, and sum of a finite number of bounded sets is again bounded (by Lemma 11 of Chapter 1), it is possible to define inductively a sequence of elements $d_n \in K^\times$ and a sequence of bounded subsets V_n of K, such that

$$d_n \to 0 \qquad \text{in } (K,T) \tag{3}$$

$$d_n \in V_0 \qquad d_1V_0 + d_1V_0 \subset V_0 \tag{4}$$

$$0,\ 1 \in V_1 \tag{5}$$

$$V_nV_n \subset V_{n+1} \qquad \text{holds for every } n \in \mathbb{N} \tag{6}$$

$$V_n + \cdots + V_n \ (n+1 \text{ times}) \subset V_{n+1} \qquad \text{for } n \in \mathbb{N} \tag{7}$$

$$d_nV_0 + d_n((V_0V_n + \cdots + V_0V_n) \ n \text{ times}) \subset d_{n+1}V_0 \tag{8}$$

holds for every $n = 2, 3, \ldots$

$$V_1 \cup V_2 \cup \cdots = K \tag{9}$$

Conditions (5), (6), and (8) imply that $\Sigma_{1=m+1}^{m+s} d_iV_0 \subset d_mV_0$ holds for every m, s $\in \mathbb{N}$, i.e., $d_NV_0 \subset d_nV_0$ if $N \geq n$.

Now we are able to construct a topology $T_L \in L_r(L)$ extending T. Note that L is the quotient field of the polynomial ring K[x]. Let

$$f(x) = x^n + b_{n-1}x^{n-1} + \cdots + b_0 \in K[x]$$

Define

$$n(f) = \min\{n \in \mathbb{N} : b_i \in V_n \text{ for every } i = 0, 1, \ldots, n-1\} \tag{10}$$

It is easy to see that if f, g $\in$ K[x] are both monic, then

$$n(fg) \leq n(f) + n(g) + \deg(f) + \deg(g) \tag{11}$$

Let $m \in \mathbb{N}$ and let $N = (n_1, n_2, \ldots)$ be an increasing sequence of natural numbers. Put

$$W_{mN} = \left\{ \sum_{i=1}^{s} \frac{1}{f_i(x)} b_i x^{q_i} \in L : f_i(x) = x^{k_i} + \sum_{j=0}^{k_i - 1} b_{ji} x^j, \right.$$

$$\left. b_i \in V_0 d_{n_i} d_{m(q_i + k_i + n(f_i))} \right\} \tag{12}$$

It can be shown (Arnautov [9]) that the set of all W_{mN} taken as a base of neighborhoods of zero defines a ring topology $\mathcal{T}_L \in \mathcal{L}_r(L)$ satisfying $\mathcal{T}_L|L = \mathcal{T}$.

The following fact plays an essential role in the proof that $\mathcal{T}$ has the desired properties. Let

$$f(x) = x^k + \sum_{i=0}^{k-1} b_i x^i \qquad n \geq k \text{ and } b'_j \in V_0 d_{n+k+n(f)} \tag{13}$$

hold for every j = 0, 1, ..., n. If

$$\frac{1}{f(x)} \sum_{i=0}^{n} b'_j x^j = \sum_{i=0}^{n-k} c_i x^i + \frac{1}{f(x)} \sum_{i=0}^{k-1} c'_i x^i$$

then $c_i \in V_0$ holds for every i = 0, 1, ..., n - k.

15

Open Problems

In this chapter we shall put together several questions concerning topological fields, posed by several authors. Moreover, we shall reproduce Mutylin's classification of complete fields (see Mutylin [4]) and describe the state of these problems.

1. (old problem) Does there exist a locally unbounded minimal field topology? The answer is not known, even in the case of the rational number field.

2. Is it true that for every $\mathcal{T} \in L_f(K)$, $\mathcal{T}$ being a proper topology, there exists a minimal field topology $\mathcal{T}$ on K, weaker than $\mathcal{T}$?

3. Describe algebraic properties of the completion $\hat{K}$ of $(K,\mathcal{T})$, where K is a field and $\mathcal{T}$ is a proper field topology on K. In particular, is it true that the completion of a field with respect to a field topology is either a field or has zero divisors?

It is known (Kowalsky [2]; Mutylin [6]) that $\hat{K}$ is a field for minimal $\mathcal{T}$. On the other hand, Zobel [1] proved that the completion of the rational number field $\mathbb{Q}$ with respect to a locally unbounded ring topology (Mutylin's example) is a proper integral domain in which the only invertible elements are those of $\mathbb{Q}^{\times}$.

Stenzel [1] has proved several interesting results dealing with the algebraic structure of the completion of a field. Recall first the following notion. Let $(A,\mathcal{T})$ be a topological ring. An element $a \in A$ is called *topologically regular* if for every net $x_\alpha \in A$, $ax_\alpha \to 0$ implies that $x_\alpha \to 0$. Stenzel has proved the following result (Stenzel [1, Chap. 5, Th. (5.2)]). Suppose that K is a field and $T \in L_r(K)$ a proper locally bounded topology such that $N_0(T) \neq 0$. Then every topologically regular element from the completion $\hat{K}$ of (K,T) is a unit in $\hat{K}$. This result implies that if (K,T) is a locally bounded topological ring, K being a field, such that every nonzero element of the completion $\hat{K}$ of (K,T) is topologically regular, then $\hat{K}$ is a topological field. Moreover, Stenzel has constructed a field K with the topology induced by a norm and such that the completion $\hat{K}$ has (algebraically) nilpotent elements.

We mention also a result of Rigo and Warner [1]. Let v be a real valuation on a field K of characteristic $p > 0$ such that $\hat{K}$ contains an element $a \notin K$ satisfying $a^p \in K$. Then there exist p topologies on K(a) extending the topology of K, all defined by norms; the completion of K for one is a field, topologized by the unique extension of v to K(a), and the completion for the others are local rings whose maximal ideals are nonzero and nilpotent.

4. (Kiltinen [2]) Under what conditions does the uniform continuity of the inverse of a polynomial of degree 3 or more imply that the topology is of type V? (See Chapter 5, Section 2.)

5. Describe all proper locally bounded topological fields which have only finitely many nonisomorphic locally bounded field extensions.

The case of minimal locally bounded field topologies can be found in Knopfmacher and Sinclair [1] and Więsław [5] (see Chapter 4, Theorem 11).

6. (Więsław [6, p. 884]) Can a nonalgebraically closed field be complete in two proper and nonequal field topologies?

For topologies induced by absolute values or maximal valuation, the answer is no (see Chapter 8). Berrondo [5], answering this question, gave an example of a nonalgebraically closed field complete with respect to two independent (nonmaximal) valuations. It is not clear whether such an example can be found with metrizable topologies (see Chapter 8, Example 2).

7. (Więsław [6, p. 883]) Are every two different minimal field topologies independent? For partial results, see Chapter 7 (also Weber [2] and Więsław [10]).

8. (Więsław [6, p. 881]) Are $\mathbb{R}$ and $\mathbb{C}$ the only complete and connected fields with a finite group of topological automorphisms?

9. (Więsław [6, p. 882]) Is $\mathbb{C}$ the only nondiscrete complete and algebraically closed field with a finite group of topological automorphisms?

10. (Heine [2]) Does every infinite field admit field topologies of type N?

11. (Heine [3]) Is it true that every uncountable field F admits $2^{|F|}$ first-countable ring topologies of type N?

12. (Kiltinen [4]) How many first countable, locally bounded field topologies are there on a field K of finite transcendence degree over its prime subfield? The number of such topologies is not greater than $2^{|K|}$ (see Chapter 2, Theorem 9, and Chapter 13, Theorem 6).

We present here a calculation of the number of the first countable, locally bounded ring topologies on a field K of finite transcendence degree over its prime subfield, given by Hans Weber (in a letter dated December 6, 1982).

If K is algebraic over a finite field, the discrete and the trivial topologies are the only locally bounded ring topologies of K. Thus the required number is 2.

Suppose that the transcendence degree of K over its prime field is finite but K is not algebraic over a finite field. This implies that K is countable. Thus every locally bounded ring topology on K satisfies the first axiom of countability, since then there is a bounded neighborhood U of zero such that $\{aU : a \in K^\times\}$ is a base of the neighborhoods of zero. Now either K is algebraic over $\mathbb{Q}$ (in this case we put $F = \mathbb{Q}$) or there is a subfield k of K and an element $x \in K$, transcendental over k, such that K is algebraic over $F = k(x)$. Let $\varphi_1, \varphi_2, \ldots$ be pairwise nonequivalent absolute values of F. Then it is easy to see that the almost-orders $\mathcal{O}(M) = \{a \in F : \forall\, n \in M, \varphi_n(a) \leq 1\}$, $M \subset \mathbb{N}$, define nonequivalent ring topologies of F. The number of such almost-orders is $2^{\aleph_0}$. Every such topology can be extended from F to K (see Chapter 14, Theorem 6). Consequently, the required number of topologies is $2^{|K|}$.

Now let K be a global field. Then every locally bounded field topology on K is the supremum of finitely many topologies induced by absolute values (Chapter 3, Theorem 12, and Chapter 9, Theorem 11). Consequently, the number of first countable locally bounded field topologies on K equals $|K| = \aleph_0$.

13. Is it possible to describe all locally bounded ring topologies on a simple transcendental extension of an infinite field, for example, on $\mathbb{Q}(X)$ or $\mathbb{R}(X)$?

14. (D. B. Šahmatov, in a letter dated February 7, 1984) Does there exist a topological field whose underlying space is heriditary normal (T_5) but not perfectly normal (T_6)?

15. (Shell [4, p. 43]) Can a simple or algebraic extension of a disconnected field be connected?

16. (Shell [4, p. 48]) Can every field with a Hausdorff ring topology be embedded in a field with a connected ring topology?

17. (A. V. Arhangelskiĭ) Does there exist in every topological field a family of open sets B_0, linearly ordered by inclusion, such that $\cap B_0 = \{0\}$?

18. (Šahmatov [3, p. 257]) Let $(K,\mathcal{T})$ be a topological field. Does there exist a linear order $\leq$ on the set K, such that $Y_x = \{y \in K : y \leq x\}$ is closed in $(K,\mathcal{T})$ for every $x \in K$?

19. (Šahmatov [3, p. 257]) Does there exist a topological space which can be topologically embedded in a topological division ring, but cannot be embedded in a topological field?

20. (Šahmatov [3, p. 257]) Assume that a topological space X can be topologically embedded in a topological field of characteristic p. Can X be topologically embedded also in a topological field of characteristic q ($p \nmid q$ primes or zero)?

21. (V. K. Belnov) Does there exist for every $n \in \mathbb{N}$ a topological field $(K_n,\mathcal{T}_n)$ whose underlying topological space has dimension n?

Other problems are stated in Tables 1 to 4, where necessary references are mentioned. We consider only proper ring topologies on fields.

Table 1. Locally Bounded Topologies on Fields

Field L	Kind of topologies on L			
	Absolute values	Valuations of rank > 1	Locally bounded field topologies	Locally bounded ring topologies
$L = \mathbb{Q}$	+ Ostrowski [3]	-	+ Mahler [2]	+ Shanks and Warner [2]; Weber [1]
$L \neq \mathbb{Q}$ $[L:\mathbb{Q}] < \infty$	+ Ostrowski [4]	-	+ Mahler [2]; Kiyek [6]	+ Weber [3]; Stenzel [1]
$L \neq \mathbb{Q}$ dim transc$[L:\mathbb{Q}]$ is finite and nonzero	?	?	? Cohen [5]	? Only the number of such topologies is known [see comments on Problem 12]
dim transc$[L:\mathbb{Q}]$ is infinite (for example, $\mathbb{R}$, $\mathbb{C}$)	? Only the number of absolute values is known: $2^{\lvert L\rvert}$	?	? Only the number of such topologies is known: $2^{\lvert L\rvert}$ Kiltinen [4]; Cohen [6]	? Only the number of such topologies is known: $2^{\lvert L\rvert}$ Heine [1]
$L = \mathbb{F}_p(X)$	+ Ostrowski [4]	-	+ Kiyek [6]; Cohen [1]; Stenzel [1]	+ Nichols and Cohen [1]; Weber [3]
$L \neq \mathbb{F}_p(X)$ $[L:\mathbb{F}_p(X)] < \infty$	+ Ostrowski [4]	-	+ Kiyek [6]; Weber [3]; Endo [5,6]	+ Nichols and Cohen [1]; Weber [3]
dim transc$[L:\mathbb{F}_p(X)] \geq 1$	? Only the number of absolute values is known: $2^{\lvert L\rvert}$ (Chapter 2, Theorem 9)	?	? Only the number of such topologies is known: $2^{\lvert L\rvert}$ Cohen [6]	?

Table 2. Structure of Complete and First-Countable Topological Fields

	Field		
Prime field[a]	Locally unbounded	Locally bounded non-absolute-valued	Absolute-valued
Locally unbounded	+ Mutylin [4]	-	+
Discrete	+ Mutylin [4]	+ Zelinsky [3]	+
$\mathbb{Q}_p$	?	+ Mutylin [2]	+
$\mathbb{R}$	+ Williamson [1]	- Gelfand-Mazur	+ Only $\mathbb{R}$ and $\mathbb{C}$ (Mazur's theorem)

[a]By the "prime field of a topological field," we mean the least closed subfield of this field.

Source: Mutylin [4].

Table 3. Structure of Complete and Non-First-Countable Topological Fields

	Field[a]					
Prime field	Locally unbounded		Locally bounded non-absolute-valued		Valued	
Locally unbounded	?	?	-	-	-	-
Discrete	+ Berrondo [6]; Stenzel [1, Chap. (4.2)]	+	?	?	+	-
$\mathbb{Q}_p$	?	?	-	-	-	-
$\mathbb{R}$	?	?	-	-	-	-

[a]The first sign corresponds to non-first-countable topological fields and the second to fields complete with respect to ring topologies without continuous division $x \mapsto x^{-1}$.

Source: Mutylin [4].

Table 4. Structure of Fields Complete with Respect to Minimal Field Topologies

Prime Field	Field: Locally unbounded	Field: Locally bounded nonvalued	Field: Valued
Locally unbounded	?	-	-
Discrete	?	- Kowalsky and Dürbaum [1]; Fleischer [1]	+
$\mathbb{Q}_p$	?	- Kowalsky and Dürbaum [1]; Fleischer [1]	+
$\mathbb{R}$	?	-	+

Source: Mutylin [4].

Appendix

Linearly Ordered Groups

DEFINITION 0 An ordered (or partially ordered) set $(X,\leq)$ is directed if for every $x, y \in X$ there exists $z \in X$ such that $x \leq z$ and $y \leq z$.

DEFINITION 1 An ordered (or partially ordered) group is a group Γ together with the relation $\leq$ satisfying the following conditions:

(i) $a \leq a$,

(ii) $a \leq b$ and $b \leq a \Rightarrow a = b$.

(iii) $a \leq b$ and $b \leq c \Rightarrow a \leq c$.

(iv) $a \leq b \Rightarrow ac \leq bc$ and $ca \leq cb$.

for every $a, b, c \in \Gamma$.

DEFINITION 2 An ordered group Γ is called *lattice ordered* (or l-group) if for every $a, b \in \Gamma$ there exist $\sup(a,b)$ and $\inf(a,b)$ (denoted also by $a \vee b$, resp. $a \wedge b$).

DEFINITION 3 An ordered group Γ is said to be *linearly ordered* if for every $a, b \in \Gamma$ either $a \leq b$ or $b \leq a$ holds.

DEFINITION 4 Let $f: \Gamma \to \Delta$ be a homomorphism from an ordered group Γ into an ordered group Δ. We say that f is an *order homomorphism* if f preserves the order, i.e., if for every $a, b \in \Gamma$, $a \leq b$ implies that $f(a) \leq f(b)$.

We define an order isomorphism similarly.

Let Γ be a linearly ordered multiplicative group with unit e. We define an absolute value in Γ in the following way:

$$|\gamma| = \gamma \text{ if } \gamma \geq e \quad \text{and} \quad |\gamma| = \gamma^{-1} \text{ otherwise}$$

DEFINITION 5 Let Γ be a linearly ordered group. A subgroup Δ of Γ is called *isolated* if the following condition is satisfied:

(i) For every $\delta \in \Delta$ and $\gamma \in \Gamma$, $|\gamma| \leq |\delta|$ implies that $\gamma \in \Delta$.

LEMMA 1 The set $G(\Gamma)$ of all isolated subgroups of a linearly ordered group Γ is linearly ordered by inclusion.

Proof: Let $\Delta_1, \Delta_2 \in G(\Gamma)$. Suppose that there exists $\delta_2 \in \Delta_2 \setminus \Delta_1$. Then $\delta_2 > \delta_1$ for every $\delta_1 \in \Delta_1$, since otherwise there would be a $\delta_0 \in \Delta_1$ with $|\delta_2| \leq |\delta_0|$ and $\delta_2 \in \Delta_1$, contrary to the assumption. Thus $\Delta_1 \subset \Delta_2$.

DEFINITION 6 The *rank* of a linearly ordered group Γ is defined as the order type of the set $G(\Gamma) \setminus \{\Gamma\}$.

In particular, Γ is of rank $0 \Leftrightarrow \Gamma = \{e\}$.

DEFINITION 7 A linearly ordered group $(\Gamma,\leq)$ is said to be *Archimedean* (and the order $\leq$ is called Archimedean) if for every a, $b \in \Gamma$, $a > e$, there exists $n \in \mathbb{N}$ satisfying $b \leq a^n$. Otherwise, we say that the order $\leq$ (and group) is *non-Archimedean*.

LEMMA 2 Let $(\Gamma,\leq)$, $\Gamma \neq \{e\}$, be a linearly ordered group. The following conditions are equivalent.

(i) Γ has rank 1.

(ii) Γ is Archimedean.

Proof: For arbitrary $a \in \Gamma$, $a \neq e$, let $\Gamma_a = \bigcup_{n \in \mathbb{N}} \{b \in \Gamma : |b| \leq |a|^n\}$. It is clear that Γ_a is the least isolated subgroup of Γ containing a. Thus the rank of Γ is $1 \Leftrightarrow$ for every $a \in \Gamma$, $a > e$, $\Gamma_a = \Gamma \Leftrightarrow$ for every $a, b \in \Gamma$, $a > e$, there exists $n \in \mathbb{N}$ satisfying $|b| \leq a^n$.

THEOREM 1 A linearly ordered group $(\Gamma,\leq)$ is Archimedean if and only if it is order isomorphic to a subgroup of the multiplicative group of positive reals $\mathbb{R}_+^\times$.

Proof: Since $\mathbb{R}_+^\times$ is an Archimedean ordered group, only the converse statement needs a proof.

Let Γ be an Archimedean ordered group. We shall define an order isomorphism of Γ into the additive reals $\mathbb{R}^+$. (Clearly $\mathbf{R}^+$ and $\mathbf{R}_+^\times$ are order isomorphic.)

Suppose first that there exists $g \in \Gamma$ such that $g > e$ and $e \leq x < g$ implies that $x = e$. Since Γ is Archimedean, for every $a \in \Gamma$ one can find $n \in \mathbb{N}$ with $g^n \leq a < g^{n+1}$, hence $e \leq g^{-n}a < g$. Thus $ag^{-n} = e$ and $a = g^n$. This shows that the mapping $g^n \mapsto n$ is an order isomorphism of $\Gamma = gp(g)$ onto $\mathbb{Z}$.

Now suppose that for every $x \in \Gamma$, $e < x$, there exists $y \in \Gamma$ with $e < y < x$. We have $y^2 \leq x$, or $(xy^{-1})^2 \leq x$; i.e., for every $x > e$ there exists $z \in \Gamma$ satisfying $e < z < x$ and $z^2 \leq x$. Indeed, $y^2 > x$ and $(xy^{-1})^2 > x$ cannot hold simultaneously: $(xy^{-1})^2 = xy^{-1}xy^{-1} > x$ gives $y^{-1}xy^{-1} > e$, or $xy^{-1} > y$, $x > y^2$, contradicting the assumption $y^2 > x$.

Now we prove that an Archimedean ordered group $(\Gamma,\leq)$ is abelian. Suppose to the contrary that there are elements $a, b \in \Gamma$, $b > e$, with $ab \neq ba$. Let, for example, $ba < ab$. Put $x = [a,b] = aba^{-1}b^{-1}$. Since $x > e$, there exists $z \in \Gamma$ satisfying $e < z < z^2 < x$. The order $\leq$ of Γ being Archimedean, there exist $m, n \in \mathbb{N}$ such that

$$z^m \leq a < z^{m+1} \quad\text{and}\quad z^n \leq b < z^{n+1} \tag{2}$$

Thus for $x = [a,b]$, $x = aba^{-1}b^{-1} < z^{m+1}z^{n+1}z^{-m}z^{-n} = z^2$, contradicting $z^2 \leq x$. This proves that Γ is abelian.

Finally, we are able to construct the desired isomorphism. Let $a \in \Gamma$, $a > e$. Put $f(a) = 1$. Define for $b \in \Gamma$,

$$L_b = \left\{\frac{m}{n} \in \mathbb{Q} : n > 0,\ a^m \leq b^n\right\} \qquad U_b = \left\{\frac{m}{n} \in \mathbb{Q} : n > 0,\ a^m > b^n\right\}$$

Since Γ is Archimedean, L_b and U_b are nonempty. Moreover, L_b and U_b define a Dedekind cut of $\mathbb{Q}$ ($L_b \subset \mathbb{Q}$; $L_b \neq \emptyset$; $p \in L_b$, $q < p \Rightarrow q \in L_b$; L_b has no greatest number). Indeed, let $p = m/n \in L_b$, $q = r/s < p$. Then $rn < sm$. The inequality $a > e$ gives $e < a^{rn} < a^{sm} = (a^m)^s \leq (b^n)^s = b^{ns}$, i.e., $(a^{-r}b^s)^n \geq e$. This implies that $a^{-r}b^s \geq e$ and $a^r \geq b^s$. Let $\beta \in \mathbb{R}^+$ be the real number defined by the Dedekind cut L_b. Put $f(b) = \beta$. The mapping $f\colon \Gamma \to \mathbb{R}^+$ is a homomorphism: $f(bc) = f(b) + f(c)$ holds for every $b, c \in \Gamma$. In fact, take $m/n \in L_b$, $r/s \in L_c$; then $a^m \leq b^n$, $a^r \leq c^s$. Thus

$$a^{ms+nr} \le (bc)^{ns} \qquad \text{i.e.,} \quad \frac{ms + nr}{ns} = \frac{m}{n} + \frac{r}{s} \in L_{bc}$$

since $f(b) = 0$ cannot hold for $b > e$ (otherwise, $a \le b^n$, or $1/n \in L_b$, contradicting the fact that L_b contains only negative numbers). Finally, it is immediate that $b \le c$ implies that $L_b \subset L_c$.

Thus we have proved that f is an order isomorphism of Γ into $\mathbb{R}^+$.

Remark There exist linearly ordered abelian groups of rank > 1. For example, the product $\Gamma_1 \times \Gamma_2$ of two Archimedean ordered groups Γ_1 and Γ_2 with lexicographic ordering

$$(\gamma_1, \gamma_2) \le (\delta_1, \delta_2) \Leftrightarrow \gamma_1 < \delta_1 \quad \text{or} \quad \gamma_1 = \delta_1 \quad \text{and} \quad \gamma_2 \le \delta_2$$

has rank greater than 1.

Notation

$\|A\|$	cardinal number of a set A
$\mathbb{N}$	{1,2,3,...}—the natural numbers
$\mathbb{Z}$	integers
$\mathbb{Q}$	rational numbers
$\mathbb{R}$	real numbers
$\mathbb{R}_+$	real nonnegative numbers
$\mathbb{R}_+^\times$	real positive numbers
$\mathbb{C}$	complex numbers
$A^\times$	$A \setminus \{0\}$, where A is a ring; in particular,
$K^\times$	multiplicative group of a field K
$A < B$	A is a subgroup of a group B
gp(A)	group generated by A
A[X]	ring of polynomials in one variable over A
$A[\{X_i\}_{i \in I}]$	ring of polynomials in $\|I\|$ variables over A; in particular,
$A[X_1,\ldots,X_n]$	ring of n variables over A
K(X)	rational function field of a variable X over a field K
A + B	$\{a + b : a \in A, b \in B\}$, where A, B are subsets of a group with commutative addition $(x,y) \mapsto x + y$
AB	$\{ab : a \in A, b \in B\}$, where A, B are subsets of a set with a binary associative operation $(x,y) \mapsto xy$
A^{-1}	$\{a^{-1} : a \in A\}$, where $A \subset K^\times$, K being a field
$\frac{A}{B}$	AB^{-1} - A, B subsets of a field K; $B \subset K^\times$
A^n	$AA\cdots A$ (n times) inductively

Bibliography

Abouabdillah, D.

1. Topologies de corps A-linéaires, *Bull. Assoc. Enseignan. Math. Sér. B* (Rabat), no. 1 (1978), 3-14.
2. Topologies de corps A-linéaires, *Colloq. Math.* 46 (1982), no. 2, 171-180.
3. Topologies de corps A-linéaires, *Pacific J. Math.* 107 (1983), no. 2, 257-266.

Adams, H. E.

1. The completion of a ring with a valuation, *Proc. Amer. Math. Soc.* 36 (1972), 21-25.
2. Filtrations and valuations on rings, *Pacific J. Math.* 43 (1972), 7-11.

Alajbegovič, J.

1. On a Cauchy sequences of a valued field, Algebr. Conf., Novi Sad 1981, Novi Sad, 1982, 39-46.

Alas, O. T.

1. Metrizable topologies on the real numbers, *Amer. Math. Monthly* 78 (1971), 773.

Albert, A. A.

1. On p-adic fields and rational division algebras, *Ann. Math.* 41 (1940), 674-693.
2. Division algebras over a function field, *Duke Math. J.* 8 (1941), 750-762.
3. Absolute valued real algebras, *Ann. Math.* 48 (1947), 495-501.
4. Absolute valued algebraic algebras, *Bull. Amer. Math. Soc.* 55 (1949), 763-768.

Alling, N. L.

1. Valuation theory of meromorphic function field over open Riemann surfaces, *Acta Math.* 110 (1965), 79-95.

Andrunakievič, V. A., and Arnautov, V. I.

1. Invertibility in topological rings, *Soviet Math. Dokl.* 7 (1966), no. 5, 1254-1257.

Anthony, J. M.

1. Topologies for quotient fields of commutative integral domains, *Pacific J. Math.* 36 (1971), 583-601.

Arapović, M.

1. Approximation theorems for fields and commutative rings, *Glas. Mat. Ser. III* 18 (38) (1983), 61-66.

Arens, R. F.

1. Linear topological division algebras, *Bull. Amer. Math. Soc.* 53 (1947), 623-630.
2. A generalization of normed rings, *Pacific J. Math.* 2 (1952), 455-471.

Arens, R. F. and Hoffman, K.

1. Algebraic extensions of normed algebras, *Proc. Amer. Math. Soc.* 7 (1956), 203-210.

Aranutov, V. I.

1. Criterion of normability of topological rings (Russian), *Algebra i Logika* 4 (1965), vyp. 4, 4-24.
2. Topologizations of the ring of integers (Russian), *Izv. Akad. Nauk. Moldav. SSR, Ser. Fiz.-Tehn. Mat. Nauk* 1 (1968), 3-15.
3. On topologization of countable rings, *Siberian Math. J.* 9 (1968), 939-946.
4. An example of an infinite ring admitting only the discrete topology (Russian), *Mat. Issled.* 5 (1970), vyp. 3 (17), 182-185.
5. Nondiscrete topologizability of infinite commutative rings (Russian), *Mat. Issled.* 5 (1970), vyp. 4 (18), 3-15.
6. Nondiscrete topologizability of countable rings, *Soviet Math. Dokl.* 11 (1970), no. 2, 423-426.
7. On topologizations of infinite modules (Russian), *Mat. Issled.* 7 (1972), vyp. 4 (26), 241-243.
8. Continuation of a topology of a commutative ring to some of its quotient rings (Russian), *Mat. Issled.*, vyp. 48 (1978), Algebra i moduli, 3-13.
9. Continuation of a locally bounded topology of a field to its simple transcendental extension (Russian), *Algebra i Logika* 20 (1981), no. 5, 511-521.

Arnautov, V. I., and Ursul, M. I.

1. Embedding of topological rings in connected ones (Russian), *Mat. Issled.*, vyp. 49 (1979), 11-15.

Arnautov, V. I., and Vizitiu, V. N.

1. Extension of a locally bounded topology on a field to algebraic extensions of the field, *Soviet Math. Dokl.* 15 (1974), no. 3, 808-812.

2. ———(Russian), *Bul. Akad. Štiince RSS Moldoven* 9 (1974), no. 2, 29-43.

Arnautov, V. I., Vodinčar, M. I., and Mihalev, A. V.

1. *An Introduction to the Theory of Topological Rings and Modules* (Russian), Akad. Nauk Moldavskoi SSR, Kishinev, "Štiinca," 1981, 1-174.

Artin, E.

1. Über die Bewertungen Algebraischer Zahlkörper, *J. Reine Angew. Math.* 167 (1932), 157-159.

2. *Theory of Algebraic Numbers*, Translated and distributed by Georg Striker, Schildweg 12, Gottingen, 1959.

Artin, E., and Whaples, G.

1. Axiomatic characterization of fields by the product formula for valuations, *Bull. Amer. Math. Soc.* 51 (1945), 469-492.

2. A note on arithmetic characterization of fields, *Bull. Amer. Math. Soc.* 52 (1946), 245-247.

Aubert, K. E.

1. Some characterizations of valuation rings, *Duke Math. J.* 21 (1954), 517-525.

2. Über Bewertungen mit halbgeordneter Wertgruppe, *Math. Ann.* 127 (1954), 8-14.

3. Characterizations des anneaux de valuation a l'aide de la théorie des r-ideaux, Sem. P. Dubreil et C. Pisot, no. 10 (1958).

Aurora, S.

1. Multiplicative norms for metric rings, *Pacific J. Math.* 7 (1957), 1279-1304.

2. On power multiplicative norms, *Amer. J. Math.* 80 (1958), 879-894.

3. The embedding of certain metric fields, *Michigan Math. J.* 7 (1960), 123-128.

4. A note on certain connected metric division rings, *Michigan Math. J.* 7 (1960), 129-132.

5. Normed fields which extend normed rings of integers, *Pacific J. Math.* 33 (1970), 21-28.

6. On normed rings with monotone multiplication, *Pacific J. Math.* 33 (1970), 15-20.

Ax, J., and Kochen, S.

1. Diophantine problems over local fields I, *Amer. J. Math.* 87 (1965), 605-630.

2. ——— II, *Amer. J. Math.* 87 (1965), 631-648.

3. ——— III, *Ann. Math.* 83 (1966), 437-456.

Ayoub, C.

1. On the submodules of a field, *Monatsh. Math.* 72 (1968), 193-199.

Bachman, G.

1. *Introduction to p-adic Numbers and Valuation Theory*, Academic Press, New York, 1964.

Baer, R.

1. Dichte, Archimedizität und Starrheit geordneter Körper, *Math. Ann.* 188 (1970), 165-205.

Baer, R., and Hasse, H.

1. Zusammenhang und Dimension topologischer Körperräume, *J. Reine Angew. Math.* 167 (1932), 40-45.

Ballier, F.

1. Über lineartopologischen Algebren, *J. Reine Angew. Math.* 195 (1956), 42-75.

Banach, S.

1. Remarques sur les groupes et les corps métriques, *Studia Math.* 10 (1948), 178-181.

Banaschewski, B.

1. Minimal topological algebras, *Math. Ann.* 211 (1974), 107-114.

Bandyopadhyay, S. P.

1. Valuations in groups and rings (Russian), *Sibirsk. Mat. Ž.* 6 (1965), vyp. 5, 1176-1180.

2. Valuations in groups and rings, *Czechoslovak Math. J.* 19 (94) (1969), 275-276.

3. Rings with valuations, *Math. Nachr.* 44 (1970), 1-18.

4. On U-semigroups, *Math. Nachr.* 44 (1970), 19-27.

Barot, J.

1. A multiplicative functional on the commutative topological field, *Arch. Math. (Brno)* 14 (1978), 191-192.

Bastos, G. A.

1. A characterization of complete subfields of complete valuated fields, *Bol. Soc. Brasil. Mat.* 6 (1975), no. 1, 53-56.

Beer, S.

1. Über die Diskrepanz von Folgen in bewerteten Körpern, *Manuscripta Math.* 1 (1969), 201-209.

Belfi, V. A., and Doran, R. S.

1. Norm and spectral characterization in Banach algebras, *Enseign. Math. (2)* 26 (1980), nos. 1-2, 103-130.

Benz, H.

1. *Über eine Bewertungstheorie der Algebren und ihre Bedeutung für die Arithmetik*, Akademie-Verlag, Berlin, 1961.
2. Über diskret fastbewertete perfekte Ringe und die Arithmetik der Hauptordnungen in Algebren, *J. Reine Angew. Math.* 209 (1962), 42-81.

Bergman, G. M.

1. A weak nullstellensatz for valuations, *Proc. Amer. Math. Soc.* 28 (1971), 32-38.

Bergman, G. M., and Waterman, A. G.

1. Connected fields of arbitrary characteristic, *J. Math. Kyoto Univ.* 5 (1966), 177-184.

Berrondo, F.

1. Prolongement des valuations discrétes aux extensions de dimension finie séparable, *C. R. Acad. Sci. Paris, Sér. A-B* 274 (1972), 151-154.
2. Critères d'irréductibilité des polynômes sur les anneaux des valuation henséliens, *C. R. Acad. Sci. Paris Sér. A-B* 276 (1973), 1269-1272.
3. Corps topologiquement henséliens, *C. R. Acad. Sci. Paris Sér. A-B* 281 (1975), 305-307.
4. Characterization topologique des corps values henseliens, Publ. Dép. Math. (Brest), 1975.
5. Un corps non algebriquement clos complet pour deux valuations independantes, *C. R. Acad. Sci. Paris Sér. A-B* 282 (1976), 675-677.
6. Corps topologiques linéaires non localement bornés et complete, *C. R. Acad. Sci. Paris Sér. A-B* 284 (1977), 843-846.

Boeme, T. K.

1. On the limits of the Gelfand-Mazur theorem. With a discussion by A. C. Cochran, Proc. Conf. Convergence Spaces, Univ. Nevada, Reno, Nevada, 1976, 1-4.

Borewicz, Z. I., and Šafarevič, I. R.

1. *The Theory of Numbers* (Russian), second printing, Izdatelstvo "Nauka," Moscow, 1972.

Bourbaki, N.

1. *Elémentes des mathématiques*, Livre III, *Topologie gènérale*, Hermann, Paris.

Bourne, S.

1. On normed semialgebras, *Studia Math.* 21 (1961), 45-54.
2. On locally compact positive half-fields, *Math. Ann.* 146 (1962), 423-426.
3. On convex topological halfalgebras, *Portugal. Math.* 24 (1965), 59-63.

Braconnier, J.

1. Groupes d'automorphismes d'un groupe localement compact, *C. R. Acad. Sci. Paris Sér. A-B* 220 (1945), 382-384.
2. Sur les modules localement compacts, *C. R. Acad. Sci. Paris Sér. A-B* 222 (1946), 527-529.
3. Sur les espaces vectoriels localement compacts, *C. R. Acad. Sci. Paris Sér. A-B* 222 (1946), 777-778.

Brown, R.

1. An approximation theorem for extended prime spots, *Canad. J. Math.* 24 (1972), 167-184.
2. Valuations, primes and irreducibility in polynomial rings and rational function fields, *Trans. Amer. Math. Soc.* 174 (1972), 451-488.
3. Tamely ramified extensions of Henselian fields, *Rocky Mountain J. Math.* 5 (1975), 543-557.

Brown, R., and Harrison, D. K.

1. Tamely ramified extensions of linearly compact fields, *J. Algebra* 15 (1970), 371-375.

Brown, R., and Warner H. D.

1. Quadratic extensions of linearly compact fields, *Trans. Amer. Math. Soc.* 163 (1972), 379-399.

Brungs, H. H.

1. Generalized discrete valuation rings, *Canad. J. Math.* 21 (1969), 1404-1408.

Buchanan, T., and Hähl, H.

1. On the kernel and the nuclei of 8-dimensional locally compact quasifields, *Arch. Math. (Basel)* 29 (1977), no. 5, 472-480.

Canals, I.

1. Generalization of the concept of algebraicity over complete valued fields, *Bol. Soc. Mat. Mexicana (2)* 13 (1968), 63-83.

Cantini, A.

1. On non-Archimedean structures of D. Klaua, *Math. Nachr.* 87 (1979), 35-41.

Cantor, D. G.

1. On the Stone-Weierstrass approximation theorem for valued fields, *Pacific J. Math.* 21 (1967), 473-478.

Carini, L.

1. Sull'estensions delle topologie (m)-adiche di un dominio fattoriale al suo campo delle frazioni, *Rend. Circ. Mat. Palermo* 25 (1976), no. 1-2, 136-144.

Carruth, P. W.

1. Generalized power series fields, *Trans. Amer. Math. Soc.* 63 (1948), 548-559.

Cashwell, E. D., and Everett, C. J.

1. Formal power series, *Pacific J. Math.* 13 (1963), 45-64.

Chabauty, C.

1. Sur la theorie des fonctions sur un corps value, *C. R. Acad. Sci. Paris* 231 (1950), 396-397, 432-434.

Chabauty, R.

1. Une generalization au théoreme de Gelfand-Mazur, *C. R. Acad. Sci. Paris Sér. A-B* 275 (1972), 519-522.

Charnow, A.

1. The automorphisms of an algebraically closed field, *Canad. Math. Bull.* 13 (1970), 95-97.

Chernoff, P. R., Rasala, R. A., and Waterhouse, W. C.

1. The Stone-Weierstrass theorem for valuable fields, *Pacific J. Math.* 27 (1969), 233-240.

Chion, J.

1. Associative ordered rings (Russian), *Dokl. Akad. Nauk. SSSR Ser. Mat.-Fiz. Nauk* 101 (1955), 1005-1007.

2. Rings normed by semigroups (Russian), *Izv. Akad. Nauk SSSR Ser. Mat.* 21 (1957), 311-328.

von Choosy, R., and Priess-Crampe, S.

1. Ordnungsvertägliche Bewertungen eines angeordneten Körpers, *Arch. Math. (Basel)* 26 (1975), no. 4, 372-387.

Coffi-Nketsia, B.-J.

1. Valuation des anneaux á l'aide des demigroupes totalement ordonnés avec zéro, Thése doct. sci. math., Fac. Sci. Paris, 1970, 253 pp.

Cohen, I. S.

1. On the non-Archimedean normed spaces, *Indag. Math.* 10 (1948), 244-249.

Cohen, I. S., and Zariski, O.

1. A fundamental inequality in the theory of extensions of valuations, *Illinois J. Math.* 1 (1957), 1-8.

Cohen, J.-A.

1. Locally bounded topologies on F(X), *Pacific J. Math.* 70 (1977), 125-132.
2. The strong approximation theorem and locally bounded topologies on F(X), *Pacific J. Math.* 87 (1980), 59-63.
3. Topologies on the ring of integers of a global field, *Pacific J. Math.* 93 (1981), no. 2, 269-276.
4. Locally bounded topologies on the ring of integers of a global field, *Canad. J. Math.* 33 (1981), no. 3, 571-584.
5. Norms on F(X), *Pacific J. Math.* 109 (1983), no. 1, 81-87.
6. On the number of locally bounded field topologies, *Proc. Amer. Math. Soc.* 90 (1984), no. 2, 207-210.
7. Extensions of valued topologies, *Pacific J. Math.* 125 (1986), no. 1, 39-44.
8. Topologies on the quotient field of a Dedekind domain, *Pacific J. Math.* 117 (1985), no. 1, 51-67.

Cohen, P. J.

1. Decision procedure for real and p-adic fields, *Comm. Pure Appl. Math.* 22 (1969), 131-151.

Cohn, P. M.

1. An invariant characterization of pseudovaluations on a field, *Proc. Cambridge Philos. Soc.* 50 (1954), 159-177.

Cohn, P. M., and Mahdavi-Hezavehi, M.

1. Extensions of valuations on skew fields, ring theory, Antwerp, 1980 (Proc. Conf., Univ. Antwerp, Antwerp, 1980), 28-41. Lecture Notes in Math., 825, Springer, Berlin, 1980.

Cohn, P. M., and Mahler, K.

1. On the composition of pseudo-valuations, *Nieuw Arch. Wisk.* 3 (1953), 161-198.

Correl, E.

1. On topologies for fields, Diss., Purdue Univ., Lafayette, Ind., 1958.
2. On a class of topologies for fields (preprint, 1958).
3. Topologies on quotient fields, *Duke Math. J.* 35 (1968), 175-178.

Correl, E., and Henriksen, M.

1. On rings of bounded continuous functions with values in division ring, *Proc. Amer. Math. Soc.* 7 (1956), 194-198.

Dankert, G.

1. Hensel's lemma for Laurent series, *Bull. Acad. Polon. Sci. Sér. Sci. Math. Astronom. Phys.* 18 (1970), 287-291.

van Dantzig, D.

1. Studien over Topologische Algebra, Diss., Amsterdam, 1931.

2. Zur topologischen Algebra I, Komplettierungstheorie, *Math. Ann.* 107 (1933), 587-626.

Dazard, J.

1. Sur les anneaux filtres de Gelfand, *C. R. Acad. Sci. Paris Sér. A-B* 262 (1966), 326-328.

De Marco, G., and Richter, M.

1. Rings of continuous functions with values in a non-Archimedean ordered field, *Rend. Sem. Mat. Univ. Padova* 45 (1971), 327-336.

Deuring, M.

1. Verzweigungstheorie bewerteter Körper, *Math. Ann.* 105 (1931), 277-307.

2. Lectures on the theory of algebraic functions of one variable, *Lecture Notes in Mathematics* 314, Springer-Verlag, New York, 1973.

Dieudonné, J.

1. Sur les fonctions continues p-adiques, *Bull. Sci. Math.* 68 (1944), 79-95.

2. Sur les corps topologiques connexes, *C. R. Acad. Sci. Paris Sér. A-B* 221 (1945), 396-398.

3. Linearly compact spaces over sfields, *Amer. J. Math.* 73 (1951), 13-19.

Dikranjan, D.

1. Minimal precompact topologies on rings, *Annuaire Univ. Sofia Fac. Math.* 67 (1972-1973), 391-397.

2. Minimal extensions of finite fields, Mathematics and education in mathematics, Proc. 8th Spring Conf. Union Bulg. Math., Sunny Beach, 1979, 208-215 (Bulgarian).

3. Extensions of minimal ring topologies, General topology and its relations to modern analysis and algebra, V, Proc. 5th Prague Topol. Symp., 1981, *Sigma Ser. Pura Math.* 3 (1983), 132-144.

Dikranjan, D., and Więsław, W.

1. Rings with only ideal topologies, *Comment. Math. Univ. St. Paul* 29 (1980), no. 2, 147-157.

Dock, S.-R.

1. Relatively complete fields, *Duke Math. J.* 24 (1957), 197-200.

Doignon, J.-P.

1. Sur les espaces projectifs topologiques, *Math. Z.* 122 (1971). 57-60.

Dubrovin, N. I.

1. Non-commutative valuation rings (Russian), *Trudy Moskov. Mat. Obšč.* 45 (1982), 265-280.

2. Non-commutative valuation rings in simple finite-dimensional algebras over fields (Russian), *Mat. Sb.* 123 (1984), no. 4, 496-509.

Dürbaum, H.

1. Über die Ganzheitsbereiche bewerteter Körper, *Math. Z.* 57 (1952), 86-93.

2. Zur Theorie der nichtkommutativen Bewertungen, *Proc. Amer. Math. Soc.* 4 (1953), 418-422.

Ellis, R.

1. A note on the continuity of the inverse, *Proc. Amer. Math. Soc.* 8 (1957), 372-373.

El-Mallah, M. L.

1. Sur les algèbres absolument valuées qui vérifient l'identité (x,x,x) = 0, *J. Algebra* 80 (1983), 314-322.

El-Mallah, M. L., and Micali, A.

1. Sur les dimensions des algèbres absolument valuées, *J. Algebra* 68 (1981), 237-246.

Endler, O.

1. Über einem Existenzsatz der Bewertungstheorie, *Math. Ann.* 150 (1963), 54-65.

2. Bewertungstheorie unter Benutzung einer Vorlesung von W. Krull, Bonner Math. Schriften, Bonn, 1963.

3. Sôbre a continuidade das raízes de um polinômio, Atas Do V Colóquio Brasileiro De Matemática, São Paulo, 1965, 59-70.

4. A note on Henselian valuation rings, *Canad. Math. Bull.* 11 (1968), 185-189.

5. Über henselsche Erweiterungen bewerteter Körper und die Fortsetzung von Bewertungen, *J. Reine Angew. Math.* 239/240 (1969), 306-320.

6. Endliche separable Körpererweiterungen mit vorgeschriebenen Bewertungsfortsetzungen I, *Abh. Math. Sem. Univ. Hamburg* 33 (1969), 80-101.

7. *Valuation Theory*, Springer-Verlag, New York, 1972.

8. On Henselizations of valued fields, *Bol. Soc. Brasil. Mat.* 4 (1973), no. 2, 97-109.

Endler, O.

9. Finite separable field extensions with prescribed extensions of valuations II, *Manuscripta Math.* 17 (1975), no. 4, 383-408.

Endler, O., and Engler, J.

1. Fields with henselian valuation rings, *Math. Z.* 152 (1977), 191-193.

Endo, M.

1. On the embedding of topological rings into quotient rings I, *J. Fac. Sci. Univ. Tokyo Sect. I-A Math.* 10 (1964), 196-214.
2. On the embedding of topological rings into quotient rings II, *Sci. Papers College Gen. Ed. Univ. Tokyo* 14 (1964), no. 1, 51-54.
3. A note on locally compact division rings, *Comment. Math. Univ. St. Paul.* 14 (1966), 57-64.
4. On the topologies of the rational number field, *Comment. Math. Univ. St. Paul.* 16 (1967), 11-20.
5. The strong approximation theorem and locally bounded topologies of algebraic function fields, *Comment. Math. Univ. St. Paul.* 30 (1981), no. 1, 77-86.
6. Corrections for "The strong approximation theorem and locally bounded topologies of algebraic function fields," *Comment. Math. Univ. St. Paul.* 31 (1982), no. 1, 61.

Engelking, R.

Outline of General Topology, North-Holland, Amsterdam, 1968.

Engler, A. J.

1. Fields with two incomparable henselian valuation rings, *Manuscripta Math.* 23 (1978), no. 4, 373-386.
2. The relative separable closure of a valued field in its completion, *Manuscripta Math.* 24 (1978), no. 1, 83-95.

Fakhruddin, S. M.

1. Linearly compact modules over Noetherian rings, *J. Algebra* 24 (1973), 544-550.

Fenstad, J. E.

1. An observation on the Gelfand-Mazur theorem, *Norske Vid. Selsk. Forh. (Trondheim)* 36 (1963), 41-45.

Figiel, T.

1. On a functional equation $f(x + y) = f(x) + f(y)$ (Polish), *Wiadom. Mat.* 11 (1969), 15-18.

Fleischer, I.

1. Sur les corps localement bornés, *C. R. Acad. Sci. Paris Sér. A-B* 237 (1953), 546-548.

Fleischer, I.

2. Sur les corps topologiques et les valuations, *C. R. Acad. Sci. Paris Sér. A-B* 236 (1953), 1320-1322.

3. Sur les espaces normés nonarchimédiens, *Indag. Math.* 16 (1954), 165-168.

4. Maximality and ultracompleteness in normed modules, *Proc. Amer. Math. Soc.* 9 (1958), 151-157.

5. Über Dualität in linear-topologischen Moduln, *Math. Z.* 72 (1959-1960), 439-445.

Fraedrich, A. M.

1. Beiträge zur Theorie der topologischen Körper, Diss., Tech. Univ. Braunschweig, 1965.

Freudenthal, H.

1. L'algèbre topologique. . ., XIIe congrès international d'histoire des sciences, Paris, 1968.

Fuchs, L.

1. The generalization of the valuation theory, *Duke Math. J.* 18 (1951), 19-26.

2. *Partially Ordered Algebraic Systems*, Pergamon Press, Elmsford, N.Y., 1963.

3. Vector spaces with valuations, *J. Algebra* 35 (1975), 23-38.

Fukawa, M.

1. An extension theorem for valuations, *J. Math. Soc. Japan* 17 (1965), 67-71.

2. On the theory of valuations, *Fac. Sci. Univ. Tokyo*, Sect. I, 12 (1966), 57-79.

Fuster, A.

1. Cuerpos maximales arquimedianos contenidos en un cuerpo ordenado no arquimediano, *Acta Salamant.* 6 (1965), no. 4, 51-60.

2. Considerations on ordered fields and maximal real fields (Spanish), *Collect. Math.* 19 (1968), 143-153.

Gelbaum, B., Kalisch, G. K., and Olmstead, J. H.

1. On the embedding of topological semigroups and integral domains, *Proc. Amer. Math. Soc.* 2 (1951), 807-821.

Gelfand, I. M.

1. Normierte Ringe, *Mat. Sb.* 9 (1941), 3-24.

Gilmer, R.

1. Extension of an order to a simple transcendental extension, *Contemp. Math.* 8 (1982), 113-118.

Goldhaber, J. K., and Wolk, E. S.

1. Maximal ideals in rings of bounded continuous functions, *Duke Math. J.* 21 (1954), 565-569.

Gould, G. G.

1. Locally unbounded topological fields and box topology on products of vector spaces, *J. London Math. Soc.* 36 (1961), 273-281.

Gräter, J.

1. Der Approximationssatz für Manisbewertungen, *Arch. Math. (Basel)* 37 (1981), 335-340.
2. Der allgemeine Approximationssatz für Manisbewertungen, *Monatsh. Math.* 93 (1982), 277-288.

Gravett, K. A. H.

1. Note on a result of Krull, *Proc. Cambridge Math. Soc.* 52 (1956), 379.

de Groot, J., and Loonstra, F.

1. Topologische Eigenschaften bewerteter Körper, *Nederl. Akad. Wetensch. Indag. Math.* 4 (1942), 658-664.

Gültekin, K. B.

1. Maximal ideals in function algebras over topological fields, *METU J. Pure Appl. Math. Sci.* 3 (1970), 1-3.

Güntzer, U., and Remmert, R.

1. Bewertungstheorie und Spektralnorm, *Abh. Math. Sem. Univ. Hamburg* 38 (1972), 32-48.

Haifawi, M.

1. Normed spaces over valued fields, *J. Reine Angew Math.* 261 (1973), 153-156.

Hanson, J. R.

1. An infinite groupoid which admits only trivial topologies, *Amer. Math. Monthly* 74 (1967), no. 5, 568-569.

Harrison, D. K., and Warner, D. K.

1. Infinite primes of fields and completions, *Pacific J. Math.* 45 (1973), 201-216.

Hartmann, P.

1. Ein Beweis des Henselschen Lemmas für maximal bewertete Körper, *Arch. Math.* 37 (1981), no. 2, 163-168.

Hasse, H.

1. Über p-adische Schiefkörper und ihre Bedeutung für die Arithmetik hyperkomplexer Zahlsysteme, *Math. Ann.* 104 (1931), 495-534.
2. *Zahlentheorie*, Akademie-Verlag, Berlin. 1963.

Hasse, H., and Schmidt, F. K.

1. Die Struktur diskret bewerteter Körper, *J. Reine Angew. Math.* 170 (1934), 4-63.

Heine, J.

1. Ringtopologien auf nichtalgebraischen Körpern, *Math. Ann.* 199 (1972), 205-211.
2. Existence of locally unbounded fields, and field topologies which are not the intersection of bounded ring topologies, *J. London Math. Soc.* 5 (1972), 481-487.
3. Ring topologies of type N on fields, *General Topology Appl.* 3 (1973), 135-148.

Heine, J., and Warner, S.

1. Ring topologies on the quotient field of a Dedekind domain, *Duke Math. J.* 40 (1973), 473-486.

Heise, W., and Sörensen, K.

1. Zur Definition topologischer projektiver Räume, *Arch. Math. (Basel)* 21 (1970), 218-220.

Hensel, K.

1. *Theorie der algebraischen Zahlen*, Erster Band, Druck und Verlag von B. B. Teubner, Leipzig and Berlin, 1908.
2. *Zahlentheorie*, G. J. Göschen'sche Verlagshandlung, Berlin and Leipzig, 1913.

Hesse, G.

1. Zur Topologisierbarkeit von Gruppen, Diss., Universität Hannover, 1979.

Heuer, G. A.

1. Continuous multiplications in $\mathbb{R}^2$, *Math. Mag.* 45 (1972), 72-77.

Hill, L.

1. Henselian valued fields with prescribed value group and residue fields, *Trans. Amer. Math. Soc.* 195 (1974), 213-222.

Hinrichs, L. A.

1. The existence of topologies on field extensions, *Trans. Amer. Math. Soc.* 113 (1964), 397-405.
2. Integer topologies, *Proc. Amer. Math. Soc.* 15 (1964), 991-995.

Hinrichs, L. A., Niven, I., and van Eynden, C. L. N.

1. Fields defined by polynomials, *Pacific J. Math.* 14 (1964), 537-545.

Hirschfeld, R. A.

1. A Mazur-Gelfand theorem for generalized division algebras, *Bull. Acad. Polon. Sci. Sér. Sci. Math. Astronom. Phys.* 27 (1979), no. 10, 763-770.

Hochster, M.

1. Existence of topologies for commutative rings with identity, *Duke Math. J.* 38 (1971), 551-554.

Hochster, M., and Kiltinen, J. O.

1. Commutative rings with identity have ring topologies, *Bull. Amer. Math. Soc.* 76 (1970), 419-420.

Holland, S. S., Jr.

1. *-valuations and ordered *-fields, *Trans. Amer. Math. Soc.* 262 (1980), 219-243.

Huckaba, J.

1. Extensions of pseudo-valuations, *Pacific J. Math.* 29 (1969), 295-302.
2. Some results on pseudo-valuations, *Duke Math. J.* 37 (1970), 1-9.

Ingleton, A. W.

1. The Hahn-Banach theorem for non-Archimedean valued fields, *Proc. Cambridge Philos. Soc.* 48 (1952), 41-45.

Iseki, K.

1. On the fundamental theorem of algebra, *J. Math. Soc. Japan* 6 (1954), 129-130.

Isidro, J. M.

1. Topologias en cuerpos, *Rev. Mat. Hisp.-Amer.* 37 (1977), 19-24.

Iwasawa, K.

1. On the rings of valuation vectors, *Ann. Math.* 57 (1953), 331-356.

Jacobson, N.

1. Totally disconnected locally compact rings, *Amer. J. Math.* 58 (1936), 433-449.
2. A note on topological fields, *Amer. J. Math.* 59 (1937), 889-894.
3. *Lectures on Abstract Algebra, Vol. III,* van Nostrand, Toronto, New York, London, 1964.

Jacobson, N., and Taussky, O.

1. Locally compact rings, *Proc. Nat. Acad. Sci. U.S.A.* 21 (1935), 106-108.

Jaeger, A.

1. Adjunction of subfield closures to ordered division rings, *Trans. Amer. Math. Soc.* 73 (1952), 35-39.

Jaffard, P.

1. Corps demie-values, *C. R. Acad. Sci. Paris Sér. A-B* 231 (1950), 1401-1403.

2. Theorie arithmetique de anneaux du type de Dedekind I, II, *Bull. Soc. Math. France* 80 (1952), 61-100; 81 (1953), 41-61.

3. La notion de valuation, *Enseignement Math.* 40 (1955), 5-26.

4. Solution d'un probleme de Krull, *Bull. Soc. Math. France* 85 (1961), 127-135.

Jančevskiǐ, V. I.

1. Division algebras over Henselian discretely valued fields (Russian), *Dokl. Akad. Nauk BSSR* 20 (1976), no. 11, 971-974.

Jauer, W.

1. Miquelsche topologische Kettenebenen, Diss., Technische Universität Hannover, 1978.

Jebli, A.

1. Topologies et valuations essentielles, *C. R. Acad. Sci. Paris Sér. A-B* 272 (1971), 1173-1174.

2. Topologies linéaires et valuations essentielles, These, Université de Bordeaux I, 1972.

3. Sur certaines topologies linéaires, *C. R. Acad. Sci. Paris Sér. A-B* 274 (1972), 444-446.

4. Corps des fractions et topologie artiniennes, *C. R. Acad. Sci. Paris Sér. A-B* 278 (1974), 973-976.

5. Topologies linéaires du corps des fractions d'un anneau noetherien, *Publ. Sem. Math. Univ. Rennes* (1974), fasc. 2, 1-44.

6. Topologies linéaires sur le corps des fractions d'un anneau de polynomes, *C. R. Acad. Sci. Paris* 280 (1975), 701-703.

Kallman, R. R.

1. Uniqueness results for the ax + b group and related algebraic objects, *Fund. Math.* 124 (1984), 255-262.

Kallman, R. R., and Simmons, F. W.

1. A theorem on planar continua and an application to automorphisms of the field of complex numbers, *Topology Appl.* 20 (1985), 251-255.

Kametani, S.

1. An elementary proof of the fundamental theorem of normed fields, *J. Math. Soc. Japan* 4 (1952), 96-99.

Kaplansky, I.

1. Maximal fields with valuations I, II, *Duke Math. J.* 9 (1942), 303-321; 12 (1945), 243-248.

Kaplansky, I.

2. Topological methods in valuation theory, *Duke Math. J.* 14 (1947), 527-541.

3. Topological rings, *Amer. J. Math.* 69 (1947), 153-183.

4. Topological rings, *Bull. Amer. Math. Soc.* 54 (1948), 809-826.

5. Polynomials in topological fields, *Bull. Amer. Math. Soc.* 54 (1948), 909-916.

6. The Weierstrass theorem in fields with valuations, *Proc. Amer. Math. Soc.* 1 (1950), 356-357.

Kaplansky, I., and Schilling, O. F. G.

1. Some remarks on relatively complete fields, *Bull. Amer. Math. Soc.* 48 (1942), 744-747.

Kapuano, I.

1. Sur les corps de nombres à une dimension distincts du corps réel, *Rev. Fac. Sci. Univ. Istanbul* (A) 11 (1946), 30-39.

Karpelevič, F. I.

1. Pseudo-norms on the ring of integers (Russian), *Uspehi Mat. Nauk* 3 (1948), vyp. 5, 174-177.

Karzel, H.

1. Beziehungen zwischen topologischen Inzidenzgruppen und topologischen Fastkörpern, Celebrazioni archimedee del seculo XX, Simposio di topologia, Gubbio, 1964.

Kasahara, S.

1. On formulations of topological linear spaces by topological semifields, *Math. Japon.* 19 (1974), 121-134.

Keller, H. A.

1. On valued, complete fields and their automorphisms, *Pacific J. Math.* 121 (1986), no. 2, 397-406.

Kertesz, A., and Szele, T.

1. On existence of non-discrete topologies on infinite abelian groups, *Publ. Math. Debrecen* 3 (1953), no. 1-2, 187-189.

Kestelman, H.

1. Automorphisms of the field of complex numbers, *Proc. London Math. Soc.* (2) 53 (1947), 1-12.

Kiltinen, J. O.

1. Inductive ring topologies, *Trans. Amer. Math. Soc.* 134 (1968), 149-169.

2. Inverses of polynomial functions in topological fields, *Proc. Amer. Math. Soc.* 24 (1970), 9-17.

Kiltenen, J. O.

3. Embedding a topological domain in a countably generated algebraic ring extension, *Duke Math. J.* 37 (1970), 647-654.

4. On the number of field topologies on an infinite field. *Proc. Amer. Math. Soc.* 40 (1973), 30-36.

5. Infinite abelian groups are highly topologizable, *Duke Math. J.* 41 (1974), 151-154.

6. On countably extending ring topologies (preprint).

Kiyek, K.

1. Bewertungssysteme in algebraischen Funktionskörpern, *Arch. Math. (Basel)* 18 (1967), 250-263.

2. Bewertungssysteme in Moduln, *Math. Z.* 112 (1969), 190-204.

3. Ein Struktursatz für Pseudobewertungen, *Math. Z.* 118 (1970). 139-156.

4. Bewertungssteme in Moduln II, *Math. Z.* 119 (1971), 219-238.

5. Homogene Pseudobewertungen, *Arch. Math. (Basel)* 22 (1971), 602-611.

6. Pseudobetragsfunktionen auf Quotientenkörpern von Dedekindringen, *J. Reine Angew. Math.* 274/275 (1975), 244-257.

Klotzek, B., and Weinert, H.-J.

1. Bemerkungen zur Theorie der bewerteten Körper, *Math. Nachr.* 39 (1969), 97-115.

Knopfmacher, J., and Sinclair, A. M.

1. Fields with few extensions, *Proc. Amer. Math. Soc.* 29 (1971), 255-258.

Kohn, S., and Newman, D. J.

1. Multiplication from other operations, *Proc. Amer. Math. Soc.* 27 (1971), 244-246.

Kolmogoroff, A.

1. Zur Begründung der projectiven Geometrie, *Ann. Math.* 33 (1932), 175-176.

Korec, I.

1. A class of Lebesgue non-measurable subfields of the field of the real numbers, *Acta Fac. Rerum Natur. Univ. Comenian. Math.* 26 (1972), 29-32.

Kowalsky, H.-J.

1. Zur topologischen Kennzeichnung von Körpern, *Math. Nachr.* 9 (1953), 261-268.

2. Beiträge zur topologischen Algebra, *Math. Nachr.* 11 (1954), 143-185.

Kowalsky, H.-J.

3. Stonesche Körper und ein Überdeckungssatz, *Math. Nachr.* 14 (1955), 57-64.

4. *Topologische Räume*, Birkhauser, Basel and Stuttgart, 1961.

Kowalsky, H.-J., and Dürbaum, H.-J.

1. Arithmetische Kennzeichnung von Körpertopologien, *J. Reine Angew. Math.* 191 (1953), 135-151.

Krull, W.

1. Allgemeine Bewertungstheorie, *J. Reine Angew. Math.* 167 (1932), 160-196.

2. Über unendliche algebraische Erweiterungen bewerteter Körper, *Rend. Circ. Mat. Palermo* 1 (1952), 164-169.

3. Über geschlossene Bewertungssysteme, *J. Reine Angew. Math.* 190 (1952), 75-92.

4. Zur Theorie der Bewertungen mit nichtarchimedisch geordneter Wertgruppe und der nichtarchimedisch geordneten Körper, Colloq. d'Algebre Superieure, Bruxelles, 1956, 45-77.

5. Über einen Existenzsatz der Bewertungstheorie, *Abh. Math. Sem. Univ. Hamburg* 23 (1959), 29-35.

6. Ordnungsfunktionen und Bewertungen von Körpern, *Math. Z.* 77 (1961), 135-148.

7. Einige Bemerkungen zur Bewertungstheorie, *An. Acad. Brasil. Ci.* 35 (1963), 475-481.

Kurepa, D.

1. On p-adic space of Hensel, *Publ. Inst. Math. (Beograd)* 5 (19) (1965), 133-135.

Kürschak, J.

1. Über Limesbildung und allgemeine Körpertheorie, *J. Reine Angew. Math.* 142 (1913), 211-253.

Larson, R. E., and Andima, S. J.

1. The lattice of topologies: A survey, *Rocky Mountain J. Math.* 5 (1975), no. 2, 177-198.

Laugwitz, D.

1. Eine nichtarchimedische Erweiterung angeordneter Körper, *Math. Nachr.* 37 (1968), 225-236.

2. Bemerkungen über Stetigkeit und angeordnete Körper, *Demonstratio Math.* 6 (1973), 191-209.

Lazard, M.

1. Bemerkungen zur Theorie der bewerteten Körper und Ringe, *Math. Nachr.* 12 (1954), 67-73.

Lenskoj, D. N.

1. A remark on Hensel's lemma, *Izv. Vysš. Učeb. Zaved.* (1960), vyp. 4 (17), 128-132 (Russian).

Levi-Civita, T.

1. Sugli infiniti ed infinitesimi actuali quali elementi analytici (1892-1893), *Opere Matematiche*, Vol. I (1954), 1-39.

Lipkina, Z. S.

1. On pseudonormability of topological rings (Russian), *Sibirsk. Mat. Ž.* 6 (1965), no. 5, 1046-1052.
2. Locally compact rings without zero divisors (Russian), *Izv. Akad. Nauk SSSR Ser. Mat.* 31 (1967), 1239-1262.

Loonstra, F.

1. Analytische Untersuchungen über bewertete Körper, Diss., N. V. Noord-Hollandsche Uitgevers Maatschappij, Amsterdam, 1941.
2. Im Kleinen kompakte nichtarchimedisch bewertete Körper, *Nederl. Akad. Wetensch. Indag. Math.* 4 (1942), 665-668.
3. Représentation d'éléments p-adiques et q-adiques, *Indag. Math.* 9 (1947), 531-539.
4. Interprétation topologique d'un théorème de K. Mahler concernant les pseudo-evaluations, *Indag. Math.* 9 (1947), 373-378.

Lorenzen, P.

1. Über die Komplettierung in der Bewertungstheorie, *Math. Z.* 59 (1953-54), 84-87.

Lucas, D.

1. A multiplication in N-space, *Proc. Amer. Math. Soc.* 74 (1979), no. 1, 1-18.

Luendemann, J. K.

1. On the embedding of compact domains, *Math. Z.* 115 (1970), 113-116.
2. On the embedding of topological domains into quotient fields, *Manuscripta Math.* 3 (1970), 213-226.

Luxemburg, W. A. J.

1. On a class of valuation fields introduced by A. Robinson, *Israel J. Math.* 25 (1976), 189-201.

de Maar, R.

1. Partially ordered fields, *Amer. Math. Monthly* 74 (1967), 418-420.

van Maaren, H.

1. Pseudo-ordered fields, *Indag. Math.* 36 (1974), 463-476.

MacLane, S.

1. A construction for absolute values in polynomial rings, *Trans. Amer. Math. Soc.* 40 (1936), 363-395.
2. A construction for prime ideals as absolute values of an algebraic field, *Duke Math. J.* 2 (1936), 492-510.
3. The uniqueness of the power series representation of certain fields with valuation, *Ann. Math.* 39 (1938), 370-382.
4. The universality of formal power series fields, *Bull. Amer. Math. Soc.* 45 (1939), 888-890.
5. Subfields and automorphism groups of p-adic fields, *Ann. Math.* 40 (1939), 423-442.
6. Note on the relative structure of p-adic fields, *Ann. Math.* 41 (1940), 751-753.

Mahler, K.

1. Über Pseudobewertungen I, *Acta Math.* 66 (1936), 79-119.
2. Über Pseudobewertungen Ia (Zerlegungssätze), *Nederl. Akad. Wetensch. Proc. Ser. A* 29 (1936), 57-65.
3. Über Pseudobewertungen II, *Acta Math.* 67 (1936), 51-80.
4. Über Pseudobewertungen III, *Acta Math.* 67 (1936), 282-328.
5. On a geometrical representation of p-adic numbers, *Ann. Math.* 41 (1940), 8-56.
6. An interpolation series for continuous functions of a p-adic variable, *J. Reine Angew. Math.* 199 (1958), 23-34.
7. Lectures on diophantine approximation, Part 1: g-adic numbers and Roth's theorem, University of Notre Dame, South Bend, Ind., 1961.

Marchisa, G.

1. Una condizione sufficiente per la continuità degli zeri dei polinomi a coefficienti in un campo topologico, *Matematiche* 32 (1977), no. 2, 316-322.

Maroscia, P.

1. Topological properties of some classes of G-domains, *Boll. Un. Mat. Ital. A (5)* 15 (1978), no. 3, 688-698.

Marshall, M. A.

1. Ramification theory for valuations of arbitrary rank, *Canad. J. Math.* 26 (1974), 908-916.

Massaza, C.

1. Sul completamento dei campi ordinati, *Univ. e Politec. Torino Rend. Sem. Mat.* 29 (1969/70), 329-348.

Massaza, C.

2. Serie di potenze a coefficienti in un campo non archimedeo, *Riv. Mat. Univ. Parma* (2) 11 (1970), 133-157.

3. On the completion of ordered fields, *Prakt. Akad. Athēnōn* 45 (1970), 174-178.

Mathiak, K.

1. Bewertungen nicht kommutativer Körper, *J. Algebra* 48 (1977), 217-235.

2. Zur Bewertungstheorie nicht kommutativer Körper, *J. Algebra* 73 (1981), no. 2, 586-600.

3. Der Approximationssatz für Bewertungen nicht kommutativer Körper, *J. Algebra* 76 (1982), 280-295.

Matignon, M., and Reversat, M.

1. Sous-corps fermés d'un corps valué, *J. Algebra* 90 (1984), no. 2, 491-515.

Matzinger, H.

1. Another characterization of classical fields, *Math. Z.* 105 (1968), 173-176.

Mazur, S.

1. Sur les anneaux linéaires, *C. R. Acad. Sci. Paris Sér. A-B* 207 (1938), 1025-1027.

McShane, N.

1. On the periodicity of homeomorphisms of the real line, *Amer. Math. Monthly* 68 (1961), 562-563.

Meisters, G. H., and Monk, J. D.

1. Constructions of the reals via ultrapowers, *Rocky Mountain J. Math.* 3(1973), 141-158.

Meloni, G. C.

1. Analisi non standard di anelli e corpi topologici, *Ist. Lombardo Accad. Sci. Lett. Rend.* A 107 (1973), 503-510; also *Collect. Math.* no. 422 (1973).

Miličić, P. M.

1. Sur les endomorphisms du corps des quaternions qui conservent les valeurs absolutes, *Mat. Vesnik* 11 (1974), no. 4, 273-275.

Misfeld, J.

1. Topologische projective Räume, *Abh. Math. Sem. Univ. Hamburg* 32 (1968), 232-263.

Misfeld, J., and Timm, J.

1. Topologische Dicksonsche Fastkörper, *Abh. Math. Sem. Univ. Hamburg* 37 (1972), 60-67.

Močkoř, J.

1. An extension of approximation theorems, *Colloq. Math.* 32 (1975), 159-165.

2. The completion of Prüfer domains, *Proc. Amer. Math. Soc.* 67 (1977), 1-11.

3. The completion of valued fields and nonstandard models, *Comment. Math. Univ. St. Paul.* 30 (1981), 1-16.

4. *Groups of Divisibility*, Reidel, Dordrecht-Boston-Lancaster, 1983.

Monna, A. F.

1. *Analyse non-archimédienne*, Springer-Verlag, Berlin-Heidelberg-New York, 1970.

Moriya, M.

1. Bewertungstheoretischer Aufbau der multiplikativen Idealtheorie, *J. Fac. Sci. Hokkaido Imp. Acad.* 8 (1940), 109-144.

2. Struktur der Divisionalgebren über diskret bewerteten perfekten Körpern, *Proc. Imp. Acad. (Tokyo)* 18 (1942), 5-11.

3. Über die Restklassenkörper bewerteter perfekter Körper, *Nagoya Math. J.* 4 (1952), 15-27.

4. Zur Theorie der halb-topologischen Gruppen and Körper, *Math. J. Okayama Univ.* 1 (1952), 109-124.

5. Charakterisierung der nicht-archimedischen Bewertungen durch Grössenordnungen, *Math. J. Okayama Univ.* 3 (1953), 29-38.

6. Zur Theorie der topologischen Körper, *Math. J. Okayama Univ.* 4 (1955), 115-134.

Moriya, M., and Schilling, O. F. G.

1. Divisionalgebren über unendlichen perfekten Körpern, *J. Fac. Sci. Hokkaido Imp. Univ.* 6 (1937), 103-111.

Mothon, B.

1. Prolongements de valuations dans un cas non commutatif, *C. R. Acad. Sci. Paris Sér. A-B* 291 (1980), no. 1, A5-A6.

Moyls, B. N.

1. The structure of valuations of rational function field k(x), *Trans. Amer. Math. Soc.* 71 (1951), 102-112.

Mutylin, A. F.

1. Imbeddability of discrete fields into connected fields, *Soviet Math.* 7 (1966), no. 3, 772-775.

2. Non-normable extensions of p-adic number fields, *Math. Notes* 2 (1967), 496-498.

3. An example of a nontrivial topologization of the field of rational numbers. Complete locally bounded fields, *Amer. Math. Soc. Transl.* 73 (1968), 159-179.

Mutylin, A. F.

4. Connected, complete, locally bounded fields. Complete not locally bounded fields, *Math. USSR-Sb.* 5 (1968), no. 3, 433-449.

5. Corrections to my article "An example of a nontrivial topologization of the field of rational numbers. Complete locally bounded fields," *Math. USSR-Izv.* 2 (1968), no. 1, 237-238.

6. Completely simple commutative rings, *Math. Notes* 5 (1969), 100-105.

Nachbin, L.

1. On strictly minimal topological division rings, *Bull. Amer. Math. Soc.* 55 (1949), 1128-1136.

Nagata, M.

1. A theorem on valuation rings and its applications, *Nagoya Math. J.* 29 (1967), 85-91.

2. A generalization of the notion of a valuation, *Amer. J. Math.* 101 (1979), 245-257.

Nagata, M., Nakayama, T., and Tuzuku, T.

1. On an existence lemma in valuation theory, *Nagoya Math. J.* 6 (1953), 59-61.

Nakano, T.

1. On the locally bounded fields, *Comment. Math. Univ. St. Paul* 9 (1961), 77-85.

2. Some examples of the locally bounded fields, *Comment. Math. Univ. St. Paul.* 10 (1962), 63-66.

3. A generalized valuation and its value groups, *Comment. Math. Univ. St. Paul.* 12 (1964), 1-22.

4. A theorem on lattice ordered groups and its applications to the valuation theory, *Math. Z.* 83 (1964), 140-146.

Narici, L., Beckenstein, E., and Bachman, G.

1. *Functional Analysis and Valuation Theory*, Marcel Dekker, New York, 1971.

Narkiewicz, W.

1. *Elementary and Analytic Theory of Algebraic Numbers*, Polish Scientific Publishers, Warszawa, 1974.

Neukirch, J.

1. Zur Verzweigungstheorie der allgemeinen Krullschen Bewertungen, *Abh. Math. Sem. Univ. Hamburg* 32 (1968), 207-215.

Neumann, B. H.

1. On ordered division rings, *Trans. Amer. Math. Soc.* 66 (1949), 202-252.

Nichols, E., and Cohen, J.-A.

1. Locally bounded topologies on global fields, *Duke Math. J.* 44 (1977), 853-862.

Nieminen, T.

1. The completion of orderable topological groups and fields, *Ann. Acad. Sci. Fenn. Ser. AI* no. 365 (1965).

Nikodým, O. M.

1. A study in algebraic extensions of linearly ordered non archimedean fields, *J. Math. Pure Appl.* 138 (1959), fasc. 1, 61-96.

Nobusawa, N.

1. On a system of valuations of algebraic function fields with several variables, *Osaka J. Math.* 10 (1958), 205-212.

Ögmundson, V.

1. Multiplication in n dimensions, *Nordisk. Mat. Tidskr.* 7 (1958), 111-116.

O'Meara, O. T.

1. *Introduction to Quadratic Forms*, Springer-Verlag, Berlin-Göttingen-Heidelberg, 1963.

Ono, T.

1. On the extension property of normed spaces over field with non-Archimedean valuations, *J. Math. Soc. Japan.* 5 (1953), 1-5.
2. A generalization of the Hahn-Banach theorem, *Nagoya Math. J.* 6 (1953), 171-176.

Ostrowski, A.

1. Über einige Fragen der allgemeinen Körpertheorie, *J. Reine Angew. Math.* 143 (1913), 255-284.
2. Über sogenannte perfekte Körper, *J. Reine Angew. Math.* 147 (1917), 191-204.
3. Über einige Lösungen der Funktionalgleichung $\varphi(x)\varphi(y) = \varphi(xy)$, *Acta Math.* 41 (1918), 271-284.
4. Untersuchungen zur arithmetischen Theorie der Körper, *Math. Z.* 39 (1935), 269-404.

Otobe, Y.

1. On locally compact fields, *Japan. J. Math.* 19 (1945), 189-202.

Palais, R.

1. The classification of real division algebras, *Amer. Math. Monthly* 75 (1968), 366-368.

Paljutin, E. A., Seese, D. G., and Taǐmanov, A. D.

1. A remark on the topologization of algebraic structures, *Rev. Roumaine Math. Pures Appl.* 26 (1981), 617-618.

Pank, M., and Roquette, P.

1. On immediate extensions of valued fields; splitting property of the ramification field and Kaplansky's hypothesis A (preprint, 1982).

Pestov, G. G.

1. The structure of ordered fields (Russian), Tomsk, 1980, Izd. Tomskovo Universiteta, 1980.

Peterson, P. H.

1. Jordan-Divisionalgebren und Bewertungen, *Math. Ann.* 202 (1973), 215-243.
2. Lokal kompakte Jordan-Divisionringe, *Abh. Math. Sem. Univ. Hamburg* 39 (1973), 164-179.
3. Composition algebras over a field with a discrete valuation, *J. Algebra* 29 (1974), 414-426.
4. Classification of locally compact Jordan division rings, *J. Algebra* 58 (1979), 350-360.

Phillips, K.

1. A note on local field duality, *Rocky Mountain J. Math.* 8 (1978), no. 3, 433-437.

Pieper, I., and Biallas, D.

1. Topologische Quasimoduln und ihre Geometrien I, *Abh. Math. Sem. Univ. Hamburg* 34 (1969/70), 170-183.

Plaumann, P., and Strambach, K.

1. Zusammenhängende Quasikörper mit Zentrum, *Arch. Math. (Basel)* 21 (1970), 455-465.

Podewski, K.-P.

1. The number of field topologies on countable fields, *Proc. Amer. Math. Soc.* 39 (1973), 33-38.
2. Transcendental extensions of field topologies on countable fields, *Proc. Amer. Math. Soc.* 39 (1973), 39-42.
3. Topologisierung algebraischer Strukturen, *Rev. Roumaine Math. Pures Appl.* 22 (1977), 1283-1290.

Pontrjagin, L. S.

1. Über stetige algebraische Körper, *Ann. Math.* 33 (1932), 163-174.
2. *Topological Groups*, Gordon and Breach, New York, 1966.

Prestel, A., and Ziegler, M.

1. Model theoretic methods in the theory of topological fields, *J. Reine Angew. Math.* 299/300 (1978), 318-341.
2. Non-axiomatizable classes of V-topological fields, *J. Reine Angew. Math.* 316 (1980), 211-214.

Prodanov, I.

1. Precompact minimal group topologies and p-adic numbers, *Annuaire Univ. Sofia Fac. Math.* 66 (1971/72), 249-266.

Prolla, J. B.

1. *Topics in Functional Analysis Over Valued Division Rings*, North-Holland Mathematics Studies 77, 1982.

van der Put, M.

1. Continuous derivatives and valued fields, *Bull. Soc. Math. France* 101 (1973), 71-112.

Radó, F.

1. A generalization of the valued vector space, *Mathematica (Cluj)* 19 (42) (1977), no. 2, 203-209.

Rangan, G.

1. Non-archimedean metrizability of topological groups, *Fund. Math.* 68 (1970), 179-182.

Rayner, F. J.

1. Gravett's proof of a theorem of Krull, *Quart. J. Math. Oxford Ser. (2)* 24 (1973), 409-410.

Rees, D.

1. Filtrations as limits of valuations, *J. London Math. Soc.* 32 (1957), 97-102.

Rhodes, C. P. L.

1. A note on primary decomposition of a pseudovaluation, *Pacific J. Math.* 47 (1973), 507-513 .
2. A relation for ideal has an analog for pseudovaluations, *J. Algebra* 51 (1978), 526-535.

Ribenboim, P.

1. Le Theoreme d'approximation pour les valuations de Krull, *Math. Z.* 68 (1957), 1-18.
2. Sur les groupes totalement ordonnes et l'arithmetique des anneaux de valuation, *Summa Brasil. Math.* 4 (1958), no. 1, 1-63.
3. Sur la theorie du prolongement des valuations de Krull, *Math. Z.* 75 (1961), 449-466.
4. An existence theorem for fields with valuations, *Trans. Amer. Math. Soc.* 105 (1962), 278-294.
5. On the decomposition theory for Krull valuations, *Illinois J. Math.* 6 (1962), 524-532.
6. On the existence of fields with few discrete valuations, *J. Reine Angew. Math.* 216 (1964), 43-49.

7. On the completion of a valuation ring, *Math. Ann.* 155 (1964), 392-396.

8. A short note on henselian fields, *Math. Ann.* 176 (1967), 83-88.

9. Higher ramification groups for rank one valuations, *Math. Ann.* 173 (1967), 253-259.

10. *Théorie des valuations*, Presses de l'Université de Montréal, Montreal, 1968.

11. Equivalent forms of Hensel's lemma, *Expositio Math.* 3 (1985), no. 1, 3-24.

Rigo, T., and Warner, S.

1. Topologies extending valuations, *Canad. J. Math.* 30 (1978), 164-169.

Rim, D. S.

1. Relatively complete fields, *Duke Math. J.* 24 (1957), 197-200.

Rodriguez-Salinas, R., and Bou, L.

1. A Hahn-Banach theorem for arbitrary vector spaces, *Boll. Un. Mat. Ital.* 10 (1974), 390-393.

Rolewicz, S.

1. On a certain class of linear metric spaces, *Bull. Acad. Polon. Sci. Sér. Sci. Math. Astronom. Phys.* 5 (1957), 471-473.

van Rooij, A. C. M., and Schikhof, W. H.

1. Non-Archimedean analysis, *Nieuw Arch. Wisk.* 19 (1971), 120-160.

Roquette, P.

1. On the prolongation of valuations, *Trans. Amer. Math. Soc.* 88 (1958), 42-57.

Rothman, N. J.

1. A note on compact rings, *Math. Z.* 91 (1966), 179-184.

Rychlik, K.

1. Zur Bewertungstheorie der algebraischen Körper, *J. Reine Angew. Math.* 153 (1924), 94-107.

Šafarevič, I. R.

1. On the normability of topological fields (Russian), *Dokl. Akad. Nauk SSSR* 40 (1943), 133-135.

Šahmatov, D. B.

1. Cardinal invariants of topological fields (Russian), *Dokl. Akad. Nauk SSSR* 271 (1983), no. 6, 1332-1336.

2. Embeddings in topological fields and construction of a field, whose space is not normal (Russian), *Comm. Math. Univ. Carolinae* 24 (1983), no. 3, 525-540.

Saltman, D. J.

1. Division algebras over discrete valued fields, *Comm. Algebra* 8 (1980), no. 18, 1749-1774.

Salzmann, H.

1. Topological planes, *Advances in Math.* 2 (1967), no. 1, 1-60.
2. Zahlbereiche; Teil I: Die reellen Zahlen; Teil II: Die rationalen Zahlen; Teil III: Die komplexen Zahlen. Vorlesung von Professor H. Salzmann gehalten an der Universität Tübingen im 1971-72.

Sangare, D.

1. Sur la composition des filtrations et des pseudo-valuations, *C. R. Acad. Sci. Paris Sér. A-B* 284 (1977), no. 5, A287-A290.

Sankaran, N.

1. Order structure in rings and fields, *Math. Student* 27 (1959), 165-172.

Sankaran, N., and Yadav, R. A.

1. G-domains and pseudovaluations, *Rend. Sem. Mat. Univ. Padova* 61 (1979), 33-38.

Satyanarayanamoorthy, P. V.

1. Nonarchimedean valuations on rational composition division algebras, *Notices Amer. Math. Soc.* 26 (1979), no. 6, A499.
2. Archimedean valuations on rational composition division algebras, *Notices Amer. Math. Soc.* 26 (1979), no. 5, A424.

Schaefer, H. H.

1. *Topological Vector Spaces*, 3rd ed., Springer-Verlag, New York-Heidelberg-Berlin, 1971.

Schiffels, G., and Stenzel, M.

1. Einbettung lokal-beschränkter topologischer Ringe in Quotientenringe, *Resultate Math.* 7 (1984), 234-248.

Schilling, O. F. G.

1. Arithmetic in fields of formal power series in several variables, *Ann. Math.* 38 (1937), 551.576.
2. Regular normal extensions over complete fields, *Trans. Amer. Math. Soc.* 47 (1940), 440-454.
3. Normal extensions of relatively complete fields, *Amer. J. Math.* 65 (1943), 309-334.
4. Automorphisms of fields of formal power series, *Bull. Amer. Math. Soc.* 50 (1944), 892-901.
5. Noncommutative valuations, *Bull. Amer. Math. Soc.* 51 (1945), 297-304.

Schilling, O. F. G.

6. *The Theory of Valuations*, Math. Surveys (1950), no. IV, Amer. Math. Soc.

Schmidt, F. K.

1. Mehrfach perfekte Körper, *Math. Ann.* 108 (1933), 1-25.

Schrot, M. D.

1. Pseudo-valuations sur les domaines de Dedekind, *Sem. Delange-Pisot-Poitou* (1966/67), no. 10, 1-7.

Schwarz, F.

1. Ein Struktursatz fur eine Klasse von Pseudobewertungen, *Arch. Math. (Basel)* 22 (1971), 597-601.

Sen, S.

1. On automorphisms of local fields, *Ann. Math.* 90 (1969), 33-46.

Shanks, M. E., and Warner, S.

1. Topologies on the rational field, *Bull. Amer. Math. Soc.* 79 (1973), 1281-1282.
2. Locally bounded topologies on the rational field, *Fund. Math.* 92 (1976), 57-61.

Shelah, S.

1. On a Kurosh problem: Jonsson groups; Frattini subgroups and untopologized groups (preprint, 1976).

Shell (Shilkret), N.

1. Non-Archimedean Banach algebras, *Duke J. Math.* 37 (1970), 315-322.
2. Non-Archimedean Gelfand theory, *Pacific J. Math.* 32 (1970), 541-550.
3. Maximal and minimal ring topologies, *Proc. Amer. Math. Soc.* 68 (1978), no. 1, 23-26.
4. Connected and disconnected fields, *Topology Appl.* 27 (1987), 37-50.

Shilkret, N., Narici, L., and Beckenstein, E.

1. Locally compact non-Archimedean spaces, *Rev. Mat. Hisp.-Amer.* 38 (1978), no. 1, 46-49.

Shu-Bun Ng, and Warner, S.

1. Continuity of positive and multiplicative functionals, *Duke Math. J.* 39 (1972), no. 2, 281-284.

Singh, S.

1. Pseudo-valuation and pseudonorm, *Rend. Sem. Mat. Univ. Padova* 51 (1974), 83-96.
2. Note on a paper of P. M. Cohn, *Proc. Cambridge Philos. Soc.* 78 (1975), 211-213.

Singh, S., and Saluja, K. S.

1. Locally bounded ring topologies on global fields, *Indian J. Pure Appl. Math.* 14 (7) (1983), 919-929.

Skornjakov, L. A.

1. Topological projective planes (Russian), *Trudy Mosk. Mat. Obšč.* 3 (1954), 347-373.

Skornjakov, L. S.

1. Elements of the lattice theory (Russian), Izdatelstvo "Nauka," Moscow, 1970.

Sörensen, K.

1. Topologische Inzidenzgruppen, *Abh. Math. Sem. Univ. Hamburg* 35 (1970), 75-82.

Soundararajan, T.

1. A covering theorem for fields, *Indag. Math.* 27 (1965), 133-135.

Staum, R.

1. The completion of an algebraically closed valued field, *Boll. Un. Mat. Ital.* 10 (1974), 553-555.

Stenzel, M.

1. Inversenbildung und Komplettierung bei topologischen Ringen und Körpern, Diss., Univ. Bielefeld, 1978.
2. ———, *Resultate Math.* 3 (1980), 264-265.

Štipelman, I. J.

1. Valuations on the skew-field of quotients of a quantum-mechanic ring (Russian), *Funkcional. Anal. i Priložen.* 7 (1973), vyp. 1, 56-64.

Stone, A. L.

1. Nonstandard analysis in topological algebra, Applications of model theory to algebra, analysis, and probability, W. A. J. Luxemburg, ed., Holt, Reinhart and Winston, Inc., New York, 1969, pp. 285-299.

Strano, R.

1. Alcune properta dei corpi henseliani, *Ann. Univ. Ferrara Sez. VII (N.S.)* 21 (1975), 69-86.
2. Sulla henselizzatione stretta di un corpo valuato, *Ann. Univ. Ferrara Sez. VII (N.S.)* 21 (1975), 117-130.

Strzelecki, E.

1. Algebras under a minimal norm, *Colloq. Math.* 11 (1963), 41-52.
2. Power-associative regular real normed algebras, *Austral. J. Math.* 6 (1966), 183-209.

Suvorov, N. M., and Ufnarovskaja, T. P.

1. An example of infinite quasi-group admitting only the discrete topology (Russian), Issledovania po teorii kvazigrup, Kishinev, 1973, 163-165.

Szambien, H.

1. Desarguessche uniforme projektive Ebenen, Diss., Univ. Hannover, 1981.

Szpiro, L.

1. Fonctions d'ordre et valuations, *Bull. Soc. Math. France* 94 (1966), 301-311.

Tabaâ, M.

1. Sur une question de compacité linéaire, Diss., Univ. Mohammed V, Rabat, 1978.
2. Corps linéarement compacts, *C. R. Acad. Sci. Paris Sér. A-B* 290 (1980), 531-532.

Tai-Chih-Chang

1. On complete fields with multiple valuations (Chinese), *Shuxue Jinzahan* 9 (1966), 401-406.

Taĭmanov, A. D.

1. An example of a semigroup admitting only the discrete topology (Russian), *Algebra i Logika* 12 (1973), no. 1, 114-116.
2. On topologizations of countable algebras (Russian), Matematičeskiĭ analiz i smežnye voprosy matematiki, Novosibirsk, 1978, 254-275.

Tamari, D.

1. Sur l'immersion d'un sémigroupe topologique dans un groupe topologique. Paris: Colloques Internationaux du Centre National de la Recherche Scientifique, Algebre et Théorie des Nombres, 1950, 217-221.

Taussky, O.

1. Analytical methods in hypercomplex systems, *Compositio Math.* 3 (1936), 181-225.

Teichmüller, O.

1. Über die Struktur diskret bewerteter Körper, *Nachr. Ges. Wiss. Göttingen I*, N.F. 1 (1936), 151-161.
2. Diskret bewertete perfekte Körper mit unvollkommenen Restklassenkörpern, *J. Reine Angew. Math.* 176 (1937), 141-152.

Theess, M.

1. Bewertete Vektorräume und pseudovollständige Körper, *Mitt. Math. Gesellsch. Hamburg*, Band X, Heft 5 (1977), 385-389.

Tornheim, L.

1. Normed fields over the real and complex fields, *Michigan Math. J.* 1 (1952), 61-68.

Urbanik, K.

1. Absolute-valued algebras with an involution, *Fund. Math.* 49 (1961), 247-258.
2. Reversibility in absolute-valued algebras, *Fund. Math.* 51 (1962/63), 131-140.
3. Remarks on ordered absolute-valued algebras, *Colloq. Math.* 11 (1963), 31-38.

Urbanik, K., and Wright, F. B.

1. Absolute valued algebras, *Bull. Acad. Polon. Sci. Sér. Sci. Math. Astronom. Phys.* 8 (1960), 285-286.
2. Absolute valued algebras, *Proc. Amer. Math. Soc.* 11 (1960), 861-866.

Uspenskiĭ, V. V.

1. Metrizable space is embeddable in a metrizable field (Russian), *Uspehi Mat. Nauk* 38 (1983), vyp. 3 (321), 171-172.

Vámos, P.

1. Multiply maximally complete fields, *J. London Math. Soc.* 12 (1975), 103-111.

Varouchakis, N. Ch.

1. Anneaux ultrametroides values, *Math. Balc.* 5 (1975), 279-293.

Viswanathan, T. M.

1. On fully ordered rings, *J. Indian Math. Soc.* (N.S.) 32 (1968), suppl. I, 381-395.
2. Hahn valuations and (locally) compact rings, *J. Algebra* 17 (1971), 94-109.
3. A characterization of Henselian valuations via the norm, *Bol. Soc. Brasil. Mat.* 4 (1973), 51-53.

Vizitiu, V. N.

1. A continuation of m-topologies of a field (Russian), *Mat. Issled.* 10 (1975), vyp. 2 (36), 64-78.
2. Topological rings with m-minimal topologies (Russian), *Mat. Issled.* 1976, vyp. 38, 97-115.
3. Uniqueness of the continuation of m-topologies of a field to an algebraic extension (Russian), *Mat. Issled.* 1979, vyp. 53, 27-42.
4. Continuation of norms on fields (Russian), Rings, Algebras, Modules, *Mat. Issled.* 1981, vyp. 62, 3-8.

Vizitiu, V. N.

5. On continuous continuation of differentiations of rings (Russian), *Mat. Issled.* 1983, vyp. 74, 3-17.

Vodinčar, M. I.

1. Criterion of normability of topological algebras (Russian), *Mat. Issled.* 2 (1967), vyp. 1, 133-140.

Volk, G.

1. Eine Maßtheoretische Charakterisierung von Körpertopologien, *Math. Nachr.* 43 (1970), 19-23.

van der Waerden, B. L.

1. Topological algebra, *Nieuw Arch. Wisk.* 23 (1975), 212-227.

Wagner, C. G.

1. Automorphisms of p-adic number fields, *Amer. Math. Monthly* 81 (1974), 51-52.

Warner, S.

1. Compact noetherian rings, *Math. Ann.* 141 (1960), 161-170.
2. Compact and finite-dimensional locally compact vector spaces, *Illinois J. Math.* 13 (1969), 383-393.
3. Normability of certain topological rings, *Proc. Amer. Math. Soc.* 33 (1972), 423-427.
4. Openly embedding local noetherian domains, *J. Reine Angew. Math.* 253 (1972), 146-151.
5. A topological characterization of complete, discretely valued fields, *Pacific J. Math.* 48 (1973), 293-298.
6. A new approach to Gelfand-Mazur theory and the extension theorem, *Michigan Math. J.* 26 (1979), 13-17.
7. Generalizations of a theorem of Mutylin, *Proc. Amer. Math. Soc.* 78 (1980), no. 3, 327-330.
8. Nonuniqueness of immediate maximal extensions of a valuation, *Math. Scand.* 56 (1985), 191-202.
9. Residual fields in valuation theory, *Math. Scand.* 56 (1985), 203-221.

Waterman, A. G., and Bergman, G. M.

1. Connected fields of arbitrary characteristic, *J. Math. Kyoto Univ.* 5 (1966), 177-184.

Weber, H.

1. Ringtopologien auf Z und Q, *Math. Z.* 155 (1977), 287-298.
2. Zu einem Problem von H. J. Kowalsky, *Abh. Braunschweig. Wiss. Gesellsch.* 29 (1978), 127-134.

Weber, H.

3. Charakterisierung der lokalbeschränkten Ringtopologien auf globalen Körpern, *Math. Ann.* 239 (1979), 193-205.

4. Topologische Charakterisierung globaler Körper und algebraischer Funktionenkörper in einer Variablen, *Math. Z.* 169 (1979), no. 2, 167-177.

5. Unabhängige Topologien, Zerlegung von Ringtopologien, *Math. Z.* 180 (1982), no. 3, 379-393.

6. Bestimmung aller Integritätsbereiche, deren einzige lokalbeschränkte Ringtopologien Idealtopologien sind, *Arch. Math. (Basel)* 41 (1983), no. 4, 328-336.

Wefelscheid, H.

1. Vervollständigung topologisch-algebraischer Strukturen, Diss., Hamburg, 1966.

2. Vervollständigung topologischer Fastkörper, *Math. Z.* 99 (1967), 279-298.

3. Zur Konstruktion bewerteter Fastkörper, *Abh. Math. Sem. Univ. Hamburg* 38 (1972), 106-117.

4. Bewertungen und Topologie in Fastkörpern, *Abh. Math. Sem. Univ. Hamburg* 39 (1973), 130-146.

Weil, A.

1. *Basic Number Theory*, Springer-Verlag, Berlin-Heidelberg-New York, 1967.

Weiss, E.

1. Boundedness in topological rings, *Pacific J. Math.* 6 (1956), 149-158.

Weiss, E., and Zierler, N.

1. Locally compact division rings, *Pacific J. Math.* 8 (1958), 369-371.

Więsław, W.

1. On some characterizations of the complex number field, *Bull. Acad. Polon. Sci. Sér. Sci. Math. Astronom. Phys.* 19 (1971), 353-354.

2. A remark on complete and connected rings, *Bull. Acad. Polon. Sci. Sér. Sci. Math. Astronom. Phys.* 19 (1971), 981-982.

3. On some characterizations of the complex number field, *Colloq. Math.* 24 (1972), 139-145.

4. A characterization of locally compact fields of zero characteristic, *Fund. Math.* 76 (1972), 149-155. Errata, *Fund. Math.* 82 (1973), 91-92.

Więsław, W.

5. Extensions of topological fields, *Colloq. Math.* 28 (1973), 203-204.

6. On topological fields, *Colloq. Math.* 29 (1974), 119-146.

7. An example of a locally unbounded complete extension of the p-adic number field, *Colloq. Math.* 30 (1974), 105-108. Errata, *Colloq. Math.* 34 (1976), 293-294.

8. A remark on complete and connected rings II, *Bull. Acad. Polon. Sci. Sér. Sci. Math. Astronom. Phys.* 22 (1974), 15-17.

9. A characterization of locally compact fields II, *Fund. Math.* 88 (1975), 121-125.

10. Independent topologies on fields, *Comment. Math. Univ. St. Paul.* 26 (1977), 87-94.

11. Locally bounded topologies on fields and rings, *Mitt. Math. Gesellsch. Hamburg*, Band X, Heft 6 (1978), 481-508.

12. A characterization of locally compact fields III, *Colloq. Math.* 39 (1978), no. 2, 313-317.

13. Locally bounded topologies on Euclidean rings, *Arch. Math. (Basel)* 31 (1978), no. 1, 33-37.

14. Embeddings of fields in complete ones, *Math. Nachr.* 93 (1979), 133-137.

15. Locally bounded topologies on some Dedekind rings, *Arch. Math. (Basel)* 33 (1979), no. 1, 41-44.

16. Topological fields, *Acta Univ. Wratislav. Mat. Fiz. Astronom.* XLIII (1982), no. 675, 219 pp.

17. Topological division rings, Summer School on Number Theory held at Chlébské, September 1983, Brno, 1985, 80-97.

Williamson, W. H.

1. On the topologizing of the field $\mathbb{C}(t)$, *Proc. Amer. Math. Soc.* 5 (1954), 729-734.

Wong, C. M.

1. Familles de valuations à caractère compact, *Canad. J. Math.* 24 (1972), 416-425.

Wright, F. B.

1. Absolute valued algebras, *Proc. Nat. Acad. Sci. U.S.A.* 39 (1953), 330-332.

Wright, M. J.

1. On the number of prolongations of a finite rank valuation, *Canad. J. Math.* 23 (1971), 553-556.

Yakabe, I.

1. On semi-valuations I, *Mem. Fac. Sci. Kyushu Univ. Ser. A* 16 (1962), 1-8.

2. On semi-valuations II, *Mem. Fac. Sci. Kyushu Univ. Ser. A* 17 (1963), 10-28.

3. Equivalence of the Krull-Müller-Jaffard theorem and Ribenboim's approximation theorem, *Mem. Fac. Sci. Kyushu Univ. Ser. A* 17 (1963), 145-152.

4. An extension theorem for semi-valuations of the first kind, *Mem. Fac. Sci. Kyushu Univ. Ser. A* 18 (1964), 50-55.

5. Approximation theorems for semi-valuations, Queen's Univ., Kingston, Ontario, preprint 1968-16.

Zariski, O.

1. Generalized semi-local rings, *Summa Brasil. Math.* 1 (1946), no. 8, 169-195.

Zariski, O., and Samuel, P.

1. *Commutative Algebra*, Vols. I, II, Springer-Verlag, New York-Heidelberg-Berlin, 1975.

Żelazko, W.

1. A theorem on B_0 division algebras, *Bull. Acad. Polon. Sci. Sér. Sci. Math. Astronom. Phys.* 8 (1960), 373-375.

2. Some remarks on topological algebras, *Studia Math.* Ser. Spec. z. 1 (1963), 141-149.

3. Metric generalizations of Banach algebras, *Rozprawy Mat.* 47 (1965).

4. On generalized topological divisors of zero in m-convex locally convex algebras, *Studia Math.* 27 (1966), 9-16.

5. *Banach Algebras*, Polish Scientific Publishers, Warszawa, 1973.

Zelinsky, D.

1. Nonassociative valuations, *Bull. Amer. Math. Soc.* 54 (1948), 175-183.

2. Topological characterization of fields with valuations, *Bull. Amer. Math. Soc.* 54 (1948), 1145-1150.

3. Complete fields from local rings, *Proc. Nat. Acad. Sci. U.S.A.* 37 (1951), 379-381.

4. Rings with ideal nuclei, *Duke Math. J.* 18 (1951), 431-442.

Zobel, R.

1. Direkte Gruppen- und Ringtopologien, Diss., Tech. Univ. Braunschweig, 1973.

2. Gruppentopologien auf nichtabelschen abzählbaren Gruppen, *Manuscripta Math.* 14 (1974), 207-216.

Author Index

Subject Index